Lecture Notes in Mathematics 2169

More information about this series at http://www.springer.com/series/304

Bo'az Klartag • Emanuel Milman
Editors

Geometric Aspects of Functional Analysis

Israel Seminar (GAFA) 2014–2016

 Springer

Editors
Bo'az Klartag
School of Mathematical Sciences
Tel Aviv University
Tel-Aviv, Israel

Emanuel Milman
Mathematics Department
Technion - Israel Institute of Technology
Haifa, Israel

ISSN 0075-8434   ISSN 1617-9692   (electronic)
Lecture Notes in Mathematics
ISBN 978-3-319-45281-4   ISBN 978-3-319-45282-1   (eBook)
DOI 10.1007/978-3-319-45282-1

Library of Congress Control Number: 2016959616

Mathematics Subject Classification (2010): 80M35, 26A51, 32-XX, 46-XX, 60-XX

Printed on acid-free paper

This Springer imprint is published by Springer Nature
The registered company is Springer International Publishing AG
The registered company address is: Gewerbestrasse 11, 6330 Cham, Switzerland

Preface

Since the mid-1980s, the following volumes containing collections of papers reflecting the activity of the Israel Seminar in Geometric Aspects of Functional Analysis have appeared:

1983–1984   Published privately by Tel Aviv University
1985–1986   Springer Lecture Notes in Mathematics, vol. 1267
1986–1987   Springer Lecture Notes in Mathematics, vol. 1317
1987–1988   Springer Lecture Notes in Mathematics, vol. 1376
1989–1990   Springer Lecture Notes in Mathematics, vol. 1469
1992–1994   Operator Theory: Advances and Applications, vol. 77, Birkhäuser
1994–1996   MSRI Publications, vol. 34, Cambridge University Press
1996–2000   Springer Lecture Notes in Mathematics, vol. 1745
2001–2002   Springer Lecture Notes in Mathematics, vol. 1807
2002–2003   Springer Lecture Notes in Mathematics, vol. 1850
2004–2005   Springer Lecture Notes in Mathematics, vol. 1910
2006–2010   Springer Lecture Notes in Mathematics, vol. 2050
2011–2013   Springer Lecture Notes in Mathematics, vol. 2116

The first six were edited by Lindenstrauss and Milman; the seventh by Ball and Milman; the subsequent four by Milman and Schechtman; the penultimate one by Klartag, Mendelson, and Milman; and the last by the present editors.

As in the previous seminar notes, the current volume reflects general trends in the study of geometric aspects of functional analysis, and many of the papers deal with different aspects of asymptotic geometric analysis, understood in a broad sense. A classical theme in the local theory of Banach spaces, which is well represented in this volume, is the identification of lower-dimensional structures, such as diameter bounds, Euclidean structure of sections, and super-Gaussian tail decay of projections, in high-dimensional objects, such as subclasses of high-dimensional convex bodies and other distributions. More recent applications of high dimensionality are represented by contributions in random matrix theory, establishing bounds on expectation of norms of matrices and their inverses, deviations about the expectation, and the restricted invertibility property. Naturally, the Gaussian measure plays a central role in many of these topics and is studied in this volume—the recent breakthrough proof of the Gaussian correlation conjecture is revisited, moment inequalities for log-concave random variables are obtained, and a Poincaré-type inequality on the boundary of convex domains on Gaussian space is derived. As expected, probabilistic tools play a significant role, and concentration results for non-Lipschitz functions and empirical multiplier processes are presented. The interplay of the theory with harmonic analysis is also well apparent in several examples on large Lie groups and the discrete cube. The classical relation to both the primal and dual Brunn–Minkowski theories is not abscent, with contributions pertaining to the stability of Brunn–Minkowski type inequalities and characterization of the radial sum. Related algebraic structures, such as constructable functions of valuations, valuations on quasi-concave functions, generalized valent functions on the complex plane, and rigidity of the chain rule, are also discussed. Other classical topics such as the theory of type and cotype are covered as well. All contributions are original research papers and were subject to the usual refereeing standards.

We are grateful to Vitali Milman for his help and guidance in preparing and editing this volume.

Tel Aviv, Israel   Bo'az Klartag
Haifa, Israel   Emanuel Milman

Contents

On Repeated Sequential Closures
of Constructible Functions in Valuations

Semyon Alesker

Abstract The space of constructible functions form a dense subspace of the space
of generalized valuations. In this note we prove a somewhat stronger property that
the sequential closure, taken sufficiently many (in fact, infinitely many) times, of the
former space is equal to the latter one. This stronger property is necessary for some
applications in Alesker (Geom Funct Anal 20(5):1073–1143, 2010).

1 Main Results

The main results of this note are Theorem 1.1 and Corollaries 1.2, 1.3 below.
Corollary 1.2 says that the taken sufficiently (infinitely) many times sequential
closure of constructible functions inside the space of generalized valuations is equal
to the whole space. Corollary 1.3 says, in particular, that if a sequentially continuous
linear operator from generalized valuations on a manifold either with or without
compact support to a Hausdorff linear topological space vanishes on constructible
functions, then it vanishes. Recall that a map between two topological spaces is
called sequentially continuous if it maps convergent sequences to convergent ones.
Notice that for non-metrizable topological spaces sequential continuity of a map
does not imply topological continuity.

The reason to write this note is to correct a mistake made by the author in [7],
where it was wrongly claimed that several operations, such as pull-back, push-
forward, and product on generalized valuations with given wave front sets are
topologically continuous, while they satisfy, in fact, only a weaker property of
sequential continuity. This property comes from the fact that operations of pull-back,
push-forward, and product on generalized functions or distributions with given wave

S. Alesker (✉)
Department of Mathematics, Tel Aviv University, Ramat Aviv, 69978 Tel Aviv, Israel
e-mail: semyon@post.tau.ac.il

© Springer International Publishing AG 2017   1
B. Klartag, E. Milman (eds.), *Geometric Aspects of Functional Analysis*,
Lecture Notes in Mathematics 2169, DOI 10.1007/978-3-319-45282-1_1

front sets are only sequentially continuous in appropriate (so called Hörmander) topology,[1] but in general they are not topologically continuous, see Sect. 3.1 in [10].

Let X be either smooth manifold or real analytic manifold. It will always be assumed to be countable at infinity, i.e. X can be presented as a countable union of compact subsets. We will denote by $V_c^\infty(X)$ and $V^\infty(X)$ the space of smooth valuations on X with and without compact support respectively. Also we denote by $V_c^{-\infty}(X)$ and $V^{-\infty}(X)$ the spaces of generalized valuations with and without compact support respectively. We refer to [7] for the definitions of all these spaces and further details (see also [6]). For a compact subset $Z \subset X$ we denote by $V_Z^{-\infty}(X)$ the space of generalized valuations with support contained in Z; clearly it is a closed subspace of $V^{-\infty}(X)$. We equip $V_Z^{-\infty}(X)$ with the weak topology induced from $V^{-\infty}(X)$. We denote by $V_c^{-\infty}(X)$ the space of all generalized compactly supported valuations. We equip it with the topology of (strict) inductive limit:

$$V_c^{-\infty}(X) = \varinjlim_{z \text{ compact}} V_Z^{-\infty}(X).$$

Since all our manifolds are countable at infinity, the limit can be made countable.

Let X be a smooth manifold (not necessarily real analytic). Let $P \subset X$ be a compact submanifold with corners. Then P defines a generalized valuation as follows. By definition, generalized valuations are continuous linear functionals on $V_c^\infty(X)$. Then P defines a linear functional $[\phi \mapsto \phi(P)]$. We denote this generalized valuation by $\Xi^\infty(\mathbb{1}_P)$. We denote by $\mathcal{F}(X)$ the span over $\mathbb{C}$ of all generalized valuations of the form $\Xi^\infty(\mathbb{1}_P)$ where $P \subset X$ is a compact submanifold with corners. Clearly $\mathcal{F}(X)$ is a subspace of $V_c^{-\infty}(X)$. Furthermore, for a closed subset $Z \subset X$ we denote by $\mathcal{F}_Z(X)$ the span over $\mathbb{C}$ of generalized valuations of the form $\Xi^\infty(\mathbb{1}_P)$ where $P \subset Z$ is a compact submanifold with corners.

Let now X be a real analytic manifold. We denote by $\mathcal{F}^{an}(X)$ the space of so called $\mathbb{C}$-valued constructible functions. Let us remind the definition of this notion following [5]. We refer to §8.2 in [12] for the definition and basic properties of subanalytic sets (see also Sect. 1.2 in [5]). An integer valued function $f: X \to \mathbb{Z}$ on a real analytic manifold is called constructible if it satisfies:

1. for every $m \in \mathbb{Z}$ the set $f^{-1}(m)$ is subanalytic;
2. the family of sets $\{f^{-1}(m)\}_{m \in \mathbb{Z}}$ is locally finite.

Now a $\mathbb{C}$-valued function $f: X \to \mathbb{C}$ is called *constructible* if f is a finite linear combination with $\mathbb{C}$-coefficients of integer valued constructible (in the above sense) functions. Furthermore, for a closed subset $Z \subset X$ we denote by $\mathcal{F}_Z^{an}(X)$ the subspace of $\mathcal{F}^{an}(X)$ consisting of functions supported in Z.

For a real analytic manifold X there is a canonical injective imbedding $\mathcal{F}^{an}(X) \to V^{-\infty}(X)$; see Sect. 8.1 in [5].

---

[1]This fact was pointed out to the author by C. Brouder in September 2013. I am very grateful to him for this remark.

In order to formulate our main results, let us remind the notion of a sequential closure of transfinite order of a set. This notion was already known to S. Banach, see p. 213 in his classical book [9]. Let $\mathcal{X}$ be a topological space. Let $A \subset \mathcal{X}$ be a subset. A sequential closure of A is defined by

$$scl(A) = \{x \in \mathcal{X} \mid \exists \text{ sequence } \{a_i\} \subset A \text{ s.t. } a_i \to x\}.$$

It is clear that $A \subset scl(A)$, and if $\mathcal{X}$ is a linear topological space and $A \subset \mathcal{X}$ is a linear subspace then $scl(A)$ is a linear subspace. A subset $A \subset \mathcal{X}$ is called sequentially closed if for any converging in $\mathcal{X}$ sequence $\{x_i\}_{i=1}^{\infty} \subset A$ its limit belongs to A; equivalently $scl(A) = A$. Clearly any closed subset is sequentially closed, but the converse is not true in general. If $\mathcal{X}$ is not metrizable, $scl(A)$ may not be closed or even sequentially closed, i.e. $scl(scl(A)) \neq scl(A)$. We can repeat the procedure of taking sequential closure any number of times, even any infinite number of times corresponding to any ordinal. More precisely, for any ordinal η one can define by transfinite induction the subset $scl^{\eta}(A)$ as follows:

- if $\eta = 0$ then $scl^0(A) = A$;
- if $\eta = \xi + 1$ then $scl^{\eta}(A) = scl(scl^{\xi}(A))$;
- if η is a limit ordinal then $scl^{\eta}(A) = \cup_{\xi < \eta} scl^{\xi}(A)$.

Furthermore there exists an ordinal η such that for any $\eta' > \eta$ one has $scl^{\eta'}(A) = scl^{\eta}(A)$. We will denote the latter subset by $scl^*(A)$. It is also clear that $scl^*(A)$ is sequentially closed, i.e. $scl(scl^*(A)) = scl^*(A)$. Clearly if $\mathcal{X}$ is a linear topological space and $A \subset \mathcal{X}$ is a linear subspace then $scl^{\eta}(A)$ is a linear subspace for any η.

Here is the main result of the note.

Theorem 1.1 *Let X be a real analytic (resp. smooth) manifold countable at infinity. Let $Z_1 \subset X$ be a compact subset. Let Z_2 be a compact neighborhood of Z_1. Then in the above notation the subspace $scl^*(\mathcal{F}_{Z_2}^{an}(X)) \subset V_c^{-\infty}(X)$ (resp. $scl^*(\mathcal{F}_{Z_2}(X)) \subset V_c^{-\infty}(X)$) contains $V_{Z_1}^{-\infty}(X)$.*

Let us deduce two immediate corollaries.

Corollary 1.2 *Let X be a real analytic (resp. smooth) manifold. Then in the above notation $scl^*(\mathcal{F}^{an}(X)) = V^{-\infty}(X)$ (resp. $scl^*(\mathcal{F}(X)) = V^{-\infty}(X)$).*

Proof of Corollary 1.2 By Theorem 1.1 $scl^*(\mathcal{F}^{an}(X))$ (resp. $scl^*(\mathcal{F}(X))$) contains $V_c^{-\infty}(X)$. But $V_c^{-\infty}(X)$ is sequentially dense in $V^{-\infty}(X)$ since X is assumed to be countable at infinity. Q.E.D.

Corollary 1.3 *Let X be a real analytic (resp. smooth) manifold countable at infinity. Let $R: V^{-\infty}(X)(or\ V_c^{-\infty}(X)) \to E$ be a linear operator into a Hausdorff topological vector space E. Assume that R is sequentially continuous, i.e. R maps convergent sequences in $V^{-\infty}(X)(or\ V_c^{-\infty}(X))$ to convergent sequences in E. Let $L \subset E$ be a sequentially closed subset. Then if $R(\mathcal{F}^{an}(X)) \subset L$ (resp. $R(\mathcal{F}(X)) \subset L)$ then the whole image of R is contained in L. In particular if $R(\mathcal{F}^{an}(X)) = 0$ (resp. $R(\mathcal{F}(X)) = 0)$ then $R \equiv 0$.*

Proof By transfinite induction $R(scl^*(\mathcal{F}^{an}(X))) \subset L$ (resp. $R(scl^*(\mathcal{F}(X)) \subset L))$. In all cases (smooth or real analytic X, compact or non-compact support) $scl^*(\mathcal{F}(X))$ is equal to the whole space. Q.E.D.

Remark 1.4 Corollary 1.3 is used in [7] in the proof of the inversion formula for the Radon transform on valuations with respect to the Euler characteristic. Khovansky and Pukhlikov [13] proved it for constructible functions, and then one extends it to valuations using sequential continuity with the use of Corollary 1.3.

2 Proofs

The two cases of real analytic and smooth manifold X are almost identical and will be treated simultaneously. Let us remind some notation.

For a finite dimensional real vector space W we denote by $\mathbb{P}_+(W)$ the so called oriented projectivization of W, i.e. the manifold of oriented lines passing through the origin. For a smooth manifold X we denote by $\mathbb{P}_X$ the oriented projectivization of the cotangent bundle T^*X, i.e. the fiber of $\mathbb{P}_X$ over X is equal to $\mathbb{P}_+(T_x^*X)$.

For either smooth or real analytic manifold X and for sufficiently nice subsets $P \subset X$ (i.e. compact submanifolds with corners or compact subanalytic subsets) we denote by $N(P)$ the normal cycle of P; in general it is a current on $\mathbb{P}_X$, see Sect. 1.5 in [7] for the easier case of compact submanifolds with corners and [11] for the case of subanalytic sets.

For a finite dimensional real vector space V we denote by $\mathcal{K}(V)$ the family of all convex compact non-empty subsets of V, and by $\mathcal{K}^\infty(V)$ the subfamily of compact convex sets with non-empty interior and infinitely smooth boundary with everywhere positive Gauss curvature.

Lemma 2.1 *Let σ be an infinitely smooth measure on a vector space V. Let $A \in \mathcal{K}^\infty(V)$. Then $K \mapsto \sigma(K + A)$ is a smooth valuation.*

Proof Consider the map

$$p: V \times \mathbb{P}_+(V^*) \times [0, 1] \to V$$

given by $(x, n, t) = x + t\nabla h_A(n)$, where $h_A: V^* \to \mathbb{R}$ is the supporting functional of A. Since h_A is 1-homogeneous, its gradient ∇h_A is 0-homogeneous, and hence can be considered as a map $\nabla h_A: \mathbb{P}_+(V^*) \to V$. Since $A \in \mathcal{K}^\infty(V)$ the latter map is infinitely smooth.

We may and will assume that $0 \in int(A)$; the general case reduces to this one by translation. In this case the restriction of p to $N(K) \times [0, 1]$ is a homeomorphism onto the closure of $(K + A) \backslash K$. Hence

$$\sigma(K + A) = \sigma(K) + \int_{N(K) \times [0,1]} p^*\sigma. \tag{1}$$

Let $q: V \times \mathbb{P}_+(V^*) \times [0, 1] \to V \times \mathbb{P}_+(V^*)$ is the obvious projection. Then by (1) we have

$$\sigma(K + A) = \sigma(K) + \int_{N(K)} q_* p^* \sigma.$$

Obviously $q_* p^* \sigma$ is a smooth $(\dim V - 1)$-form on $V \times \mathbb{P}_+(V^*)$. This proves the lemma. Q.E.D.

Lemma 2.2 *Let $Z_1 \subset V$ be a compact set, $A \in \mathcal{K}^\infty(V)$, and Z_2 be a compact neighborhood of $Z_1 - A$. Let σ be a smooth measure with $\mathrm{supp}(\sigma) \subset Z_1$. Then*

(1) one has

$$\sigma(\bullet + A) = \int_V \mathbb{1}_{x-A} \cdot d\sigma(x) \qquad (2)$$

   as generalized valuations;
(2) $\sigma(\bullet + A)$ belongs to $V_{Z_2}^\infty(V) \cap scl^*(\mathcal{F}_{Z_2}(V))$.
(3) If, in addition, A is subanalytic then $\sigma(\bullet + A)$ belongs to $V_{Z_2}^\infty(V) \cap scl^*(\mathcal{F}_{Z_2}^{an}(V))$.

Proof First notice that $\sigma(\bullet + A)$ is a smooth valuation by Lemma 2.1.

To prove part (1) it suffices to apply both sides to an arbitrary smooth compactly supported valuation and to prove that the result is the same. It suffices to apply them to such valuations of the form $\eta = \omega(\bullet + B)$ where ω is a smooth compactly supported measure and $B \in \mathcal{K}^\infty(V)$, since linear combinations of such valuations are dense in $V^\infty(V)$ (this easily follows from Corollary 3.1.7 in [3]).

Apply the left hand side of (2) to η:

$$< \sigma(\bullet + A), \eta > \; = \; (\sigma \boxtimes \omega)(\Delta(V) + (A \times B))$$

$$\overset{Fubini}{=} \int d\sigma(x)\omega(x - A + B),$$

where $\Delta: V \to V \times V$ is the diagonal imbedding given by $\Delta(x) = (x, x)$.

Apply the right hand side of (2) to η:

$$< \int d\sigma(x)\mathbb{1}_{x-A}, \eta > = \int d\sigma(x) \cdot \eta(x - A) = \int d\sigma(x) \cdot \omega(x - A + B)$$

$$= \; < \sigma(\bullet + A), \eta > .$$

Thus part (1) is proved.

Let us prove part (3); part (2) can be proven along exactly the same lines. It remains to show that $\sigma(\bullet + A) \in scl^*(\mathcal{F}_{Z_2}^{an}(V))$. We use the equality (2) and replace the integral in the right hand side by a Riemann sum corresponding to a subdivision of V whose diameter will tend to 0. Let us show that these Riemann sums converge to the integral in the weak topology on $V^{-\infty}(V)$. Let $\{C_i^N\}_{N=1}^\infty$ be a sequence of subdivisions whose diameter tends to 0 as $N \to \infty$. Choose a point $x_i^N \in C_i^N$. Apply the corresponding Riemann sum to η:

$$< \sum_i \sigma(C_i^N) \mathbb{1}_{x_i^N - A}, \eta > = \sum_i \sigma(C_i^N) \eta(x_i^N - A) \underset{N \to \infty}{\to} \int d\sigma(x) \eta(x - A)$$

$$= < \int d\sigma(x) \mathbb{1}_{x-A}, \eta >,$$

where we have used the fact that continuous scalar valued functions are Riemann integrable, to the function $[x \mapsto \eta(x - A)]$. It only remains to notice that the i-th Riemann sum belongs to $\mathcal{F}_{Z_2}^{an}(V)$ for $i \gg 1$.Q.E.D.

Lemma 2.3 *Let $Z_1 \subset V$ be a compact set. Let $A_1, A_2 \in \mathcal{K}^\infty(V)$. Let Z_2 be a compact neighborhood of $Z_1 - A_1$. Let σ_1, σ_2 be smooth measures on V such that $supp(\sigma_1) \subset Z_1$. Let $\phi_i := \sigma_i(\bullet + A_i)$, $i = 1, 2$. Then*

$$\phi_1 \cdot \phi_2 \in V_{Z_2}^\infty(V) \cap scl^*(\mathcal{F}_{Z_2}(V)).$$

If, in addition, A_1, A_2 are subanalytic then $\phi_1 \cdot \phi_2 \in V_{Z_2}^\infty(V) \cap scl^*(\mathcal{F}_{Z_2}^{an}(V))$.

Proof We consider the subanalytic case only since the two cases are essentially the same. First let us show that in the space $V^{-\infty}(V)$ (or, equivalently, in $V_c^{-\infty}(V)$) one has

$$\phi_1 \cdot \phi_2 = \int \int d\sigma_1(x) d\sigma_2(y) \mathbb{1}_{(x-A_1) \cap (y-A_2)}, \tag{3}$$

where the integral is understood in the sense the limit of Riemann sums converging in the weak topology. Again we have to show that if we apply the two sides on the same $\eta \in V_c^\infty(V)$ then we get the same result. It suffices to choose η of the form $\eta = \omega(\bullet + C)$, where $C \in \mathcal{K}^\infty(V)$, and ω is a smooth compactly supported measure on V. Applying the right hand side of (3) on such η we get

$$< \int \int d\sigma_1(x) d\sigma_2(y) \mathbb{1}_{(x-A_1) \cap (y-A_2)}, \eta > \tag{4}$$

$$= \int \int d\sigma_1(x) d\sigma_2(y) \eta((x - A_1) \cap (y - A_2)) \tag{5}$$

$$= \int \int d\sigma_1(x) d\sigma_2(y) \omega \left([(x - A_1) \cap (y - A_2)] + C \right). \tag{6}$$

Now let us apply to η the left hand side of (3) (in the computation Δ is the diagonal map $V \to V \times V \times V$ given by $x \mapsto (x, x, x)$):

$$< \phi_1 \cdot \phi_2, \eta > = (\phi_1 \cdot \phi_2 \cdot \eta)(V) \tag{7}$$

$$= (\sigma_1 \boxtimes \sigma_2 \boxtimes \omega)(\Delta(V) + (A_1 \times A_2 \times C)) \tag{8}$$

$$\overset{Fubini}{=} \int \int d\sigma_1(x) d\sigma_2(y)\omega([(x-A_1) \cap (y-A_2)] + C) = (6). \tag{9}$$

Thus equality (3) is proven. It remains to show that the right hand side of (3) belongs to $scl^*(\mathcal{F}_{Z_2}^{an}(V))$. To do that, we will approximate the double integral by Riemann sums belonging to $\mathcal{F}_{Z_2}^{an}(V)$ which converge to the double integral in the weak topology on $V^{-\infty}(V)$.

Consider a sequence $\{C_i^N\}_{N=1}^\infty$ of subdivisions of V with diameter tending to 0 as $N \to \infty$. Choose a point $x_i^N \in C_i^N$. For the corresponding Riemann sum

$$< \sum_{i,j} \sigma_1(C_i^N)\sigma_2(C_j^N) \mathbb{1}_{(x_i^N-A_1)\cap(x_j^N-A_2)}, \eta > \tag{10}$$

$$= \sum_{i,j} \sigma_1(C_i^N)\sigma_2(C_j^N)\eta((x_i^N - A_1) \cap (x_j^N - A_2)). \tag{11}$$

Similarly for the double integral we have

$$< \int \int d\sigma_1(x) d\sigma_2(y) \mathbb{1}_{(x-A_1)\cap(y-A_2)}, \eta > \tag{12}$$

$$= \int \int d\sigma_1(x) d\sigma_2(y) \cdot \eta((x - A_1) \cap (y - A_2)). \tag{13}$$

We see that (11) is a Riemann sum for (13), and we have to show that the former converges to the latter. In other words we have to show that the function $V \times V \to \mathbb{R}$ given by $(x, y) \mapsto \eta((x - A_1) \cap (y - A_2))$ is Riemann integrable.

Notice that the above function does not have to be continuous. But obviously this function is bounded. By the Lebesgue criterion of Riemann integrability (see e.g. [15], Sect. 11.1) it suffices to show that the above function is continuous almost everywhere. For that it suffices to prove that the function $\Xi: V \times V \to \mathcal{K}(V) \cup \{\emptyset\}$ given by $\Xi(x, y) = (x - A_1) \cap (y - A_2)$ is continuous almost everywhere in the Hausdorff metric on $\mathcal{K}(V)$.

To prove the last statement let us consider a closed convex set $M \subset V \times V \times V$ defined by

$$M := \{(x, y, z) \mid x - z \in A_1, y - z \in A_2\}.$$

Let $q: V \times V \times V \to V \times V$ be the projection onto the first two copies of V. Then clearly

$$q^{-1}(x, y) \cap M = (x - A_1) \cap (y - A_2) = \Xi(x, y),$$

and the restriction of q to M is proper. Applying Theorem 1.8.8 of [14],[2] it follows that Ξ is continuous outside of the boundary of the set

$$q(M) = \{(x, y)|(x - A_1) \cap (y - A_2) \neq \emptyset\} = \{(x, y)|x - y \in A_1 - A_2\}.$$

From this description it is clear that $q(M)$ is a closed convex set. Its boundary always has Lebesgue measure zero. Finally let us notice that the N-th Riemann sum belongs to $\mathcal{F}_{Z_2}^{an}(V)$ for $N \gg 1$. Q.E.D.

In [4] we have defined a canonical filtration on $V^\infty(X^n)$ by closed subspaces

$$V^\infty(X^n) = W_0^\infty(X^n) \supset W_1^\infty(X^n) \supset \cdots \supset W_n^\infty(X^n).$$

Here for a closed subset $Z \subset X$ we will also denote by $W_{i,Z}^\infty := W_i^\infty \cap V_Z^\infty(X)$.

Lemma 2.4 *Let V be an n-dimensional real vector space. Let $Z_1 \subset V$ be a compact domain with infinitely smooth boundary, and Z_2 be a compact neighborhood of Z_1. The image of $W_{i,Z_1}^\infty \cap scl^*(\mathcal{F}_{Z_2}^{an}(V))$ (resp. $W_{i,Z_1}^\infty \cap scl^*(\mathcal{F}_{Z_2}(V))$) in $W_{i,Z_1}^\infty / W_{i+1,Z_1}^\infty \simeq C_{Z_1}^\infty(V, Val_i^\infty(V))$[3] is a dense linear subspace.*

Proof Let us consider the real analytic case only; the smooth case is very similar. Let us fix $A \in \mathcal{K}^\infty(V)$. Let $\tilde{\phi}_\varepsilon(K) := vol(K + \varepsilon A)$. Define

$$\phi(K) := \frac{i!}{n!} \frac{d^{n-i}}{d\varepsilon^{n-i}}\Big|_{\varepsilon=0} \tilde{\phi}_\varepsilon(K) = V(K[i], A[n - i]).$$

Clearly $\phi \in W_i^\infty$. Now let $\psi = \omega(\bullet - B)$, where $B \in \mathcal{K}^\infty(V)$ such that $0 \in B$, and ω is a smooth compactly supported measure on V such that $supp(\omega) + B \subset Z_1$. Thus $supp(\psi) \subset Z_1$. By Lemma 2.3 for small $\varepsilon > 0$

$$\psi \cdot \tilde{\phi}_\varepsilon \in V_{Z_1}^\infty(V) \cap scl^*(\mathcal{F}_{Z_2}^{an}(V)).$$

Hence also $\psi \cdot \phi \in V_{Z_1}^\infty(V) \cap scl^*(\mathcal{F}_{Z_2}^{an}(V))$. But since $\phi \in W_i^\infty$ and $V^\infty(V) \cdot W_i^\infty \subset W_i^\infty$, we deduce that

$$\phi \cdot \psi \in W_{i,Z_1}^\infty \cap scl^*(\mathcal{F}_{Z_2}^{an}(V)). \tag{14}$$

---

[2]This theorem says that if $K, L \in \mathcal{K}(V)$ cannot be separated by a hyperplane (i.e. there is no hyperplane such that K and L are contained in different closed subspaces defined by the hyperplane) and if convex compact sets $K_i \to K$ and $L_i \to L$ in the Hausdorff metric as $i \to \infty$, then $K_i \cap L_i \to K \cap L$.

[3]This canonical isomorphism was proved in Lemma 5.1.3(1) of [5]. $C_{Z_1}^\infty(V, Val_i^\infty(V))$ denotes the space of smooth functions on V with support in Z_1 and with valued in the Fréchet space $Val_i^\infty(V)$.

Next let us compute the image of this valuation in $C_{Z_1}^\infty(V, Val_i^\infty(V))$. We have

$$(\phi \cdot \psi)(K) = \frac{i!}{n!} \frac{d^{n-i}}{d\varepsilon^{n-i}}\big|_{\varepsilon=0}(\omega \boxtimes vol)(\Delta(K) + (B \times \varepsilon A)).$$

For any valuation $\xi \in W_i^\infty, K \in \mathcal{K}(V), x \in V$ one has

$$\xi(x + \lambda K) = O(\lambda^i) \text{ as } \lambda \to +0;$$

this was the definition of W_i^∞ in [3], beginning of Sect. 3 (where it was denoted by W_i).

For such a ξ, its image $\bar{\xi}$ in $W_i^\infty/W_{i+1}^\infty = C^\infty(V, Val_i^\infty(V))$ is computed as follows:

$$(\bar{\xi}(x))(K) = \lim_{\lambda \to +0} \frac{1}{\lambda^i}\xi(x + \lambda K).$$

The limit necessarily exists and the map $\bar{\xi}$ takes values in $Val_i^\infty(V)$.

For $\xi = \phi \cdot \psi$ as above we have

$$\overline{\phi \cdot \psi}(x)(K) = \frac{1}{n!} \frac{\partial^n}{\partial\lambda^i \partial\varepsilon^{n-i}}\big|_{\lambda=\varepsilon=0}(\omega \boxtimes vol)(\Delta(\lambda K + x) + (B \times \varepsilon A)) \quad (15)$$

$$= \frac{1}{n!} \frac{\partial^n}{\partial\lambda^i \partial\varepsilon^{n-i}}\big|_{\lambda=\varepsilon=0}(\omega \boxtimes vol)(\Delta(\lambda K) + ((x + B) \times (x + \varepsilon A))) \quad (16)$$

$$= \frac{1}{n!} \frac{\partial^n}{\partial\lambda^i \partial\varepsilon^{n-i}}\big|_{\lambda=\varepsilon=0}((T_{-x})_*\omega \boxtimes vol)(\Delta(\lambda K) + (B \times \varepsilon A)), \quad (17)$$

where $(T_{-x})_*$ denotes the push-forward on measures under the shift by $-x$, namely $[y \mapsto y - x]$. We will need a lemma.

Lemma 2.5 *Let σ be a smooth measure on an n-dimensional vector space V. Let $A, B, K \in \mathcal{K}(V)$. Define the function of $(\lambda, \varepsilon) \in [0, \infty)^2$ by*

$$F_\sigma(\lambda, \varepsilon) := (\sigma \boxtimes vol)(\Delta(\lambda K) + (B \times \varepsilon A)).$$

Then the following holds:

(1) $F_\sigma \in C^\infty([0, \infty)^2)$.
(2) For $0 \le i \le n$ define

$$h_{\sigma,i}(\lambda) := \frac{i!}{n!} \frac{\partial^{n-i}}{\partial\varepsilon^{n-i}}\big|_{\varepsilon=0}F_\sigma(\lambda, \varepsilon).$$

Then $h_{\sigma,i}(\lambda) = O(\lambda^i)$ as $\lambda \to +0$.

(3) $\lim_{\lambda \to +0} \frac{h_{\sigma,i}(\lambda)}{\lambda^i} = \sigma(B) \cdot V(K[i], A[n-i]).$

Let us postpone the proof of Lemma 2.5 and finish the proof of Lemma 2.4. By Lemma 2.5 and (17) we have

$$\overline{\phi \cdot \psi}(x)(K) = ((T_{-x})_* \omega)(B) \cdot V(K[i], A[n-i]).$$

To summarize, we have proven so far the following: any smooth $Val_i^\infty(V)$-valued function on V of the form

$$x \mapsto \omega(x+B) \cdot V(\bullet[i], A[n-i])$$

belongs to the image of $W_{i,Z_1}^\infty \cap scl^*(\mathcal{F}_{Z_2}^{an}(V))$ in $W_{i,Z_1}^\infty / W_{i+1,Z_1}^\infty$, where $A, B \in \mathcal{K}^\infty(V)$ such that $0 \in B$, and ω is a smooth measure on V such that $supp(\omega) + B \subset Z_1$. Now let us show that the closure of such functions in the usual Fréchet topology on $C_{Z_1}^\infty(V, Val_i^\infty(V))$ is equal to the whole space.

Let B be the unit Euclidean ball in V. For any $l \in \mathbb{N}$ the function $\frac{\omega(x+\frac{1}{l}B)}{vol(\frac{1}{l}B)} \cdot V(\bullet[i], A[n-i])$ belongs to the image of $W_i^\infty \cap scl^*(\mathcal{F}_{Z_2}^{an}(V))$. However obviously

$$\frac{\omega(x+\frac{1}{l}B)}{vol(\frac{1}{l}B)} \to \frac{\omega}{vol}(x) \text{ in } C_{Z_1}^\infty(V) \text{ as } l \to \infty.$$

This implies that for any smooth function $h: V \to \mathbb{C}$, any subanalytic $A \in \mathcal{K}^\infty(V)$ with $supp(h) \subset int(Z_1)$ the function

$$[x \mapsto h(x) \cdot V(\bullet[i], A[n-i])] \tag{18}$$

belongs to the closure of the image of $W_{i,Z_1}^\infty \cap scl^*(\mathcal{F}_{Z_2}^{an}(V))$. Since i-homogeneous mixed volumes are dense in $Val_i^\infty(V)$ by Alesker [2] we deduce that for any $h \in C_{Z_1}^\infty(V)$ and any $\mu \in Val_i^\infty(V)$ the $Val_i^\infty(V)$-valued function $h \otimes \mu$ lies in the closure of the image of $W_{i,Z_1}^\infty \cap scl^*(\mathcal{F}_{Z_2}^{an}(V))$ (here it is the only place where we have used that the boundary of Z_1 is smooth). But linear combinations of such elements are dense in $C_{Z_1}^\infty(V, Val_i^\infty(V))$. Q.E.D.

Proof of Lemma 2.5

(1) This was proved in [1] in a more general form.
(2) We have

$$F_\sigma(\lambda, \varepsilon) = \int_{y \in \lambda K + \varepsilon A} \sigma([\lambda K \cap (y - \varepsilon A)] + B) dvol(y). \tag{19}$$

Obviously there exists a constant C such that for any $\lambda, \varepsilon \in [0,1]$

$$|\sigma([\lambda K \cap (y - \varepsilon A)] + B)| \leq C.$$

Hence for $\lambda, \varepsilon \in [0, 1]$ one has

$$|F_\sigma(\lambda, \varepsilon)| \leq Cvol(\lambda K + \varepsilon A) = \sum_{j=0}^{n} C_j \varepsilon^j \lambda^{n-j},$$

where C_j are some constants. This implies that the Taylor expansion of F_σ at $(0, 0)$ does not contain monomials $\varepsilon^a \lambda^b$ with $a + b < n$. This implies part (2) of the lemma.

(3) It was shown (in a more general form) in [1] that if a sequence $\{\sigma_N\} \subset C^\infty$ converges to σ in C^∞ (i.e. uniformly on compact subsets of V with all derivatives) then

$$F_{\sigma_N} \to F_\sigma \text{ in } C^\infty([0, \infty)^2) \text{ as } N \to \infty.$$

Hence to prove part (3) of the lemma it suffices to assume that σ has a polynomial density on V. We may and will assume that

$$\sigma = P \cdot dvol,$$

where P is a *homogeneous* polynomial of certain degree d. Define the function

$$\Phi(\lambda, \varepsilon, \delta) := (\sigma \boxtimes vol)(\Delta(\lambda K) + (\delta B \times \varepsilon A)), \ \lambda, \varepsilon \geq 0.$$

By [13] (see also [1]) this function Φ is a polynomial in $\lambda, \varepsilon, \delta \geq 0$. Obviously it is homogeneous of degree $d + 2n$. Let us write it

$$\Phi(\lambda, \varepsilon, \delta) = \sum_{p,q,r} \Phi_{pqr} \lambda^p \varepsilon^q \delta^r,$$

where p, q, r must satisfy

$$p + q + r = d + 2n. \tag{20}$$

Furthermore $F_\sigma(\lambda, \varepsilon) = \Phi(\lambda, \varepsilon, 1)$ is a polynomial, hence let us write it

$$F_\sigma(\lambda, \varepsilon) = \sum_{p,q} F_{pq} \lambda^p \varepsilon^q.$$

For the quantity we have to compute we clearly have

$$\lim_{\lambda \to +0} \frac{h_{\sigma,i}(\lambda)}{\lambda^i} = \binom{n}{i}^{-1} F_{i,n-i}. \tag{21}$$

The identity $F_\sigma(\lambda, \varepsilon) = \Phi(\lambda, \varepsilon, 1)$ immediately implies

$$F_{i,n-i} = \Phi_{i,n-i,d+n}. \tag{22}$$

To compute the last expression, let us write

$$\Phi(\lambda, \varepsilon, \delta) = \int_{y \in \lambda K + \varepsilon A} \sigma([\lambda K \cap (y - \varepsilon A)] + \delta B) dvol(y)$$

$$= \int_{y \in \lambda K + \varepsilon A} \left(\delta^{d+n} \sigma(B) + (\text{ lower degree terms in } \delta) \right) dvol(y)$$

$$= \sigma(B) \cdot vol(\lambda K + \varepsilon A) \cdot \delta^{d+n} + (\text{ lower degree terms in } \delta).$$

This immediately implies that

$$\Phi_{i,n-i,d+n} = \binom{n}{i} \sigma(B) V(K[i], A[n-i]). \tag{23}$$

Lemma follows from (21)–(23). Q.E.D.

Lemma 2.6 *Let X be a smooth manifold and $Z \subset X$ be a compact subset. Let $Z' \subset X$ be a compact neighborhood of Z. Then for any element $\psi \in V_Z^{-\infty}(X)$ there exists a sequence of elements from $V_{Z'}^\infty(X)$ converging to ψ in the topology of $V_c^{-\infty}(X)$ (or equivalently in the weak topology on $V^{-\infty}(X)$).*

Proof

Step 1. Let us prove the statement for $X = \mathbb{R}^n$. For this let us choose a sequence $\{\mu_i\}$ of smooth non-negative compactly supported measures on the Lie group $Aff(\mathbb{R}^n)$ of affine transformations of $\mathbb{R}^n$ such that $\int_{Aff(\mathbb{R}^n)} \mu_i = 1$ and $supp\{\mu_i\} \to \{id\}$ in Hausdorff metric on $Aff(\mathbb{R}^n)$. Define

$$\psi_i := \int_{g \in Aff(\mathbb{R}^n)} g^*(\psi) \cdot d\mu_i(g).$$

In the proof of Lemma 8.2 in [8] it was shown that $\psi_i \in V^\infty(X)$ for all i and $\psi_i \to \psi$ in $V^{-\infty}(X)$. It is also clear that for $i \gg 1$ one has $supp(\psi_i) \subset Z'$. This implies the lemma for $X = \mathbb{R}^n$.

Step 2. Assume now that X is a general smooth manifold. Let us choose a finite open covering $\{V_\alpha\}_\alpha$ of Z and open subsets U_α such that $V_\alpha \subset U_\alpha \subset Z'$, the closures $\bar{U}_\alpha$ are compact and are contained in the interior of Z', and there exist diffeomorphisms $U_\alpha \tilde{\to} \mathbb{R}^n$.

Let us choose a partition of unity in valuations subordinate to the covering $\{V_\alpha\}_\alpha \cup \{X \backslash Z\}$; we denote it by $\{\xi_\alpha\}_\alpha \cup \{\xi\}$ where $supp(\xi_\alpha) \subset V_\alpha$, $supp(\xi) \subset X \backslash Z$, and $\sum_\alpha \xi_\alpha + \xi = \chi$, where χ is the Euler characteristic. Such partition of unity exists by Alesker [5], Proposition 6.2.1. Since $supp(\psi) \subset Z$ we have

$$\psi = \sum_\alpha \psi \cdot \xi_\alpha.$$

Since $\psi \cdot \xi_\alpha$ has compact support contained in $V_\alpha \subset U_\alpha$ and $U_\alpha \simeq \mathbb{R}^n$, Step 1 implies that there exists a sequence $\{\psi_{\alpha,i}\}_i \subset V_{\bar{V}_\alpha}^\infty(X)$ converging to $\psi \cdot \xi_\alpha$ in the topology of $V^{-\infty}(X)$ as $i \to \infty$. Then the sequence

$$\psi_i := \sum_\alpha \psi_{\alpha,i}$$

satisfies the proposition. Q.E.D.

Proof of Theorem 1.1 We consider only the real analytic case; the smooth case is almost the same. We have to show that $V_{Z_1}^{-\infty}(X) \subset scl^*(\mathcal{F}_{Z_2}^{an}(X))$ for a real analytic manifold X. Let us fix a compact neighborhood Z_1' of Z_1 with infinitely smooth boundary contained in the interior of Z_2. By Lemma 2.6 for every element of $V_{Z_1}^{-\infty}(X)$ there exists a sequence of elements of $V_{Z_1'}^\infty(X)$ converging to this element in the weak topology on $V^{-\infty}(X)$. Hence it suffices to show that

$$V_{Z_1'}^\infty(X) \subset scl^*(\mathcal{F}_{Z_2}^{an}(X)). \tag{24}$$

Notice that $V_{Z_1'}^\infty(X) \cap scl^*(\mathcal{F}_{Z_2}^{an}(X))$ is a closed subspace of $V_{Z_1'}^\infty(X)$ since $V_{Z_1'}^\infty(X)$ is metrizable.

First let us prove (24) for X being a vector space. If this is not true then there exists a unique integer $0 \leq i \leq n$ such that $W_{i+1,Z_1'}^\infty \cap scl^*(\mathcal{F}_{Z_2}^{an}(X)) = W_{i+1,Z_1'}^\infty$ and $W_{i,Z_1'}^\infty \cap scl^*(\mathcal{F}_{Z_2}^{an}(X)) \neq W_{i,Z_1'}^\infty$. In this case the image of $W_{i,Z_1'}^\infty \cap scl^*(\mathcal{F}_{Z_2}^{an}(X))$ in $W_{i,Z_1'}^\infty / W_{i+1,Z_1'}^\infty$ is a closed subspace. However by Lemma 2.4 this image is dense. Hence $W_{i,Z_1'}^\infty \cap scl^*(\mathcal{F}_{Z_2}^{an}(X)) = W_{i,Z_1'}^\infty$ which is a contradiction. This proves (24) when X is a vector space.

Let us prove (24) for a general manifold X. Let us fix a finite open covering $\{U_\alpha\}$ of Z_1' such that the closures $\bar{U}_\alpha \subset int(Z_2)$ and each U_α is real analytically diffeomorphic to $\mathbb{R}^n$. By Proposition 6.2.1 of [5] one can construct a partition of unity in valuations subordinate to this covering, namely there exist valuations $\{\phi_\alpha\}$ such that $supp(\phi_\alpha) \subset U_\alpha$ and in a neighborhood of Z_1' one has $\sum_\alpha \phi_\alpha = \chi$ (here χ is the Euler characteristic). Any $\psi \in V_{Z_1'}^\infty(X)$ can be written $\psi = \sum_\alpha \phi_\alpha \cdot \psi$. Let us choose compact sets $Z_{\alpha,2} \subset U_\alpha$ such that $supp(\phi_\alpha) \subset int(Z_{\alpha,2})$; hence $Z_{\alpha,2} \subset Z_2$. Since $supp(\phi_\alpha \cdot \psi) \subset int(Z_{\alpha,2})$, by what we have shown for a vector space, we have

$$\phi_\alpha \cdot \psi \in scl^*(\mathcal{F}_{Z_{\alpha,2}}^{an}(U_\alpha)).$$

But obviously the extension by zero gives the natural closed imbedding $scl^*(\mathcal{F}_{Z_{\alpha,2}}^{an}(U_\alpha)) \subset scl^*(\mathcal{F}_{Z_2}^{an}(X))$. Hence $\psi \in scl^*(\mathcal{F}_{Z_2}^{an}(X))$. Q.E.D.

Acknowledgements I am very grateful to the anonymous referee for the careful reading of the first version of the paper and for numerous remarks.

Partially supported by ISF grant 1447/12.

References

1. S. Alesker, Integrals of smooth and analytic functions over Minkowski's sums of convex sets. Convex geometric analysis (Berkeley, CA, 1996), pp, 1–15. Math. Sci. Res. Inst. Publ, vol. 34 (Cambridge University Press, Cambridge, 1999)
2. S. Alesker, Description of translation invariant valuations on convex sets with solution of P. McMullen's conjecture. Geom. Funct. Anal. **11**(2), 244–272 (2001)
3. S. Alesker, Theory of valuations on manifolds. I. Linear spaces. Isr. J. Math. **156**, 311–339 (2006)
4. S. Alesker, Theory of valuations on manifolds. II. Adv. Math. **207**(1), 420–454 (2006)
5. S. Alesker, Theory of valuations on manifolds. IV. New properties of the multiplicative structure, in *Geometric Aspects of Functional Analysis*. Lecture Notes in Mathematics, vol. 1910 (Springer, Berlin, 2007), pp. 1–44
6. S. Alesker, Theory of valuations on manifolds: a survey. Geom. Funct. Anal. **17**(4), 1321–1341 (2007)
7. S. Alesker, Valuations on manifolds and integral geometry. Geom. Funct. Anal. **20**(5), 1073–1143 (2010)
8. S. Alesker, A. Bernig, The product on smooth and generalized valuations. Am. J. Math. **134**(2), 507–560 (2012). Also:arXiv:0904.1347
9. S. Banach, Théorie des operations linéaires. Warszawa (1932)
10. C. Brouder, N.V. Dang, F. Hélein, Continuity of the fundamental operations on distributions having a specified wave front set (with a counterexample by Semyon Alesker). Studia Math. **232**(3), 201–226 (2016)
11. J.H.G. Fu, Curvature measures of subanalytic sets. Am. J. Math. **116**(4), 819–880 (1994)
12. M. Kashiwara, P. Schapira, Sheaves on manifolds. With a chapter in French by Christian Houzel. Corrected reprint of the 1990 original. Grundlehren der Mathematischen Wissenschaften [Fundamental Principles of Mathematical Sciences], vol. 292 (Springer, Berlin, 1994)
13. A.G. Khovanski, A.V. Pukhlikov, Finitely additive measures of virtual polyhedra. (Russian) Algebra i Analiz **4**(2), 161–185 (1992). Translation in St. Petersburg Math. J. **4**(2), 337–356 (1993)
14. R. Schneider, *Convex Bodies: The Brunn-Minkowski Theory*. Encyclopedia of Mathematics and Its Applications, vol. 44 (Cambridge University Press, Cambridge, 1993)
15. V.A. Zorich, *Mathematical Analysis. II*. Universitext (Springer, Berlin, 2004). Translated from the 2002 fourth Russian edition by Roger Cooke

Orbit Point of View on Some Results of Asymptotic Theory; Orbit Type and Cotype

Limor Ben-Efraim, Vitali Milman, and Alexander Segal

Abstract We develop an orbit point of view on the notations of type and cotype and extend Kwapien's theorem to this setting. We show that such approach provides an exact equality in the latter theorem. In addition, we discuss several well known theorems and reformulate them using the orbit point of view.

1 Introduction

Let $X = (\mathbb{R}^n, ||\cdot||)$ be an n-dimensional normed space. For a given integer k define by $\alpha(k), \beta(k)$ the smallest possible constants, satisfying

$$\left(E \left\| \sum_{i=1}^{k} \gamma_i x_i \right\|^2 \right)^{1/2} \leq \alpha(k) \left(\sum_{i=1}^{k} ||x_i||^2 \right)^{1/2}$$

and

$$\left(E \left\| \sum_{i=1}^{k} \gamma_i x_i \right\|^2 \right)^{1/2} \geq \beta^{-1}(k) \left(\sum_{i=1}^{k} ||x_i||^2 \right)^{1/2}$$

for any $\{x_i\}_1^k \subset X$ and γ_i independent normalized Gaussian random variables. We say that X has type 2 α where $\alpha = \sup_k \alpha(k)$. Similarly we say that X has cotype 2 constant β where $\beta = \sup_k \beta(k)$. By a result of Tomczak-Jaegermann (see [11]), it is known that $\alpha \leq 2\alpha(n)$ and $\beta \leq 2\beta(n)$. Thus, up to a universal constant we may

L. Ben-Efraim (✉)
Tel Aviv, Israel
e-mail: limor.benefraim@gmail.com

V. Milman
Tel Aviv University, Tel Aviv, Israel

A. Segal
Afeka College of Engineering, Tel Aviv, Israel

© Springer International Publishing AG 2017
B. Klartag, E. Milman (eds.), *Geometric Aspects of Functional Analysis*,
Lecture Notes in Mathematics 2169, DOI 10.1007/978-3-319-45282-1_2

15

always deal with n-tuples in the definition of type and cotype for n-dimensional spaces. Both notions play an important role in the study of Banach spaces and local theory.

Remark 1.1 In this note we consider only Gaussian type and cotype constants, and we do not deal with Rademacher type and cotype (see [6, 7, 11]).

Before we discuss a few examples, recall that given two n dimensional normed spaces X, Y, the Banach Mazur distance between X, Y is

$$d(X, Y) = \sup\{\|T\|\|T^{-1}\| : T : X \to Y \text{ is an isomorphism}\}.$$

Whenever Y is a Euclidean space, we will denote $d(X, Y)$ by d_X. The next theorem, due to Kwapien, provides an upper bound for d_X through type 2 and cotype 2 constants.

Theorem 1.2 (Kwapien [4]) *Let X be a (finite or infinite) Banach space. Then, X is isomorphic to a Hilbert space if and only if it has a finite type 2 and a finite cotype 2 constants. Moreover, in this case we have $d_X \leq \alpha\beta$, where α is the type 2 constant and β is the cotype 2 constant of X.*

It can be shown that the bound in Theorem 1.2 is not optimal. That is, we can find a space X such that $\alpha\beta$ is of order n, which is clearly not optimal since d_X is always bounded by $\sqrt{n}$ (John's Theorem). In this note we present a new point of view on the above result, which provides us an equality instead of an upper bound in Theorem 1.2. To this end, we present the notion of orbits in normed spaces.

Definition 1.3 Let $x = (x_1, \ldots x_k) \subset X$. We say that a k-tuple $y = (y_1, \ldots y_k)$ belongs to the orbit set of x if there exists $U = (u_{ij}) \in O(k)$ such that

$$y_i = \sum_{j=1}^{k} u_{ij}x_j.$$

The set of all such k-tuples will be denoted by $O(x) = \{Ux : U \in O(k)\}$ and called the orbit of x.

Using this notion, we may define the Gaussian type 2 and cotype 2 of an orbit x as the smallest constants $\alpha(x), \beta(x)$ such that

$$\left(E \left\| \sum_{i=1}^{k} \gamma_i y_i \right\|^2 \right)^{1/2} \leq \alpha(x) \left(\sum_{i=1}^{k} \|y_i\|^2 \right)^{1/2}$$

and

$$\left(E \left\| \sum_{i=1}^{k} \gamma_i y_i \right\|^2 \right)^{1/2} \geq \beta^{-1}(x) \left(\sum_{i=1}^{k} \|y_i\|^2 \right)^{1/2}$$

for all $y \in O(x)$. Clearly, $\alpha(x) = \alpha(y)$ and $\beta(x) = \beta(y)$ for all $y \in O(x)$, so the constants are well defined. Denote

$$g(x, \gamma) = E \left\| \sum_{i=1}^{k} \gamma_i x_i \right\|^2 ,$$

where $\gamma = \{\gamma_i\}_1^k$. Due to the rotation invariance of the standard Gaussian measure we have that if $y \in O(x)$ then $g(x, \gamma) = g(y, \gamma')$, where $\gamma' = \{\gamma_i'\}_1^k$ are independent Gaussian variables, which are also independent of γ (see e.g. [9, Chap. 2, p. 13]). Hence,

$$\alpha(x)\beta(x) = \inf \left\{ \left(\frac{\sum_{i=1}^{k} ||y_i||^2}{\sum_{i=1}^{k} ||z_i||^2} \right)^{1/2} : y, z \in O(x) \right\} \tag{1}$$

Using the notion of orbits it is possible to write the exact formula for d_X in Theorem 1.2:

Theorem 1.4 *For any n dimensional normed space X we have*

$$d_X = \sup\{\alpha(x)\beta(x) | x = (x_1, \ldots x_k), k = 1, 2, \ldots\}.$$

Moreover,

$$d_X \le 4 \sup\{\alpha(x)\beta(x) : x = (x_1, \ldots x_n)\}.$$

Of course, the first formula is correct for infinite dimensional spaces as well.

Remark 1.5 The question of the exact formula for d_X was also considered in the Master Thesis of Limor Ben-Efraim, under the supervision of V. Milman (not published).

Remark 1.6 It was noted by Pivovarov (private communication, 2016) that Theorem 1.4 easily implies that $d_X \le 4\sqrt{n}$.

In the spirit of Theorem 1.4, it is possible to reformulate several well known theorems regarding embeddings of l_1^k and l_∞^k in X, such as Alon-Milman's theorem (see [1]) and Elton's theorem (see [2]). However, since those theorems involve Rademacher averaging instead of Gaussian averaging, the results will not be precise, as those averages are not equivalent in the general case.

However, the following two theorems may be reformulated in an exact way:

Theorem 1.7 (Figiel-Lindenstrauss-Milman [3]) *Let X be an n dimensional normed space with the unit ball K. Let $x = (x_1, x_2, \ldots x_n)$ be an orbit with cotype 2 constant $\beta(x)$. If x is the orthogonal basis of the maximal volume ellipsoid of K then X contains a subspace of dimension $k = cn\beta(x)^{-2}$ that is 2-isomorphic to l_2^k, for some universal constant $c > 0$.*

Theorem 1.8 ([6, Theorem 9.7]) *Let X be an n-dimensional normed space and let $x = (x_1, \ldots, x_k) \subset X$ be a k-tuple for some $k \leq n$. If $O(x)$ has a 2-type constant α, then the space $E = \mathrm{span}\{x_i\}_1^k$ contains a space of dimension $m = \lceil c\alpha^2 \rceil$ which is 2-isomorphic to l_2^m, for some absolute constant $c > 0$.*

It may be an interesting question to analyze the Maurey-Pisier lemma for equivalence of Rademacher and Gaussian averages (see [8, Proposition 3.2]) in this context. However, one should consider a general orbit of cotype q which is not done in this note.

2 Proof of the Extended Kwapien Theorem

Proof Before we proceed with the proof of Theorem 1.4, let us recall a few definitions and facts.

Definition 2.1 An operator $u : X \to Y$ factors through a Hilbert space if there is a Hilbert space H and operators $B : X \to H$ and $A : H \to Y$ such that $u = AB$. Denote by $\Gamma_2(X, Y)$ the space of all such operators, equipped with the norm

$$\gamma_2(u) = \inf\{\|A\|\|B\|\}$$

where the infimum is taken over all factorizations of u.

A well known theorem by Lindenstrauss and Pelczynski (see [5], [7, Theorem 2.4], [11, Proposition 13.11]) provides a necessary and sufficient condition when an operator u belongs to $\Gamma_2(X, Y)$:

Theorem 2.2 *$u : X \to Y$ belongs to $\Gamma_2(X, Y)$ if and only if there exists a constant C such that for all n and all $n \times n$ orthogonal matrices (a_{ij}) we have,*

$$\left(\sum_{i=1}^n \left\| \sum_{j=1}^n a_{ij} u x_j \right\|^2 \right)^{1/2} \leq C \left(\sum_{i=1}^n \|x_i\|^2 \right)^{1/2}$$

for all $x_{,1} \ldots x_n \in X$. Moreover, $\gamma_2(u)$ coincides with the smallest possible constant C satisfying the above inequality.

Let $x = (x_j)$ be a k-tuple of elements of X and let $(a_{ij}) \in O(k)$. By the definition of Gaussian orbit cotype of x we have

$$\beta(x)^{-1} \left(\sum_{i=1}^k \left\| \sum_{j=1}^k a_{ij} x_j \right\|^2 \right)^{1/2} \leq \left(E \left\| \sum_{i=1}^k \gamma_i \sum_{j=1}^k a_{ij} x_j \right\|^2 \right)^{1/2}. \tag{2}$$

By the definition of Gaussian orbit type we have

$$g(x, \gamma)^{1/2} \leq \alpha(x) \left(\sum_{i=1}^{k} \|x_i\|^2 \right)^{1/2}. \tag{3}$$

However, since $g(x, \gamma) = g(y, \gamma)$ where

$$y_i = \sum_{j=1}^{k} a_{ij} x_j,$$

we get that

$$\left(\sum_{i=1}^{k} \left\| \sum_{j=1}^{k} a_{ij} x_j \right\|^2 \right)^{1/2} \leq \alpha(x) \beta(x) \left(\sum_{i=1}^{k} \|x_i\|^2 \right)^{1/2}. \tag{4}$$

Thus, the condition of Theorem 2.2 is satisfied with the constant

$$C = \sup_{x} \{ \alpha(x) \beta(x) \}.$$

Clearly, $\beta(x)$ and $\alpha(x)$ are the smallest possible numbers satisfying (2) and (3). Therefore, $\sup_x \{\alpha(x)\beta(x)\}$ is the smallest possible number satisfying (4) for each positive k and each k-frame x. Thus,

$$\gamma_2(Id) = \sup\{\alpha(x)\beta(x)\},$$

However, $\gamma_2(Id) = d_X$ (by definition), so the first part of the proof of Theorem 1.4 is finished. □

Remark 2.3 In the case where $\dim X = \dim Y = n$, one may consider only $n \times n$ orthogonal matrices and the best constant C in Theorem 2.2 is equivalent to $\gamma_2(u)$ up to a factor of 4. This was noted independently by Tomczak-Jaegermann and Pisier (private communication, 2000). Since the result was not published we will provide a different argument which is due to Tomczak-Jaegermann.

To this end, we recall several facts regarding absolutely summing operators (see [7, 11]).

Definition 2.4 Let X and Y be Banach spaces. An operator $u : X \to Y$ is called 2-summing operator if there exists a constant C such that for all finite sequences $\{x_i\} \subset X$:

$$\left(\sum_{i=1}^{k} \|ux_i\|^2 \right)^{1/2} \leq C \sup_{\xi \in X^*, \|\xi\| \leq 1} \left(\sum_{i=1}^{k} |\xi(x_i)|^2 \right)^{1/2}.$$

The smallest possible C satisfying the above is denoted by $\pi_2(u)$ and is called the 2-summing norm of u.

Now we will define a similar concept for an orbit and see how it relates to the definition above. From now on, unless stated otherwise, it is assumed that X is an n-dimensional normed space.

Definition 2.5 Given an operator $u : l_2^k \to X$, denote

$$\pi_2^{(k)}(u) = \sup \left(\sum_{i=1}^{k} \|uf_i\|^2 \right)^{1/2},$$

$$\delta_2^{(k)}(u) = \inf \left(\sum_{i=1}^{k} \|uf_i\|^2 \right)^{1/2},$$

where $\{f_i\}_1^k$ runs over all orthonormal bases of l_2^k.

Given an orbit $x = \{x_1, \ldots x_k\} \subset X$ we will denote $\pi_2^{(k)}(x) = \pi_2^{(k)}(u)$, $\delta_2^{(k)}(x) = \delta^{(k)}(u)$ where u is defined by

$$ue_i = x_i, \qquad 1 \leq i \leq k.$$

Remark 2.6 The standard definition of $\pi_2^{(k)}(u)$ slightly differs from definition above. It is defined as the smallest possible constant satisfying

$$\left(\sum_{i=1}^{k} \|ux_i\|^2 \right)^{1/2} \leq C \sup_{\xi \in X^*, \|\xi\| \leq 1} \left(\sum_{i=1}^{k} |\xi(x_i)|^2 \right)^{1/2},$$

for all $x_1, \ldots x_k \in X$.

By a theorem of Tomczak-Jaegermann [10] we have that for any operator $u : l_2^k \to X$ of rank n:

$$\pi_2^{(n)}(u) \leq \pi_2(u) \leq 2\pi_2^{(n)}(u). \tag{5}$$

Since the proof of (5) constructs an orthonormal basis (e_j) of l_2^n that satisfies

$$\left(\sum_{i=1}^{n} \|ue_i\|^2\right)^{1/2} \geq \frac{1}{2}\pi_2(u),$$

we get that inequality (5) holds for our definition of $\pi_2^{(n)}(u)$ as well.

An easy consequence of the above is the following lemma:

Lemma 2.7 *For each $k \geq n$ and $x = (x_1, \ldots x_k) \subset X$ there exists $y \in O(x)$ and a subset $y' \subset y$ of cardinality n such that*

$$\pi_2^{(k)}(y) \leq 2\pi_2^{(n)}(y').$$

Proof Let $u : l_2^k \to X$ be the operator defined by $ue_i = x_i$, and denote $E = ker(u)^\perp$. Denote by $P : l_2^k \to E$ the orthogonal projection such that $u = u|_E P$. Let $f_1 \ldots f_n \in E$ and $f_{n+1} \ldots f_k \in E^\perp$ be another orthonormal basis of l_2^k and denote by $y_i = uf_i$. Clearly,

$$\pi_2^{(k)}(x) \leq \pi_2(u) = \pi_2(u|_E) \leq 2\pi_2^{(n)}(y')$$

where $y' = (y_1, \ldots y_n)$.   □

Since $\delta_2^{(k)}$ is not necessarily convex, denote by $\hat{\delta}_2^{(k)}$ the largest convex function that is smaller than $\delta_2^{(k)}$. The norms $\pi_2^{(k)}$ and $\hat{\delta}_2^{(k)}$ are dual norms on $L(l_2^k, X)$ and $L(l_2^k, X^*)$. That is

$$\pi_2^{(k)}(u) = \sup\{|trace(uv)| : v^* \in L(l_2^k, X^*), \hat{\delta}_2^{(k)}(v^*) \leq 1\}.$$

The proof of this fact is similar to the proof presented in [11, Proposition 9.9], for the norms π_2 and δ_2.

By a standard duality argument we get the following corollary.

Corollary 2.8 *Let $u : l_2^k \to X$ be an operator, where $k \geq n$. Let $E = (ker\,u)^\perp$ with $\dim E = n$, and let P be the orthogonal projection $P : l_2^k \to E$. Define $\tilde{u} : E \to X$ such that $u = \tilde{u}P$. Then we have*

$$\hat{\delta}_2^{(n)}(\tilde{u}) \leq 2\hat{\delta}_2^{(k)}(u).$$

Now, we may prove the key lemma required for our goal.

Lemma 2.9 *If C satisfies*

$$\forall x = (x_1, \ldots x_n) \subset X, \quad \pi_2^{(n)}(x) \leq Cv_2^{(n)}(x), \tag{6}$$

then, for all $k > n$,

$$\forall x = (x_1, \ldots x_k) \subset X, \quad \pi_2^{(k)}(x) \le 4C v_2^{(k)}(x), \tag{7}$$

Proof Denote by X^m the space of all m-tuples of X. Take $x \in X^n$ and consider $u : l_2^m \to X$ an operator defined by $ue_i = x_i$. Clearly, by (6) and the convexity of $\hat{\delta}_2^n$ and $\pi_2^{(n)}$

$$\pi_2^{(n)}(x) \le C \hat{\delta}_2^{(n)}(u).$$

Given $k \ge n$ take $x = (x_1, \ldots x_k)$, $y \in O(x)$ and define operator u as above. As before, denote $E = (\ker u)^\perp$ and by $P : l_2^k \to E$ the orthogonal projection. Define $\tilde{u} : E \to X$ such that $u = \tilde{u}P$. Let $f_1 \ldots f_n \in E$ and $f_{n+1} \ldots f_k \in E^\perp$ be some orthonormal basis of l_2^k. Denote $y_i = uf_i$ and $y = (y_1, \ldots y_k)$, $y' = (y_1, \ldots y_n)$. Then,

$$\pi_2^{(k)}(x) = \pi_2^{(k)}(y) \le 2\pi_2^{(n)}(y')$$

and

$$\hat{\delta}_2^{(n)}(\tilde{u}) \le 2\delta_2^{(k)}(u) = v_2^{(k)}(y) = 2v_2^{(k)}(x).$$

Thus,

$$\pi_2^{(k)}(x) \le 2\pi_2^{(n)}(y') \le 2C\hat{\delta}_2^{(n)}(\tilde{u}) \le 4C v_2^{(k)}(x).$$

$\square$

Now we may finish the second part of main theorem. Let $x = (x_1, \ldots x_k)$. Notice that by (1), for each k

$$\alpha(x)\beta(x) = \frac{\pi_2^{(k)}(x)}{v_2^{(k)}(x)}.$$

Applying Lemma 2.9 we get

$$\sup_{x \in X^k} \alpha(x)\beta(x) = \sup_{x \in X^k} \frac{\pi_2^{(k)}(x)}{v_2^{(k)}(x)} \le 4 \sup_{x \in X^n} \frac{\pi_2^{(n)}(x)}{v_2^{(n)}(x)}$$

$$= 4 \sup_{x \in X^n} \{\alpha(x)\beta(x)\}$$

and the proof is complete.

Acknowledgements The authors would like to thank Nicole Tomczak-Jaegermann for very useful discussions.

References

1. N. Alon, V.D. Milman, Embedding of l_∞^k in finite-dimensional Banach spaces. Isr. J. Math. **45**(4), 265–280 (1983)
2. J. Elton, Sign-embeddings of l_1. Trans. Am. Math. Soc. **279**(1), 113–124 (1983)
3. T. Figiel, J. Lindenstrauss, V.D. Milman, The dimension of almost spherical sections of convex bodies. Bull. Am. Math. Soc. **82**(4), 575–578 (1976)
4. S. Kwapien, Isomorphic characterizations of inner product spaces by orthogonal series with vector coefficients. Stud. Math. **44**, 583–595 (1972)
5. J. Lindenstrauss, A. Pelczynski, Absolutely summing operators in L_p spaces and their applications. Stud. Math. **29**, 275–326 (1968)
6. V.D. Milman, G. Schechtman, *Asymptotic Theory of Finite Dimensional Normed Spaces* (Springer, Berlin, 1986)
7. G. Pisier, *Factorization of Linear Operators and Geometry of Banach Spaces*. Conference Board of Mathematical Sciences Number, vol. 60 (American Mathematical Society, 1986)
8. G. Pisier, *Probabilistic methods in the geometry of Banach spaces*, in *Probability and Analysis* (Varenna, 1985). Lecture Notes in Mathematics, vol. 1206 (Springer, Berlin, 1986), pp. 167–241
9. G. Pisier, *The Volume of Convex Bodies and Banach Space Geometry*. Cambridge Tracts in Mathematics, vol. 94 (Cambridge University Press, Cambridge, 1999)
10. N. Tomczak-Jaegermann, Computing 2-summing norm with few vectors. Ark. Mat. **17**, 273–277 (1979)
11. N. Tomczak-Jaegermann, *Banach-Mazur Distances and Finite-Dimensional Operator Ideals* (Longman Scientific and Technical, Essex, 1989)

Concentration Properties of Restricted Measures with Applications to Non-Lipschitz Functions

Sergey G. Bobkov, Piotr Nayar, and Prasad Tetali

Abstract We show that, for any metric probability space (M, d, μ) with a subgaussian constant $\sigma^2(\mu)$ and any Borel measurable set $A \subset M$, we have $\sigma^2(\mu_A) \le c \log (e/\mu(A)) \, \sigma^2(\mu)$, where μ_A is a normalized restriction of μ to the set A and c is a universal constant. As a consequence, we deduce concentration inequalities for non-Lipschitz functions.

2010 Mathematics Subject Classification. Primary 60Gxx

1 Introduction

It is known that many high-dimensional probability distributions μ on the Euclidean space $\mathbb{R}^n$ (and other metric spaces, including graphs) possess strong concentration properties. In a functional language, this may informally be stated as the assertion that any sufficiently smooth function f on $\mathbb{R}^n$, e.g., having a bounded Lipschitz semi-norm, is almost a constant on almost all space. There are several ways to quantify such a property. One natural approach proposed by Alon et al. [2] associates with a given metric probability space (M, d, μ) its *spread constant*,

$$ s^2(\mu) = \sup \operatorname{Var}_\mu(f) = \sup \int (f - m)^2 \, d\mu, $$

S.G. Bobkov
School of Mathematics, University of Minnesota, Minneapolis, MN 55455, USA

P. Nayar (✉)
Institute of Mathematics and its Applications, Minneapolis, MN 55455, USA
e-mail: nayar@mimuw.edu.pl

P. Tetali
School of Mathematics and School of Computer Science, Georgia Institute of Technology, Atlanta, GA 30332, USA

© Springer International Publishing AG 2017
B. Klartag, E. Milman (eds.), *Geometric Aspects of Functional Analysis*,
Lecture Notes in Mathematics 2169, DOI 10.1007/978-3-319-45282-1_3

where $m = \int f \, d\mu$, and the sup is taken over all functions f on M with $\|f\|_{\mathrm{Lip}} \leq 1$. More information is contained in the so-called subgaussian constant $\sigma^2 = \sigma^2(\mu)$ which is defined as the infimum over all σ^2 such that

$$\int e^{tf} \, d\mu \leq e^{\sigma^2 t^2/2}, \qquad \text{for all } t \in \mathbb{R}, \tag{1}$$

in the class $\mathcal{L}_0$ of all f on M with $m = 0$ and $\|f\|_{\mathrm{Lip}} \leq 1$ (cf. [8]). Describing the diameter of $\mathcal{L}_0$ in the Orlicz space $L^{\psi_2}(\mu)$ for the Young function $\psi_2(t) = e^{t^2} - 1$ (within universal factors), the quantity $\sigma^2(\mu)$ appears as a parameter in a subgaussian concentration inequality for the class of all Borel subsets of M. As an equivalent approach, it may also be introduced via the transport-entropy inequality connecting the classical Kantorovich distance and the relative entropy from an arbitrary probability measure on M to the measure μ (cf. [7]).

While in general $s^2 \leq \sigma^2$, the latter characteristic allows one to control subgaussian tails under the probability measure μ uniformly in the entire class of Lipschitz functions on M. More generally, when $\|f\|_{\mathrm{Lip}} \leq L$, (1) yields

$$\mu\{|f - m| \geq t\} \leq 2e^{-t^2/(2\sigma^2 L^2)}, \qquad t > 0. \tag{2}$$

Classical and well-known examples include the standard Gaussian measure on $M = \mathbb{R}^n$ in which case $s^2 = \sigma^2 = 1$, and the normalized Lebesgue measure on the unit sphere $M = S^{n-1}$ with $s^2 = \sigma^2 = \frac{1}{n-1}$. The last example was a starting point in the study of the concentration of measure phenomena, a fruitful direction initiated in the early 1970s by V.D. Milman.

Other examples come often after verification that μ satisfies certain Sobolev-type inequalities such as Poincaré-type inequalities

$$\lambda_1 \mathrm{Var}_\mu(u) \leq \int |\nabla u|^2 \, d\mu,$$

and logarithmic Sobolev inequalities

$$\rho \, \mathrm{Ent}_\mu(u^2) = \rho \left[\int u^2 \log u^2 \, d\mu - \int u^2 \, d\mu \, \log \int u^2 \, d\mu \right] \leq 2 \int |\nabla u|^2 \, d\mu,$$

where u may be any locally Lipschitz function on M, and the constants $\lambda_1 > 0$ and $\rho > 0$ do not depend on u. Here the modulus of the gradient may be understood in the generalized sense as the function

$$|\nabla u(x)| = \limsup_{y \to x} \frac{|u(x) - u(y)|}{d(x, y)}, \qquad x \in M$$

(this is the so-called "continuous setting"), while in the discrete spaces, e.g., graphs, we deal with other naturally defined gradients. In both cases, one has respectively the well-known upper bounds

$$s^2(\mu) \leq \frac{1}{\lambda_1}, \qquad \sigma^2(\mu) \leq \frac{1}{\rho}. \tag{3}$$

For example, $\lambda_1 = \rho = n - 1$ on the unit sphere (best possible values, [17]), which can be used to make a corresponding statement about the spread and Gaussian constants.

One of the purposes of this note is to give new examples by involving the family of the normalized restricted measures

$$\mu_A(B) = \frac{\mu(A \cap B)}{\mu(A)}, \qquad B \subset M \text{ (Borel)},$$

where a Borel measurable set $A \subset M$ is fixed and has a positive measure. As an example, returning to the standard Gaussian measure μ on $\mathbb{R}^n$, it is known that $\sigma^2(\mu_A) \leq 1$ for any convex body $A \subset \mathbb{R}^n$. This remarkable property, discovered by Bakry and Ledoux [3] in a sharper form of a Gaussian-type isoperimetric inequality, has nowadays several proofs and generalizations, cf. [5, 6]. Of course, in general, the set A may have a rather disordered structure, for instance, to be disconnected. And then there is no hope for validity of a Poincaré-type inequality for the measure μ_A. Nevertheless, it turns out that the concentration property of μ_A is inherited from μ, unless the measure of A is too small. In particular, we have the following observation about abstract metric probability spaces.

Theorem 1.1 *For any measurable set $A \subset M$ with $\mu(A) > 0$, the subgaussian constant $\sigma^2(\mu_A)$ of the normalized restricted measure satisfies*

$$\sigma^2(\mu_A) \leq c \log\left(\frac{e}{\mu(A)}\right) \sigma^2(\mu), \tag{4}$$

where c is an absolute constant.

Although this assertion is technically simple, we will describe two approaches: one is direct and refers to estimates on the ψ_2-norms over the restricted measures, and the other one uses a general comparison result due to Barthe and Milman on the concentration functions [4].

One may further generalize Theorem 1.1 by defining the subgaussian constant $\sigma_{\mathcal{F}}^2(\mu)$ within a given fixed subclass $\mathcal{F}$ of functions on M, by using the same bound (1) on the Laplace transform. This is motivated by a possible different level of concentration for different classes; indeed, in case of $M = \mathbb{R}^n$, the concentration property may considerably be strengthened for the class $\mathcal{F}$ of all convex Lipschitz functions. In particular, one result of Talagrand [18, 19] provides a dimension-free bound $\sigma_{\mathcal{F}}^2(\mu) \leq C$ for an arbitrary product probability measure μ on the n-dimensional cube $[-1, 1]^n$. Hence, a more general version of Theorem 1.1 yields the bound

$$\sigma_{\mathcal{F}}^2(\mu_A) \leq c \log\left(\frac{e}{\mu(A)}\right)$$

with some absolute constant c, which holds for any Borel subset A of $[-1, 1]^n$ (cf. Sect. 6 below).

According to the very definition, the quantities $\sigma^2(\mu)$ and $\sigma^2(\mu_A)$ might seem to be responsible for deviations of only Lipschitz functions f on M and A, respectively. However, the inequality (4) may also be used to control deviations of non-Lipschitz f — on large parts of the space and under certain regularity hypotheses. Assume, for example, $\int |\nabla f| \, d\mu \leq 1$ (which is kind of a normalization condition) and consider

$$A = \{x \in M : |\nabla f(x)| \leq L\}. \tag{5}$$

If $L \geq 2$, this set has the measure $\mu(A) \geq 1 - \frac{1}{L} \geq \frac{1}{2}$, and hence, $\sigma^2(\mu_A) \leq c\sigma^2(\mu)$ with some absolute constant c. If we assume that f has a Lipschitz semi-norm $\leq L$ on A, then, according to (2),

$$\mu_A\{x \in A : |f - m| \geq t\} \leq 2e^{-t^2/(c\sigma^2(\mu)L^2)}, \qquad t > 0, \tag{6}$$

where m is the mean of f with respect to μ_A. It is in this sense one may say that f is almost a constant on the set A.

This also yields a corresponding deviation bound on the whole space,

$$\mu\{x \in M : |f - m| \geq t\} \leq 2e^{-t^2/c\sigma^2(\mu)L^2} + \frac{1}{L}.$$

Stronger integrability conditions posed on $|\nabla f|$ can considerably sharpen the conclusion. By a similar argument, Theorem 1.1 yields, for example, the following exponential bound, known in the presence of a logarithmic Sobolev inequality for the space (M, d, μ), and with σ^2 replaced by $1/\rho$ (cf. [7]).

Corollary 1.2 *Let f be a locally Lipschitz function on M with Lipschitz semi-norms $\leq L$ on the sets (5). If $\int e^{|\nabla f|^2} \, d\mu \leq 2$, then f is μ-integrable, and moreover,*

$$\mu\{x \in M : |f - m| \geq t\} \leq 2e^{-t/c\sigma(\mu)}, \qquad t > 0,$$

where m is the μ-mean of f and c is an absolute constant.

Equivalently (up to an absolute factor), we have a Sobolev-type inequality

$$\|f - m\|_{\psi_1} \leq c\sigma(\mu) \|\nabla f\|_{\psi_2},$$

connecting the ψ_1-norm of $f - m$ with the ψ_2-norm of the modulus of the gradient of f. We prove a more general version of this corollary in Sect. 6 (cf. Theorem 6.1).

As will be explained in the same section, similar assertions may also be made about convex f and product measures μ on $M = [-1, 1]^n$, thus extending Talagrand's theorem to the class of non-Lipschitz functions.

In view of the right bound in (3) and (4), the spread and subgaussian constants for restricted measures can be controlled in terms of the logarithmic Sobolev constant ρ via

$$s^2(\mu_A) \leq \sigma^2(\mu_A) \leq c \log\left(\frac{e}{\mu(A)}\right)\frac{1}{\rho}.$$

However, it may happen that $\rho = 0$ and $\sigma^2(\mu) = \infty$, while $\lambda_1 > 0$ (e.g., for the product exponential distribution on $\mathbb{R}^n$). Then one may wonder whether one can estimate the spread constant of a restricted measure in terms of the spectral gap. In that case there is a bound similar to (4).

Theorem 1.3 *Assume the metric probability space (M, d, μ) satisfies a Poincaré-type inequality with $\lambda_1 > 0$. For any $A \subset M$ with $\mu(A) > 0$, with some absolute constant c*

$$s^2(\mu_A) \leq c \log^2\left(\frac{e}{\mu(A)}\right)\frac{1}{\lambda_1}. \tag{7}$$

It should be mentioned that the logarithmic terms in (4) and (7) may not be removed and are actually asymptotically optimal as functions of $\mu(A)$, as $\mu(A)$ is getting small, see Sect. 7.

Our contribution below is organized into sections as follows:

2. Bounds on ψ_α-Norms for Restricted Measures.
3. Proof of Theorem 1.1. Transport-Entropy Formulation.
4. Proof of Theorem 1.3. Spectral Gap.
5. Examples.
6. Deviations for Non-Lipschitz Functions.
7. Optimality.
8. Appendix.

2 Bounds on ψ_α-Norms for Restricted Measures

A measurable function f on the probability space (M, μ) is said to have a finite ψ_α-norm, $\alpha \geq 1$, if for some $r > 0$,

$$\int e^{(|f|/r)^\alpha} \, d\mu \leq 2.$$

The infimum over all such r represents the ψ_α-norm $\|f\|_{\psi_\alpha}$ or $\|f\|_{L^{\psi_\alpha}(\mu)}$, which is just the Orlicz norm associated with the Young function $\psi_\alpha(t) = e^{|t|^\alpha} - 1$.

We are mostly interested in the particular cases $\alpha = 1$ and $\alpha = 2$. In this section we recall well-known relations between the ψ_1 and ψ_2-norms and the usual L^p-norms $\|f\|_p = \|f\|_{L^p(\mu)} = (\int |f|^p \, d\mu)^{1/p}$. For the readers' convenience, we include the proof in the Appendix.

Lemma 2.1 *We have*

$$\sup_{p \geq 1} \frac{\|f\|_p}{\sqrt{p}} \leq \|f\|_{L^{\psi_2}(\mu)} \leq 4 \sup_{p \geq 1} \frac{\|f\|_p}{\sqrt{p}}, \tag{8}$$

$$\sup_{p \geq 1} \frac{\|f\|_p}{p} \leq \|f\|_{L^{\psi_1}(\mu)} \leq 6 \sup_{p \geq 1} \frac{\|f\|_p}{p}. \tag{9}$$

Given a measurable subset A of M with $\mu(A) > 0$, we consider the normalized restricted measure μ_A on M, i.e.,

$$\mu_A(B) = \frac{\mu(A \cap B)}{\mu(A)}, \qquad B \subset M.$$

Our basic tool leading to Theorem 1.1 will be the following assertion.

Proposition 2.2 *For any measurable function f on M,*

$$\|f\|_{L^{\psi_2}(\mu_A)} \leq 4e \, \log^{1/2}\left(\frac{e}{\mu(A)}\right) \|f\|_{L^{\psi_2}(\mu)}. \tag{10}$$

Proof Assume that $\|f\|_{L^{\psi_2}(\mu)} = 1$ and fix $p \geq 1$. By the left inequality in (8), for any $q \geq 1$,

$$q^{q/2} \geq \int |f|^q \, d\mu \geq \mu(A) \int |f|^q \, d\mu_A,$$

so

$$\frac{\|f\|_{L^q(\mu_A)}}{\sqrt{q}} \leq \left(\frac{1}{\mu(A)}\right)^{1/q}.$$

But by the right inequality in (8),

$$\|f\|_{\psi_2} \leq 4 \sup_{q \geq 1} \frac{\|f\|_q}{\sqrt{q}} \leq 4\sqrt{p} \sup_{q \geq p} \frac{\|f\|_q}{\sqrt{q}}.$$

Applying it on the space (M, μ_A), we then get

$$\|f\|_{L^{\psi_2}(\mu_A)} \le 4\sqrt{p} \sup_{q \ge p} \frac{\|f\|_{L^q(\mu_A)}}{\sqrt{q}}$$

$$\le 4\sqrt{p} \sup_{q \ge p} \left(\frac{1}{\mu(A)}\right)^{1/q} = 4\sqrt{p} \left(\frac{1}{\mu(A)}\right)^{1/p}.$$

The obtained inequality,

$$\|f\|_{L^{\psi_2}(\mu_A)} \le 4\sqrt{p} \left(\frac{1}{\mu(A)}\right)^{1/p},$$

holds true for any $p \ge 1$ and therefore may be optimized over p. Choosing $p = \log \frac{e}{\mu(A)}$, we arrive at (10). $\qquad\square$

A possible weak point in the bound (10) is that the means of f are not involved. For example, in applications, if f were defined only on A and had μ_A-mean zero, we might need to find an extension of f to the whole space M keeping the mean zero with respect to μ. In fact, this should not create any difficulty, since one may work with the symmetrization of f.

More precisely, we may apply Proposition 2.2 on the product space $(M \times M, \mu \otimes \mu)$ to the product sets $A \times A$ and functions of the form $f(x) - f(y)$. Then we get

$$\|f(x) - f(y)\|_{L^{\psi_2}(\mu_A \otimes \mu_A)} \le 4e \log^{1/2}\left(\frac{e}{\mu(A)^2}\right) \|f(x) - f(y)\|_{L^{\psi_2}(\mu \otimes \mu)}.$$

Since $\log\left(\frac{e}{\mu(A)^2}\right) \le 2\log\left(\frac{e}{\mu(A)}\right)$, we arrive at:

Corollary 2.3 *For any measurable function f on M,*

$$\|f(x) - f(y)\|_{L^{\psi_2}(\mu_A \otimes \mu_A)} \le 4e\sqrt{2} \log^{1/2}\left(\frac{e}{\mu(A)}\right) \|f(x) - f(y)\|_{L^{\psi_2}(\mu \otimes \mu)}.$$

Let us now derive an analog of Proposition 2.2 for the ψ_1-norm, using similar arguments. Assume that $\|f\|_{L^{\psi_1}(\mu)} = 1$ and fix $p \ge 1$. By the left inequality in (9), for any $q \ge 1$,

$$q^q \ge \int |f|^q \, d\mu \ge \mu(A) \int |f|^q \, d\mu_A,$$

so

$$\frac{\|f\|_{L^q(\mu_A)}}{q} \le \left(\frac{1}{\mu(A)}\right)^{1/q}.$$

But, by the inequality (9),

$$\|f\|_{L^{\psi_1}} \leq 6 \sup_{q \geq 1} \frac{\|f\|_q}{q} \leq 6p \sup_{q \geq p} \frac{\|f\|_q}{q}.$$

Applying it on the space (M, μ_A), we get

$$\|f\|_{L^{\psi_1}(\mu_A)} \leq 6p \sup_{q \geq p} \frac{\|f\|_{L^q(\mu_A)}}{q}$$

$$\leq 6p \sup_{q \geq p} \left(\frac{1}{\mu(A)}\right)^{1/q} = 6p \left(\frac{1}{\mu(A)}\right)^{1/p}.$$

The obtained inequality,

$$\|f\|_{L^{\psi_1}(\mu_A)} \leq 6p \left(\frac{1}{\mu(A)}\right)^{1/p},$$

holds true for any $p \geq 1$ and therefore may be optimized over p. Choosing $p = \log \frac{e}{\mu(A)}$, we arrive at:

Proposition 2.4 *For any measurable function f on M, we have*

$$\|f\|_{L^{\psi_1}(\mu_A)} \leq 6e \log \left(\frac{e}{\mu(A)}\right) \|f\|_{L^{\psi_1}(\mu)}.$$

Similarly to Corollary 2.3 one may write down this relation on the product probability space $(M \times M, \mu \otimes \mu)$ with the functions of the form $\tilde{f}(x, y) = f(x) - f(y)$ and the product sets $\tilde{A} = A \times A$. Then we get

$$\|f(x) - f(y)\|_{L^{\psi_1}(\mu_A \otimes \mu_A)} \leq 12 e \log \left(\frac{e}{\mu(A)}\right) \|f(x) - f(y)\|_{L^{\psi_1}(\mu \otimes \mu)}. \tag{11}$$

3 Proof of Theorem 1.1: Transport-Entropy Formulation

The finiteness of the subgaussian constant for a given metric probability space (M, d, μ) means that ψ_2-norms of Lipschitz functions on M with mean zero are uniformly bounded. Equivalently, for any (for all) $x_0 \in M$, we have that, for some $\lambda > 0$,

$$\int e^{d(x,x_0)^2/\lambda^2} d\mu(x) < \infty.$$

The definition (1) of $\sigma^2(\mu)$ inspires to consider another norm-like quantity

$$\sigma_f^2 = \sup_{t \neq 0}\left[\frac{1}{t^2/2}\log\int e^{tf}\,d\mu\right].$$

Here is a well-known relation (with explicit numerical constants) which holds in the setting of an abstract probability space (M, μ). Once again, we include a proof in the Appendix for completeness.

Lemma 3.1 *If f has mean zero and finite ψ_2-norm, then*

$$\frac{1}{\sqrt{6}}\|f\|_{\psi_2}^2 \leq \sigma_f^2 \leq 4\,\|f\|_{\psi_2}^2.$$

One can now relate the subgaussian constant of the restricted measure to the subgaussian constant of the original measure. Let now (M, d, μ) be a metric probability space. First, Lemma 3.1 immediately yields an equivalent description in terms of ψ_2-norms, namely

$$\frac{1}{\sqrt{6}}\sup_f \|f\|_{\psi_2}^2 \leq \sigma^2(\mu) \leq 4\sup_f \|f\|_{\psi_2}^2, \tag{12}$$

where the supremum is running over all $f : M \to \mathbb{R}$ with μ-mean zero and $\|f\|_{\mathrm{Lip}} \leq 1$. Here, one can get rid of the mean zero assumption by considering functions of the form $f(x) - f(y)$ on the product space $(M \times M, \mu \otimes \mu, d_1)$, where d_1 is the l_1-type metric given by $d_1((x_1, y_1), (x_2, y_2)) = d(x_1, x_2) + d(y_1, y_2)$. If f has mean zero, then, by Jensen's inequality,

$$\iint e^{(f(x) - f(y))^2/r^2}\,d\mu(x)\,d\mu(y) \geq \int e^{f(x)^2/r^2}\,d\mu(x),$$

which implies that

$$\|f(x) - f(y)\|_{L^{\psi_2}(\mu \otimes \mu)} \geq \|f\|_{L^{\psi_2}(\mu)}.$$

On the other hand, by the triangle inequality,

$$\|f(x) - f(y)\|_{L^{\psi_2}(\mu \otimes \mu)} \leq 2\,\|f\|_{L^{\psi_2}(\mu)}.$$

Hence, we arrive at another, more flexible relation, where the mean zero assumption may be removed.

Lemma 3.2 *We have*

$$\frac{1}{4\sqrt{6}} \sup_f \|f(x) - f(y)\|^2_{L^{\psi_2}(\mu \otimes \mu)} \le \sigma^2(\mu) \le 4 \sup_f \|f(x) - f(y)\|^2_{L^{\psi_2}(\mu \otimes \mu)},$$

where the supremum is running over all functions f on M with $\|f\|_{\mathrm{Lip}} \le 1$.

Proof of Theorem 1.1 We are prepared to make last steps for the proof of the inequality (4). We use the well-known Kirszbraun's theorem: Any function $f : A \to \mathbb{R}$ with Lipschitz semi-norm $\|f\|_{\mathrm{Lip}} \le 1$ on A admits a Lipschitz extension to the whole space [10, 14]. Namely, one may put

$$\tilde{f}(x) = \inf_{a \in A} \big[f(a) + d(a, x) \big], \quad x \in M.$$

Applying first Corollary 2.3 and then the left inequality of Lemma 3.2 to $\tilde{f}$, we get

$$\begin{aligned}
\|f(x) - f(y)\|^2_{L^{\psi_2}(\mu_A \otimes \mu_A)} &= \|\tilde{f}(x) - \tilde{f}(y)\|^2_{L^{\psi_2}(\mu_A \otimes \mu_A)} \\
&\le \left(4e\sqrt{2}\right)^2 \log\left(\frac{e}{\mu(A)}\right) \|\tilde{f}(x) - \tilde{f}(y)\|^2_{L^{\psi_2}(\mu \otimes \mu)} \\
&\le \left(4e\sqrt{2}\right)^2 \log\left(\frac{e}{\mu(A)}\right) \cdot \left(4\sqrt{6}\right)^2 \sigma^2(\mu).
\end{aligned}$$

Another application of Lemma 3.2 — in the space (A, d, μ_A) (now the right inequality) yields

$$\sigma^2(\mu_A) \le 4 \cdot \left(4e\sqrt{2}\right)^2 \log\left(\frac{e}{\mu(A)}\right) \cdot \left(4\sqrt{6}\right)^2 \sigma^2(\mu).$$

This is exactly (4) with constant $c = 4 \cdot (4e\sqrt{2})^2 (4\sqrt{6})^2 = 3 \cdot 2^{12} e^2 = 90,796.72\dots$ □

Remark 3.3 Let us also record the following natural generalization of Theorem 1.1, which is obtained along the same arguments. Given a collection $\mathcal{F}$ of (integrable) functions on the probability space (M, μ), define $\sigma_{\mathcal{F}}^2(\mu)$ as the infimum over all σ^2 such that

$$\int e^{t(f-m)} d\mu \le e^{\sigma^2 t^2 / 2}, \quad \text{for all } t \in \mathbb{R},$$

for any $f \in \mathcal{F}$, where $m = \int f \, d\mu$. Then with the same constant c as in Theorem 1.1, for any measurable $A \subset M$, $\mu(A) > 0$, we have

$$\sigma_{\mathcal{F}_A}^2(\mu_A) \le c \log\left(\frac{e}{\mu(A)}\right) \sigma_{\mathcal{F}}^2(\mu),$$

where $\mathcal{F}_A$ denotes the collection of restrictions of functions f from $\mathcal{F}$ to the set A.

Let us now mention an interesting connection of the subgaussian constants with the Kantorovich distances

$$W_1(\mu, v) = \inf \iint d(x, y)\, \pi(x, y)$$

and the relative entropies

$$D(v\|\mu) = \int \log \frac{dv}{d\mu}\, dv$$

(called also Kullback-Leibler's distances or informational divergences). Here, v is a probability measure on M, which is absolutely continuous with respect to μ (for short, $v \ll \mu$), and the infimum in the definition of W_1 is running over all probability measures π on the product space $M \times M$ with marginal distributions μ and v, i.e., such that

$$\pi(B \times M) = \mu(B), \quad \pi(M \times B) = v(B) \qquad (\text{Borel } B \subset M).$$

As was shown in [7], if (M, d) is a Polish space (complete separable), the subgaussian constant $\sigma^2 = \sigma^2(\mu)$ may be described as an optimal value in the transport-entropy inequality

$$W_1(\mu, v) \le \sqrt{2\sigma^2 D(v\|\mu)}. \tag{13}$$

Hence, we obtain from the inequality (4) a similar relation for measures v supported on given subsets of M.

Corollary 3.4 *Given a Borel probability measure μ on a Polish space (M, d) and a closed set A in M such that $\mu(A) > 0$, for any Borel probability measure v supported on A,*

$$W_1^2(\mu_A, v) \le c\sigma^2(\mu) \log\left(\frac{e}{\mu(A)}\right) D(v\|\mu_A),$$

where c is an absolute constant.

This assertion is actually equivalent to Theorem 1.1. Note that, for v supported on A, there is an identity $D(v\|\mu_A) = \log \mu(A) + D(v\|\mu)$. In particular, $D(v\|\mu_A) \le D(v\|\mu)$, so the relative entropies decrease when turning to restricted measures.

For another (almost equivalent) description of the subgaussian constant, introduce the concentration function

$$\mathcal{K}_\mu(r) = \sup \left[1 - \mu(A^r)\right] \qquad (r > 0),$$

where $A^r = \{x \in M : d(x,a) < r$ for some $a \in A\}$ denotes an open r-neighbourhood of A for the metric d, and the sup is running over all Borel sets $A \subset M$ of measure $\mu(A) \geq \frac{1}{2}$. As is well-known, the transport-entropy inequality (13) gives rise to a concentration inequality on (M, d, μ) of a subgaussian type (K. Marton's argument), but this can also be seen by a direct application of (1). Indeed, for any function f on M with $\|f\|_{\mathrm{Lip}} \leq 1$, it implies

$$\iint e^{t(f(x)-f(y))} \, d\mu(x) \, d\mu(y) \leq e^{\sigma^2 t^2}, \qquad t \in \mathbb{R},$$

and, by Chebyshev's inequality, we have a deviation bound

$$(\mu \otimes \mu)\{(x,y) \in M \times M : f(x) - f(y) \geq r\} \leq e^{-r^2/4\sigma^2}, \qquad r \geq 0.$$

In particular, one may apply it to the distance functions $f(x) = d(A, x) = \inf_{a \in A} d(a, x)$. Assuming that $\mu(A) \geq \frac{1}{2}$, the measure on the left-hand side is greater than or equal to $\frac{1}{2}(1 - \mu(A^r))$, so that we obtain a concentration inequality

$$1 - \mu(A^r) \leq 2e^{-r^2/4\sigma^2}.$$

Therefore,

$$\mathcal{K}_\mu(r) \leq \min\left\{\frac{1}{2}, 2e^{-r^2/4\sigma^2}\right\} \leq e^{-r^2/8\sigma^2}.$$

To argue in the opposite direction, suppose that the concentration function admits a bound of the form $\mathcal{K}_\mu(r) \leq e^{-r^2/b^2}$ for all $r > 0$ with some constant $b > 0$. Given a function f on M with $\|f\|_{\mathrm{Lip}} \leq 1$, let m be a median of f under μ. Then the set $A = \{f \leq m\}$ has measure $\mu(A) \geq \frac{1}{2}$, and by the Lipschitz property, $A^r \subset \{f < m + r\}$ for all $r > 0$. Hence, by the concentration hypothesis,

$$\mu\{f - m \geq r\} \leq \mathcal{K}_\mu(r) \leq e^{-r^2/b^2}.$$

A similar deviation bound also holds for the function $-f$ with its median $-m$, so that

$$\mu\{|f - m| \geq r\} \leq 2e^{-r^2/b^2}, \qquad r > 0.$$

This is sufficient to properly estimate ψ_2-norm of $f - m$ on (M, μ). Namely, for any $\lambda < 1/b^2$,

$$\int e^{\lambda|f-m|^2} \, d\mu = 1 + 2\lambda \int_0^\infty r e^{\lambda r^2} \mu\{|f - m| \geq r\} \, dr$$

$$\leq 1 + 2\lambda \int_0^\infty r e^{\lambda r^2} e^{-r^2/b^2} \, dr = 1 + \frac{\lambda}{\frac{1}{b^2} - \lambda} = 2,$$

where in the last equality the value $\lambda = \frac{1}{2b^2}$ is chosen. Thus, $\int e^{|f-m|^2/(2b^2)} \, d\mu \le 2$, which means that $\|f - m\|_{\psi_2} \le \sqrt{2}\, b$. The latter gives $\|f(x) - f(y)\|_{L^{\psi_2}(\mu \otimes \mu)} \le 2\sqrt{2}\, b$. Taking the supremum over f, it remains to apply Lemma 3.2, and then we get $\sigma^2(\mu) \le 32\, b^2$.

Let us summarize.

Proposition 3.5 *Let $b = b(\mu)$ be an optimal value such that the concentration function of the space (M, d, μ) satisfies a subgaussian bound $\mathcal{K}_\mu(r) \le e^{-r^2/b^2} (r > 0)$. Then*

$$\frac{1}{8}\, b^2(\mu) \le \sigma^2(\mu) \le 32\, b^2(\mu).$$

Once this description of the subgaussian constant is recognized, one may give another proof of Theorem 1.1, by relating the concentration function $\mathcal{K}_{\mu_A}$ to $\mathcal{K}_\mu$. In this connection, let us state below as a lemma one general observation due to Barthe and Milman (cf. [4], Lemma 2.1, p. 585).

Lemma 3.6 *Let a Borel probability measure ν on M be absolutely continuous with respect to μ and have density p. Suppose that, for some right-continuous, non-increasing function $R : (0, 1/4] \to (0, \infty)$, such that $\beta(\varepsilon) = \varepsilon/R(\varepsilon)$ is increasing, we have*

$$\nu\{x \in M : p(x) > R(\varepsilon)\} \le \varepsilon \qquad \left(0 < \varepsilon \le \frac{1}{4}\right).$$

Then

$$K_\nu(r) \le 2\beta^{-1}\big(K_\mu(r/2)\big), \quad \text{for all } r \ge 2K_\mu^{-1}(\beta(1/4)).$$

Here β^{-1} denotes the inverse function, and $K_\mu^{-1}(\varepsilon) = \inf\{r > 0 : K_\mu(r) < \varepsilon\}$.

The 2nd Proof of Theorem 1.1 The normalized restricted measure $\nu = \mu_A$ has density $p = \frac{1}{\mu(A)}\, 1_A$ (thus taking only one non-zero value), and an optimal choice of R is the constant function $R(\varepsilon) = \frac{1}{\mu(A)}$. Hence, Lemma 3.6 yields the relation

$$K_{\mu_A}(r) \le \frac{2}{\mu(A)}\, K_\mu(r/2), \qquad \text{for } r \ge 2K_\mu^{-1}(\mu(A)/4).$$

In particular, if $\mathcal{K}_\mu(r) \le e^{-r^2/b^2}$, then

$$K_{\mu_A}(r) \le \frac{2}{\mu(A)}\, e^{-r^2/(4b^2)}, \qquad \text{for } r \ge 2b\sqrt{\log(4/\mu(A))}.$$

Necessarily $K_{\mu_A}(r) \le \frac{1}{2}$, so the last relation may be extended to the whole positive half-axis. Moreover, at the expense of a factor in the exponent, one can remove the factor $\frac{2}{\mu(A)}$; more precisely, we get $K_{\mu_A}(r) \le e^{-r^2/\tilde{b}^2}$ with $\tilde{b}^2 = \frac{4b^2}{\log 2} \log \frac{4}{\mu(A)}$, that is,

$$b^2(\mu_A) \le \frac{4}{\log 2} \log \frac{4}{\mu(A)} b^2(\mu).$$

It remains to apply the two-sided bound of Proposition 3.5.   □

4 Proof of Theorem 1.3: Spectral Gap

Theorem 1.1 insures, in particular, that, for any function f on the metric probability space (M, d, μ) with Lipschitz semi-norm $\|f\|_{\mathrm{Lip}} \le 1$,

$$\mathrm{Var}_{\mu_A}(f) \le c \log \left(\frac{e}{\mu(A)} \right) \sigma^2(\mu)$$

up to some absolute constant c. In fact, in order to reach a similar concentration property of the restricted measures, it is enough to start with a Poincaré-type inequality on M,

$$\lambda_1 \mathrm{Var}_\mu(f) \le \int |\nabla f|^2 \, d\mu.$$

Under this hypothesis, a well-known theorem due to Gromov-Milman and Borovkov-Utev asserts that mean zero Lipschitz functions f have bounded ψ_1-norm. One may use a variant of this theorem proposed by Aida and Strook [1], who showed that

$$\int e^{\sqrt{\lambda_1} f} \, d\mu \le K_0 = 1.720102 \ldots \qquad (\|f\|_{\mathrm{Lip}} \le 1).$$

Hence

$$\int e^{\sqrt{\lambda_1} |f|} \, d\mu \le 2K_0 \quad \text{and} \quad \int e^{\frac{1}{2}\sqrt{\lambda_1} |f|} \, d\mu \le \sqrt{2K_0} < 2,$$

thus implying that $\|f\|_{\psi_1} \le \frac{2}{\sqrt{\lambda_1}}$. In addition,

$$\iint e^{\sqrt{\lambda_1}(f(x)-f(y))} \, d\mu(x)d\mu(y) \le K_0^2 \quad \text{and} \quad \iint e^{\sqrt{\lambda_1}|f(x)-f(y)|} \, d\mu(x)d\mu(y) \le 2K_0^2 < 6.$$

From this,

$$\int e^{\frac{1}{3}\sqrt{\lambda_1}\,|f(x)-f(y)|}\,d\mu(x)d\mu(y) < 6^{1/3} < 2,$$

which means that $\|f(x)-f(y)\|_{\psi_1} \le \frac{3}{\sqrt{\lambda_1}}$ with respect to the product measure $\mu \otimes \mu$ on the product space $M \times M$. This inequality is translation invariant, so the mean zero assumption may be removed. Thus, we arrive at:

Lemma 4.1 *Under the Poincaré-type inequality with spectral gap $\lambda_1 > 0$, for any mean zero function f on (M, d, μ) with $\|f\|_{\mathrm{Lip}} \le 1$,*

$$\|f\|_{\psi_1} \le \frac{2}{\sqrt{\lambda_1}}.$$

Moreover, for any f with $\|f\|_{\mathrm{Lip}} \le 1$,

$$\|f(x) - f(y)\|_{L^{\psi_1}(\mu \otimes \mu)} \le \frac{3}{\sqrt{\lambda_1}}. \tag{14}$$

This is a version of the concentration of measure phenomenon (with exponential integrability) in presence of a Poincaré-type inequality. Our goal is therefore to extend this property to the normalized restricted measures μ_A. This can be achieved by virtue of the inequality (11) which when combined with (14) yields an upper bound

$$\|f(x) - f(y)\|_{L^{\psi_1}(\mu_A \otimes \mu_A)} \le 36\,e \log\left(\frac{e}{\mu(A)}\right)\frac{1}{\sqrt{\lambda_1}}.$$

Moreover, if f has μ_A-mean zero, the left norm dominates $\|f\|_{L^{\psi_1}(\mu_A)}$ (by Jensen's inequality). We can summarize, taking into account once again Kirszbraun's theorem, as we did in the proof of Theorem 1.1.

Proposition 4.2 *Assume the metric probability space (M, d, μ) satisfies a Poincaré-type inequality with constant $\lambda_1 > 0$. Given a measurable set $A \subset M$ with $\mu(A) > 0$, for any function $f : A \to \mathbb{R}$ with μ_A-mean zero and such that $\|f\|_{\mathrm{Lip}} \le 1$ on A,*

$$\|f\|_{L^{\psi_1}(\mu_A)} \le 36\,e \log\left(\frac{e}{\mu(A)}\right)\frac{1}{\sqrt{\lambda_1}}.$$

Theorem 1.3 is now easily obtained with constant $c = 2\,(36e)^2$ by noting that L^2-norms are dominated by L^{ψ_1}-norms. More precisely, since $e^{|t|} - 1 \ge \frac{1}{2}t^2$, one has $\|f\|_{\psi_1}^2 \ge \frac{1}{2}\|f\|_2^2$.

Remark 4.3 Related stability results are known for various classes of probability distributions on the Euclidean spaces $M = \mathbb{R}^n$ (and even in a more general situation, where μ_A is replaced by an absolutely continuous measure with respect to μ). See, in particular, the works by Milman [15, 16] on convex bodies and log-concave measures.

5 Examples

Theorems 1.1 and 1.3 involve a lot of interesting examples. Here are a few obvious cases.

1. The standard Gaussian measure $\mu = \gamma$ on $\mathbb{R}^n$ satisfies a logarithmic Sobolev inequality on $M = \mathbb{R}^n$ with a dimension-free constant $\rho = 1$. Hence, from Theorem 1.1 we get:

Corollary 5.1 *For any measurable set $A \subset \mathbb{R}^n$ with $\gamma(A) > 0$, the subgaussian constant $\sigma^2(\gamma_A)$ of the normalized restricted measure γ_A satisfies*

$$\sigma^2(\gamma_A) \leq c \log\left(\frac{e}{\gamma(A)}\right),$$

where c is an absolute constant.

As it was already mentioned, if A is convex, there is a sharper bound $\sigma^2(\gamma_A) \leq 1$. However, it may not hold without convexity assumption. Nevertheless, if $\gamma(A)$ is bounded away from zero, we obtain a more universal principle.

Clearly, Corollary 5.1 extends to all product measures $\mu = \nu^n$ on $\mathbb{R}^n$ such that ν satisfies a logarithmic Sobolev inequality on the real line, and with constants c depending on ρ, only. A characterization of the property $\rho > 0$ in terms of the distribution function of the measure ν and the density of its absolutely continuous component may be found in [7].

2. Consider a uniform distribution ν on the shell

$$A_\varepsilon = \{x \in \mathbb{R}^n : 1 - \varepsilon \leq |x| \leq 1\}, \qquad 0 \leq \varepsilon \leq 1 \ (n \geq 2).$$

Corollary 5.2 *The subgaussian constant of ν satisfies $\sigma^2(\nu) \leq \frac{c}{n}$, up to some absolute constant c.*

In other words, mean zero Lipschitz functions f on A_ε are such that $\sqrt{n}f$ are subgaussian with universal constant factor. This property is well-known in the extreme cases—on the unit Euclidean ball $A = B_n (\varepsilon = 1)$ and on the unit sphere $A = S^{n-1} (\varepsilon = 0)$.

Let μ denote the normalized Lebesgue measure on B_n. In the case $\varepsilon \geq \frac{1}{n}$, the shell A_ε represents the part of B_n of measure

$$\mu(A_\varepsilon) = 1 - \left(1 - \frac{1}{n}\right)^n \geq 1 - \frac{1}{e}.$$

Since the logarithmic Sobolev constant of the unit ball is of order $\frac{1}{n}$, and therefore $\sigma^2(\mu) \leq \frac{c}{n}$, the assertion of Corollary 5.2 immediately follows from Theorem 1.1.

In case $\varepsilon \leq \frac{1}{n}$, the assertion follows from a similar concentration property of the uniform distribution σ_{n-1} on the unit sphere. Indeed, with every Lipschitz function f on A_ε one may associate its restriction to S^{n-1}, which is also Lipschitz (with respect to the Euclidean distance). We have $|f(r\theta) - f(\theta)| \leq |r - 1| \leq \varepsilon \leq \frac{1}{n}$, for any $r \in [1 - \varepsilon, 1]$ and $\theta \in S^{n-1}$. Hence,

$$|f(r'\theta') - f(r\theta)| \leq |f(\theta') - f(\theta)| + |f(r'\theta) - f(\theta')| + |f(r\theta) - f(\theta)|$$

$$\leq |f(\theta') - f(\theta)| + \frac{2}{n},$$

whenever $r, r' \in [1 - \varepsilon, 1]$ and $\theta, \theta' \in S^{n-1}$, which implies

$$|f(r'\theta') - f(r\theta)|^2 \leq 2|f(\theta') - f(\theta)|^2 + \frac{8}{n^2}.$$

But the map $(r, \theta) \to \theta$ pushes forward v onto σ_{n-1}, so, we obtain that, for any $c > 0$,

$$\iint \exp\{cn|f(r'\theta') - f(r\theta)|^2\}\, dv(r', \theta')\, dv(r, \theta)$$

$$\leq e^{8/n} \iint \exp\{2cn|f(\theta') - f(\theta)|^2\}\, d\sigma_{n-1}(\theta')\, d\sigma_{n-1}(\theta).$$

Here, for a certain numerical constant $c > 0$, the right-hand side is bounded by a universal constant. This constant can be replaced with 2 using Jensen's inequality. The assertion follows from Lemma 3.2.

3. The two-sided product exponential measure μ on $\mathbb{R}^n$ with density 2^{-n} $e^{-(|x_1|+\cdots+|x_n|)}$ satisfies a Poincaré-type inequality on $M = \mathbb{R}^n$ with a dimension-free constant $\lambda_1 = 1/4$. Hence, from Proposition 4.2 we get:

Corollary 5.3 *For any measurable set $A \subset \mathbb{R}^n$ with $\mu(A) > 0$, and for any function $f : A \to \mathbb{R}$ with μ_A-mean zero and $\|f\|_{\mathrm{Lip}} \leq 1$, we have*

$$\|f\|_{L^{\psi_1}(\mu_A)} \leq c \log\left(\frac{e}{\mu(A)}\right),$$

where c is an absolute constant. In particular,

$$s^2(\mu_A) \leq c \log^2\left(\frac{e}{\mu(A)}\right).$$

Clearly, Corollary 5.3 extends to all product measures $\mu = \nu^n$ on $\mathbb{R}^n$ such that ν satisfies a Poincaré-type inequality on the real line, and with constants c depending on λ_1, only. A characterization of the property $\lambda_1 > 0$ may also be given in terms of the distribution function of ν and the density of its absolutely continuous component (cf. [7]).

4a. Let us take the metric probability space $(\{0, 1\}^n, d_n, \mu)$, where d_n is the Hamming distance, that is, $d_n(x, y) = \sharp\{i : x_i \neq y_i\}$, equipped with the uniform measure μ. For this particular space, Marton established the transport-entropy inequality (13) with an optimal constant $\sigma^2 = \frac{n}{4}$, cf. [12]. Using the relation (13) as an equivalent definition of the subgaussian constant, we obtain from Theorem 1.1:

Corollary 5.4 *For any non-empty set* $A \subset \{0, 1\}^n$, *the subgaussian constant* $\sigma^2(\mu_A)$ *of the normalized restricted measure* μ_A *satisfies, up to an absolute constant c,*

$$\sigma^2(\mu_A) \leq cn \log\left(\frac{e}{\mu(A)}\right). \tag{15}$$

4b. Let us now assume that A is *monotone*, i.e., A satisfies the condition

$$(x_1, \ldots, x_n) \in A \implies (y_1, \ldots, y_n) \in A, \text{ whenever } y_i \geq x_i, \ i = 1, \ldots, n.$$

Recall that the discrete cube can be equipped with a natural graph structure: there is an edge between x and y whenever they are of Hamming distance $d_n(x, y) = 1$. For monotone sets A, the graph metric d_A on the subgraph of A is equal to the restriction of d_n to $A \times A$. Indeed, we have:

$$d_n(x, y) \leq d_A(x, y) \leq d_A(x, x \wedge y) + d_A(y, x \wedge y) = d_n(x, x \wedge y) + d_n(y, x \wedge y) = d_n(x, y),$$

where $x \wedge y = (x_1 \wedge y_1, \ldots, x_n \wedge y_n)$. Thus,

$$s^2(\mu_A, d_A) \leq \sigma^2(\mu_A, d_A) \leq cn \log\left(\frac{e}{\mu(A)}\right).$$

This can be compared with what follows from a recent result of Ding and Mossel (see [9]). The authors proved that the conductance (Cheeger constant) of (A, μ_A) satisfies $\phi(A) \geq \frac{\mu(A)}{16n}$. However, this type of isoperimetric results may not imply sharp concentration bounds. Indeed, by using Cheeger inequality, the above inequality leads to $\lambda_1 \geq c\mu(A)^2/n^2$ and $s^2(\mu_A, d_A) \leq 1/\lambda_1 \leq cn^2/\mu(A)^2$, which is even worse than the trivial estimate $s^2(\mu_A, d_A) \leq \frac{1}{2}\operatorname{diam}(A)^2 \leq n^2/2$.

5. Let (M, d, μ) be a (separable) metric probability space with finite subgaussian constant $\sigma^2(\mu)$. The previous example can be naturally generalized to the product space (M^n, μ^n), when it is equipped with the ℓ^1-type metric

$$d_n(x, y) = \sum_{i=1}^{n} d(x_i, y_i), \qquad x = (x_1, \ldots, x_n), \ y = (y_1, \ldots, y_n) \in M^n.$$

This can be done with the help of the following elementary observation.

Proposition 5.5 *The subgaussian constant of the space (M^n, d_n, μ^n) is related to the subgaussian constant of (M, d, μ) by the equality $\sigma^2(\mu^n) = n\sigma^2(\mu)$.*

Indeed, one may argue by induction on n. Let f be a function on M^n. The Lipschitz property $\|f\|_{\mathrm{Lip}} \leq 1$ with respect to d_n is equivalent to the assertion that f is coordinatewise Lipschitz, that is, any function of the form $x_i \to f(x)$ has a Lipschitz semi-norm ≤ 1 on M for all fixed coordinates $x_j \in M$ ($j \neq i$). Hence, in this case, for all $t \in \mathbb{R}$,

$$\int_M e^{tf(x)} d\mu(x_n) \leq \exp\left\{t\int_M f(x) d\mu(x_n) + \frac{\sigma^2 t^2}{2}\right\},$$

where $\sigma^2 = \sigma^2(\mu)$. Here the function $(x_1, \ldots, x_{n-1}) \to \int_M f(x) d\mu(x_n)$ is also coordinatewise Lipschitz. Integrating the above inequality with respect to $d\mu^{n-1}(x_1, \ldots, x_{n-1})$ and applying the induction hypothesis, we thus get

$$\int_{M^n} e^{tf(x)} d\mu^n(x) \leq \exp\left\{t\int_{M^n} f(x) d\mu^n(x) + n\frac{\sigma^2 t^2}{2}\right\}.$$

But this means that $\sigma^2(\mu^n) \leq n\sigma^2(\mu)$.

For an opposite bound, it is sufficient to test (1) for (M^n, d_n, μ^n) in the class of all coordinatewise Lipschitz functions of the form $f(x) = u(x_1) + \cdots + u(x_n)$ with μ-mean zero functions u on M such that $\|u\|_{\mathrm{Lip}} \leq 1$.

Corollary 5.6 *For any Borel set $A \subset M^n$ such that $\mu^n(A) > 0$, the subgaussian constant of the normalized restricted measure μ_A^n with respect to the ℓ^1-type metric d_n satisfies*

$$\sigma^2(\mu_A^n) \leq cn\sigma^2(\mu) \log\left(\frac{e}{\mu^n(A)}\right),$$

where c is an absolute constant.

For example, if μ is a probability measure on $M = \mathbb{R}$ such that $\int_{-\infty}^{\infty} e^{x^2/\lambda^2} d\mu(x) \leq 2$ ($\lambda > 0$), then for the restricted product measures we have

$$\sigma^2(\mu_A^n) \leq cn\lambda^2 \log\left(\frac{e}{\mu^n(A)}\right) \qquad (16)$$

with respect to the ℓ^1-norm $\|x\|_1 = |x_1| + \cdots + |x_n|$ on $\mathbb{R}^n$.

Indeed, by the integral hypothesis on μ, for any f on $\mathbb{R}$ with $\|f\|_{\mathrm{Lip}} \leq 1$,

$$\int_{-\infty}^{\infty}\int_{-\infty}^{\infty} e^{(f(x)-f(y))^2/2\lambda^2} d\mu(x)d\mu(y) \leq \int_{-\infty}^{\infty}\int_{-\infty}^{\infty} e^{(x-y)^2/2\lambda^2} d\mu(x)d\mu(y)$$

$$\leq \int_{-\infty}^{\infty}\int_{-\infty}^{\infty} e^{(x^2+y^2)/\lambda^2} d\mu(x)d\mu(y) \leq 4.$$

Hence, if f has μ-mean zero, by Jensen's inequality,

$$\int_{-\infty}^{\infty} e^{f(x)^2/4\lambda^2} d\mu(x) \leq \int_{-\infty}^{\infty}\int_{-\infty}^{\infty} e^{(f(x)-f(y))^2/4\lambda^2} d\mu(x)d\mu(y) \leq 2,$$

meaning that $\|f\|_{L^{\psi_2}(\mu)} \leq 2\lambda$. By Lemma 3.1, cf. (12), it follows that $\sigma^2(\mu) \leq 16\lambda^2$, so, (16) holds true by an application of Corollary 5.6.

6 Deviations for Non-Lipschitz Functions

Let us now turn to the interesting question on the relationship between the distribution of a locally Lipschitz function and the distribution of its modulus of the gradient. We still keep the setting of a metric probability space (M, d, μ) and assume it has a finite subgaussian constant $\sigma^2 = \sigma^2(\mu)(\sigma \geq 0)$.

Let us say that a continuous function f on M is locally Lipschitz, if $|\nabla f(x)|$ is finite for all $x \in M$. Recall that we consider the sets

$$A = \{x \in M : |\nabla f(x)| \leq L\}, \qquad L > 0. \qquad (17)$$

First we state a more general version of Corollary 1.2.

Theorem 6.1 *Assume that a locally Lipschitz function f on M has Lipschitz semi-norms $\leq L$ on the sets of the form (17). If $\mu\{|\nabla f| \geq L_0\} \leq \frac{1}{2}$, then for all $t > 0$,*

$$(\mu \otimes \mu)\{|f(x) - f(y)| \geq t\} \leq 2 \inf_{L \geq L_0}\left[e^{-t^2/c\sigma^2 L^2} + \mu\{|\nabla f| > L\}\right], \qquad (18)$$

where c is an absolute constant.

Proof Although the argument is already mentioned in Sect. 1, let us replace (6) with a slightly different bound. First note that the Lipschitz semi-norm of f with respect to the metric d in M is the same as its Lipschitz semi-norm with respect to the metric on the set A induced from M (which is true for any non-empty subset of M). Hence, we are in position to apply Theorem 1.1, and then the definition (1) for the normalized restriction μ_A yields a subgaussian bound

$$\iint e^{t(f(x)-f(y))}\, d\mu_A(x) d\mu_A(y) \leq e^{c\sigma^2 L^2 t^2/2}, \quad \text{for all } t \in \mathbb{R},$$

where A is defined in (17) with $L \geq L_0$, and where c is universal constant. From this, for any $t > 0$,

$$(\mu_A \otimes \mu_A)\left\{(x,y) \in A \times A : |f(x)-f(y)| \geq t\right\} \leq 2e^{-t^2/(2c\sigma^2 L^2)},$$

and therefore

$$(\mu \otimes \mu)\left\{(x,y) \in A \times A : |f(x)-f(y)| \geq t\right\} \leq 2e^{-t^2/(2c\sigma^2 L^2)}.$$

The product measure of the complement of $A \times A$ does not exceed $2\mu\{|\nabla f(x)| > L\}$, and we obtain (18). $\qquad\square$

If $\int e^{|\nabla f|^2}\, d\mu \leq 2$, we have, by Chebyshev's inequality, $\mu\{|\nabla f| \geq L\} \leq 2e^{-L^2}$, so one may take $L_0 = \sqrt{\log 4}$. Theorem 6.1 then gives that, for any $L^2 \geq \log 4$,

$$(\mu \otimes \mu)\{|f(x)-f(y)| \geq t\} \leq 2\,e^{-t^2/c\sigma^2 L^2} + 4e^{-L^2}.$$

For $t \geq 2\sigma$ one may choose here $L^2 = \frac{t}{\sigma}$, leading to

$$(\mu \otimes \mu)\{|f(x)-f(y)| \geq t\} \leq 6\,e^{-t/c\sigma},$$

for some absolute constant $c > 1$. In case $0 \leq t \leq 2\sigma$, this inequality is fulfilled automatically, so it holds for all $t \geq 0$. As a result, with some absolute constant C,

$$\|f(x)-f(y)\|_{\psi_1} \leq C\sigma,$$

which is an equivalent way to state the inequality of Corollary 1.2.

As we have already mentioned, with the same arguments inequalities like (18) can be derived on the basis of subgaussian constants defined for different classes of functions. For example, one may consider the subgaussian constant $\sigma_{\mathcal{F}}^2(\mu)$ for the class $\mathcal{F}$ of all convex Lipschitz functions f on the Euclidean space $M = \mathbb{R}^n$ (which we equip with the Euclidean distance). Note that $|\nabla f(x)|$ is everywhere finite in the n-space, when f is convex. Keeping in mind Remark 3.3, what we need is the following analog of Kirszbraun's theorem:

Lemma 6.2 *Let f be a convex function on $\mathbb{R}^n$. For any $L > 0$, there exists a convex function g on $\mathbb{R}^n$ such that $f = g$ on the set $A = \{x : |\nabla f(x)| \leq L\}$ and $|\nabla g| \leq L$ on $\mathbb{R}^n$.*

Accepting for a moment this lemma without proof, we get:

Theorem 6.3 *Assume that a convex function f on $\mathbb{R}^n$ satisfies $\mu\{|\nabla f| \geq L_0\} \leq \frac{1}{2}$. Then for all $t > 0$,*

$$(\mu \otimes \mu)\{|f(x) - f(y)| \geq t\} \leq 2 \inf_{L \geq L_0} \left[e^{-t^2/c\sigma^2 L^2} + \mu\{|\nabla f| > L\} \right],$$

where $\sigma^2 = \sigma_{\mathcal{F}}^2(\mu)$ and c is an absolute constant.

For illustration, let $\mu = \mu_1 \otimes \cdots \otimes \mu_n$ be an arbitrary product probability measure on the cube $[-1, 1]^n$. If f is convex and Lipschitz on $\mathbb{R}^n$, thus with $|\nabla f| \leq 1$, then

$$(\mu \otimes \mu)\{|f(x) - f(y)| \geq t\} \leq 2e^{-t^2/c}. \tag{19}$$

This is one of the forms of Talagrand's concentration phenomenon for the family of convex sets/functions (cf. [11, 13, 18, 19]). That is, the subgaussian constants $\sigma_{\mathcal{F}}^2(\mu)$ are bounded for the class $\mathcal{F}$ of convex Lipschitz f and product measures μ on the cube. Hence, using Theorem 6.3, Talagrand's deviation inequality (19) admits a natural extension to the class of non-Lipschitz convex functions:

Corollary 6.4 *Let μ be a product probability measure on the cube, and let f be a convex function on $\mathbb{R}^n$. If $\mu\{|\nabla f| \geq L_0\} \leq \frac{1}{2}$, then for all $t > 0$,*

$$(\mu \otimes \mu)\{|f(x) - f(y)| \geq t\} \leq 2 \inf_{L \geq L_0} \left[e^{-t^2/cL^2} + \mu\{|\nabla f| > L\} \right],$$

where c is an absolute constant.

In particular, we have a statement similar to Corollary 1.2 — for this family of functions, namely

$$\|f - m\|_{L^{\psi_1}(\mu)} \leq c \, \|\nabla f\|_{L^{\psi_2}(\mu)},$$

where m is the μ-mean of f.

Proof of Lemma 6.2 An affine function $l_{a,v}(x) = a + \langle x, v \rangle$ ($v \in \mathbb{R}^n$, $a \in \mathbb{R}$) may be called to be a tangent function to f, if $f \geq l$ on $\mathbb{R}^n$ and $f(x) = l_{a,v}(x)$ for at least one point x. It is well-known that

$$f(x) = \sup\{l_{a,v}(x) : l_{a,v} \in \mathcal{L}\},$$

where $\mathcal{L}$ denotes the collection of all tangent functions $l_{a,v}$. Put,

$$g(x) = \sup\{l_{a,v}(x) : l_{a,v} \in \mathcal{L}, \ |v| \leq L\}.$$

By the construction, $g \leq f$ on $\mathbb{R}^n$ and, moreover,

$$\|g\|_{\text{Lip}} \leq \sup\{\|l_{a,v}\|_{\text{Lip}} : l_{a,v} \in \mathcal{L}, \ |v| \leq L\}$$
$$= \sup\{|v| : l_{a,v} \in \mathcal{L}, \ |v| \leq L\} \leq L.$$

It remains to show that $g = f$ on the set $A = \{|\nabla f| \leq L\}$. Let $x \in A$ and let $l_{a,v}$ be tangent to f and such that $l_{a,v}(x) = f(x)$. This implies that $f(y) - f(x) \geq \langle y - x, v \rangle$ for all $y \in \mathbb{R}^n$ and hence

$$|\nabla f(x)| = \limsup_{y \to x} \frac{|f(y) - f(x)|}{|y - x|} \geq \limsup_{y \to x} \frac{\langle y - x, v \rangle}{|y - x|} = v.$$

Thus, $|v| \leq L$, so that $g(x) \geq l_{a,v}(x) = f(x)$. $\qquad\qquad\square$

7 Optimality

Here we show that the logarithmic dependence in $\mu(A)$ in Theorems 1.1 and 1.3 is optimal, up to the universal constant c. We provide several examples.

Example 1 Let us return to Example 4, Sect. 5, of the discrete hypercube $M = \{0, 1\}^n$, which we equip with the Hamming distance d_n and the uniform measure μ. Let us test the inequality (15) of Corollary 5.4 on the set $A \subset \{-1, 1\}^n$ consisting of $n + 1$ points

$$(0, 0, 0, \ldots, 0), \ (1, 0, 0, \ldots, 0), \ (1, 1, 0, \ldots, 0), \ \ldots, \ (1, 1, 1, \ldots, 1).$$

We have $\mu(A) = (n + 1)/2^n \geq 1/2^n$. The function $f : A \to \mathbb{R}$, defined by

$$f(x) = \sharp\{i : x_i = 1\} - \frac{n}{2},$$

has a Lipschitz semi-norm $\|f\|_{\text{Lip}} \leq 1$ with respect to d and the μ_A-mean zero. Moreover, $\int f^2 \, d\mu_A = \frac{n(n+2)}{12}$. Expanding the inequality $\int e^{tf} \, d\mu_A \leq e^{\sigma^2(\mu_A)t^2/2}$ at the origin yields $\int f^2 \, d\mu_A \leq \sigma^2(\mu_A)$. Hence, recalling that $\sigma^2(\mu) \leq \frac{n}{4}$, we get

$$\sigma^2(\mu_A) \geq \int f^2 \, d\mu_A \geq \frac{n^2}{12}$$

$$\geq \frac{n}{3} \sigma^2(\mu) \geq \frac{1}{3\log 2} \sigma^2(\mu) \log\left(\frac{1}{\mu(A)}\right).$$

This example shows the optimality of (15) in the regime $\mu(A) \to 0$.

Example 2 Let γ_n be the standard Gaussian measure on $\mathbb{R}^n$ of dimension $n \geq 2$. We have $\sigma^2(\gamma_n) = 1$. Consider the normalized measure γ_{A_R} on the set

$$A_R = \{(x_1, x_2, \ldots, x_n) \in \mathbb{R}^n : x_1^2 + x_2^2 \geq R^2\}, \qquad R \geq 0.$$

Using the property that the function $\frac{1}{2}(x_1^2 + x_2^2)$ has a standard exponential distribution under the measure γ_n, we find that $\gamma_n(A_R) = e^{-R^2/2}$. Moreover,

$$s^2(\gamma_{A_R}) \geq \mathrm{Var}_{\gamma_{A_R}}(x_1) = \int x_1^2 \, d\gamma_{A_R}(x) = \frac{1}{2}\int (x_1^2 + x_2^2) \, d\gamma_{A_R}(x)$$

$$= \frac{1}{e^{-R^2/2}} \int_{R^2/2}^{\infty} r e^{-r} \, dr = \frac{R^2}{2} + 1 = \log\left(\frac{e}{\gamma_n(A_R)}\right).$$

Therefore,

$$\sigma^2(\gamma_{A_R}) \geq s^2(\gamma_{A_R}) \geq \log\left(\frac{e}{\gamma_n(A_R)}\right),$$

showing that the inequality (4) of Theorem 1.1 is optimal, up to the universal constant, for any value of $\gamma_n(A) \in [0, 1]$.

Example 3 A similar conclusion can be made about the uniform probability measure μ on the Euclidean ball $B(0, \sqrt{n})$ of radius $\sqrt{n}$, centred at the origin (asymptotically for growing dimension n). To see this, it is sufficient to consider the cylinders

$$A_\varepsilon = \{(x_1, y) \in \mathbb{R} \times \mathbb{R}^{n-1} : |x_1| \leq \sqrt{n - \varepsilon^2} \text{ and } |y| \leq \varepsilon\}, \qquad 0 < \varepsilon \leq \sqrt{n},$$

and the function $f(x) = x_1$. We leave to the readers corresponding computations.

Example 4 Let μ be the two-sided exponential measure on $\mathbb{R}$ with density $\frac{1}{2} e^{-|x|}$. In this case $\sigma^2(\mu) = \infty$, but, as easy to see, $2 \leq s^2(\mu) \leq 4$ (recall that $\lambda_1(\mu) = \frac{1}{4}$). We are going to test optimality of the inequality (7) on the sets $A_R = \{x \in \mathbb{R} : |x| \geq R\}$ ($R \geq 0$). Clearly, $\mu(A_R) = e^{-R}$, and we find that

$$s^2(\mu_{A_R}) \geq \mathrm{Var}_{\mu_{A_R}}(x) = \int_{-\infty}^{\infty} x^2 \, d\mu_{A_R}(x) = \frac{1}{e^{-R}} \int_{R}^{\infty} r^2 e^{-r} \, dr$$

$$= R^2 + 2R + 2 \geq (R+1)^2 = \log^2\left(\frac{e}{\mu(A_R)}\right).$$

Therefore,

$$s^2(\mu_{A_R}) \geq \log^2\left(\frac{e}{\mu(A_R)}\right),$$

showing that the inequality (7) is optimal, up to the universal constant, for any value of $\mu(A) \in (0, 1]$.

Acknowledgements The authors gratefully acknowledge the support and hospitality of the Institute for Mathematics and its Applications, and the University of Minnesota, Minneapolis, where much of this work was conducted. The second named author would like to acknowledge the hospitality of the Georgia Institute of Technology, Atlanta, during the period 02/8-13/2015.

We would like to thank anonymous referee for pointing out to us several relevant references. S.G. Bobkov's research supported in part by the Humboldt Foundation, NSF and BSF grants. P. Nayar's research supported in part by NCN grant DEC-2012/05/B/ST1/00412. P. Tetali's research supported in part by NSF DMS-1407657.

Appendix

Proof of Lemma 2.1 Using the homogeneity, in order to derive the right-hand side inequality in (8), we may assume that $\sup_{p \geq 1} \frac{\|f\|_p}{\sqrt{p}} \leq 1$. Then $\int |f|^p \, d\mu \leq p^{p/2}$ for all $p \geq 1$, and by Chebyshev's inequality,

$$1 - F(t) \equiv \mu\{|f| \geq t\} \leq \left(\frac{\sqrt{p}}{t}\right)^p, \quad \text{for all } t > 0.$$

If $t \geq 2$, choose here $p = \frac{1}{4} t^2$, in which case $1 - F(t) \leq 2^{-\frac{1}{4} t^2}$. Integrating by parts, we have, for any $0 < \varepsilon < \frac{\log 2}{4}$,

$$\int e^{\varepsilon f^2} \, d\mu = -\int_0^\infty e^{\varepsilon t^2} \, d(1 - F(t))$$

$$= 1 + 2\varepsilon \int_0^2 t e^{\varepsilon t^2} (1 - F(t)) \, dt + 2\varepsilon \int_2^\infty t e^{\varepsilon t^2} (1 - F(t)) \, dt$$

$$\leq 1 + 2\varepsilon \int_0^2 t e^{\varepsilon t^2} \, dt + 2\varepsilon \int_2^\infty t e^{\varepsilon t^2} e^{-\frac{\log 2}{4} t^2} \, dt$$

$$= e^{4\varepsilon} + \frac{\varepsilon}{\frac{\log 2}{4} - \varepsilon} e^{-(\log 2 - 4\varepsilon)} = e^{4\varepsilon} \left(1 + \frac{\varepsilon}{2(\frac{\log 2}{4} - \varepsilon)}\right).$$

If $\varepsilon \leq \frac{\log 2}{8}$, the latter expression does not exceed $\frac{3}{2} e^{4\varepsilon}$ which does not exceed 2 for $\varepsilon \leq \frac{\log(4/3)}{4}$. Both inequalities are fulfilled for $\varepsilon = \frac{\log 2}{10}$, and with this value $\int e^{\varepsilon f^2} \, d\mu \leq 2$. Hence

$$\|f\|_{L^{\psi_2}(\mu)} \leq \frac{1}{\sqrt{\varepsilon}} = \sqrt{\frac{10}{\log 2}} < 4,$$

which yields the right inequality in (8). Conversely, if $\|f\|_{L^{\psi_2}(\mu)} = 1$, then $\int e^{\frac{3}{4}f^2} d\mu \leq 2$. Since $u(t) = t^p e^{-\frac{3}{4}t^2}$ is maximized in $t > 0$ at $t_0 = \sqrt{\frac{2p}{3}}$, we get

$$\|f\|_p^p = \int u(|f|)e^{f^2} d\mu \leq u(t_0) \cdot 2 = 2 \left(\sqrt{\frac{2p}{3e}} \right)^p.$$

Hence, $\frac{\|f\|_p}{\sqrt{p}} \leq 2^{1/p} \sqrt{\frac{2}{3e}} < 1$, which yields the left inequality.

Now, let us turn to (9) and assume that $\sup_{p\geq 1} \frac{\|f\|_p}{p} = 1$. Then $\int |f|^p d\mu \leq p^p$ for all $p \geq 1$, and by Chebyshev's inequality, for all $t > 0$,

$$1 - F(t) \equiv \mu\{|f| \geq t\} \leq \left(\frac{p}{t} \right)^p.$$

If $t \geq 2$, we may choose here $p = \frac{1}{2}t$ in which case $1 - F(t) \leq 2^{-\frac{1}{2}t}$, while for $1 \leq t < 2$ we choose $p = 1$, so that $1 - F(t) \leq \frac{1}{t}$. Arguing as before, we have, for any $0 < \varepsilon < \frac{\log 2}{2}$,

$$\int e^{\varepsilon|f|} d\mu = 1 + \varepsilon \int_0^1 e^{\varepsilon t} (1 - F(t)) \, dt + \varepsilon \int_1^2 e^{\varepsilon t} (1 - F(t)) \, dt + \varepsilon \int_2^\infty e^{\varepsilon t} (1 - F(t)) \, dt$$

$$\leq 1 + \varepsilon \int_0^1 e^{\varepsilon t} \, dt + \varepsilon \int_1^2 \frac{e^{\varepsilon t}}{t} \, dt + \varepsilon \int_2^\infty e^{\varepsilon t} e^{-\frac{\log 2}{2} t} \, dt.$$

The pre-last integral can be bounded by $\int_1^2 \frac{e^{2\varepsilon}}{t} \, dt = e^{2\varepsilon} \log 2$, so

$$\int e^{\varepsilon|f|} d\mu \leq e^\varepsilon + \varepsilon e^{2\varepsilon} \log 2 + \frac{\varepsilon}{\frac{\log 2}{2} - \varepsilon} e^{-2(\frac{\log 2}{2} - \varepsilon)}.$$

For $\varepsilon = \frac{1}{6}$, the latter expression is equal to $1.98903902\ldots$, and thus $\int e^{\varepsilon|f|} d\mu < 2$. Hence

$$\|f\|_{L^{\psi_1}(\mu)} \leq \frac{1}{\varepsilon} = 6.$$

Conversely, if $\|f\|_{L^{\psi_1}(\mu)} = 1$, then $\int e^{\frac{3}{4}|f|} d\mu = 2$. Since $u(t) = t^p e^{-\frac{3}{4}t}$ is maximized at $t_0 = p$, we get

$$\|f\|_p^p = \int u(|f|)e^{\frac{3}{4}|f|} d\mu \leq u(t_0) \cdot 2 = 2 \left(\frac{4p}{3e} \right)^p.$$

Hence, $\frac{\|f\|_p}{p} \leq 2^{1/p} \frac{4}{3e} < 1$, which yields the left inequality. $\square$

Proof of Lemma 3.1 First assume that $\|f\|_{\psi_2} = 1$, in particular $\int e^{\frac{7}{8}f^2} d\mu \leq 2$. The function

$$u(t) = \log \int e^{tf} d\mu$$

is smooth, convex, with $u(0) = 0$ and

$$u'(t) = \frac{\int f e^{tf} d\mu}{\int e^{tf} d\mu}.$$

In particular, $u'(0) = 0$. Note that, by Jensen's inequality, $\int e^{tf} d\mu \geq 1$, so $u(t) \geq 0$. Further differentiation gives

$$u''(t) = \frac{\int f^2 e^{tf} d\mu - (\int f e^{tf} d\mu)^2}{(\int e^{tf} d\mu)^2} \leq \int f^2 e^{tf} d\mu.$$

Using $tf \leq \frac{t^2+f^2}{2}$ and the elementary inequality $x e^{-\frac{3}{8}x} \leq \frac{8}{3}e^{-1}$, we get, for $|t| \leq 1$,

$$\int f^2 e^{tf} d\mu \leq \int f^2 e^{\frac{t^2+f^2}{2}} d\mu$$

$$= e^{t^2/2} \int f^2 e^{-\frac{3}{8}f^2} e^{\frac{7}{8}f^2} d\mu \leq e^{1/2} \frac{8}{3}e^{-1} \int e^{\frac{7}{8}f^2} d\mu \leq 4.$$

Thus, $u''(t) \leq 4$, and by Taylor's formula, $u(t) \leq 2t^2$.
    On the hand, for $|t| \geq 1$, by Cauchy's inequality,

$$\int e^{tf} d\mu \leq \int e^{\frac{t^2+f^2}{2}} d\mu = e^{t^2/2} \int e^{f^2/2} d\mu$$

$$\leq e^{t^2/2} \left(\int e^{\frac{7}{8}f^2} d\mu\right)^{4/7} = 2^{4/7} e^{t^2/2} \leq e^{(\frac{1}{2}+\frac{4}{7}\log 2)t^2} \leq e^{t^2}.$$

Hence, in this case $u(t) \leq t^2$. Thus,

$$\sigma_f^2 = \sup_{t \neq 0} \frac{u(t)}{t^2/2} \leq 4,$$

proving the right inequality of Lemma 3.1.
    For the left inequality, let $\sigma_f^2 = 1$. Then $\int e^{tf} d\mu \leq e^{t^2/2}$ for all $t \in \mathbb{R}$, which implies

$$1 - F(t) \equiv \mu\{|f| \geq t\} \leq 2e^{-t^2/2}, \qquad t \geq 0.$$

From this, integrating by parts, we have, for any $0 < \varepsilon < \frac{1}{2}$,

$$\int e^{\varepsilon f^2}\, d\mu = \int_0^\infty e^{\varepsilon t^2}\, dF(t) = -\int_0^\infty e^{\varepsilon t^2}\, d(1 - F(t))$$

$$= 1 + 2\varepsilon \int_0^\infty t e^{\varepsilon t^2}\, (1 - F(t))\, dt$$

$$\leq 1 + 4\varepsilon \int_0^\infty t e^{\varepsilon t^2}\, e^{-t^2/2}\, dt = 1 + \frac{2\varepsilon}{\frac{1}{2} - \varepsilon}.$$

The last expression is equal to 2 for $\varepsilon = \frac{1}{6}$, which means that $\|f\|_{\psi_2} \leq \sqrt{6}$.   □

References

1. S. Aida, D. Strook, Moment estimates derived from Poincaré and logarithmic Sobolev inequalities. Math. Res. Lett. **1**, 75–86 (1994)
2. N. Alon, R. Boppana, J. Spencer, An asymptotic isoperimetric inequality. Geom. Funct. Anal. **8**, 411–436 (1998)
3. D. Bakry, M. Ledoux, Lévy–Gromov's isoperimetric inequality for an infinite dimensional diffusion generator. Invent. Math. **123**, 259–281 (1996)
4. F. Barthe, E. Milman, Transference principles for log-Sobolev and spectral-gap with applications to conservative spin systems. Commun. Math. Phys. **323**(2), 575–625 (2013)
5. S.G. Bobkov, Localization proof of the isoperimetric Bakry-Ledoux inequality and some applications. Teor. Veroyatnost. i Primenen. **47**(2), 340–346 (2002) Translation in: Theory Probab. Appl. **47**(2), 308–314 (2003)
6. S.G. Bobkov, Perturbations in the Gaussian isoperimetric inequality. J. Math. Sci. **166**(3), 225–238 (2010). New York. Translated from: Problems in Mathematical Analysis, vol. 45 (2010), pp. 3–14
7. S.G. Bobkov, F. Götze, Exponential integrability and transportation cost related to logarithmic Sobolev inequalities. J. Funct. Anal. **163**(1), 1–28 (1999)
8. S.G. Bobkov, C. Houdré, P. Tetali, The subgaussian constant and concentration inequalities. Isr. J. Math. **156**, 255–283 (2006)
9. J. Ding, E. Mossel, Mixing under monotone censoring. Electron. Commun. Probab. **19**, 1–6 (2014)
10. M.D. Kirszbraun, Über die zusammenziehende und Lipschitzsche Transformationen. Fund. Math. **22**, (1934) 77–108.
11. M. Ledoux, *The Concentration of Measure Phenomenon*. Mathematical Surveys and Monographs, vol. 89 (American Mathematical Society, Providence, RI, 2001)
12. K. Marton, Bounding $\bar{d}$-distance by informational divergence: a method to prove measure concentration. Ann. Probab. **24**(2), 857–866 (1996)
13. B. Maurey, Some deviation inequalities. Geom. Funct. Anal. **1**, 188–197 (1991)
14. E.J. McShane, Extension of range of functions. Bull. Am. Math. Soc. **40**(12), 837–842 (1934)
15. E. Milman, On the role of convexity in isoperimetry, spectral gap and concentration. Invent. Math. **177**(1), 1–43 (2009)
16. E. Milman, Properties of isoperimetric, functional and transport-entropy inequalities via concentration. Probab. Theory Relat. Fields **152**(3–4), 475–507 (2012)

17. C.E. Mueller, F.B. Weissler, Hypercontractivity for the heat semigroup for ultraspherical polynomials and on the n-sphere. J. Funct. Anal. **48**, 252–283 (1992)
18. M. Talagrand, An isoperimetric theorem on the cube and the Khinchine–Kahane inequalities. Proc. Am. Math. Soc. **104**, 905–909 (1988)
19. M. Talagrand, Concentration of measure and isoperimetric inequalities in product spaces. Publ. Math. I.H.E.S. **81**, 73–205 (1995)

On Random Walks in Large Compact Lie Groups

Jean Bourgain

Abstract Let G be the group $SO(d)$ or $SU(d)$ with d large. How long does it take for a random walk on G to approximate uniform measure? It is shown that in certain natural examples an ε-approximation is achieved in time $\left(d \log \frac{1}{\varepsilon}\right)^C$.

1 Introduction

In order to put the problem considered in this Note in perspective, we first recall some other relatively recent results around spectral gaps and generation in Lie groups.

It was shown in [5] (resp. [6]) that if Λ is a symmetric finite subset of $SU(2)$ (resp. $SU(d)$) consisting of algebraic elements, such that the countable group $\Gamma = \langle \Lambda \rangle$ generated by Λ is dense, then the corresponding averaging operators

$$Tf = \frac{1}{|\Lambda|} \sum_{g \in \Lambda} f \circ g \tag{1}$$

acting on $L^2(G)$, has a uniform spectral gap (only depending on Λ). This result was generalized in [2] to simple compact Lie groups.

It is not known if the assumption for Λ to be algebraic is needed, and one may conjecture that it is not. Short of providing uniform spectral gaps, Varju [12] established the following property which is the most relevant statement for what follows.

Proposition 1 *Let G be a compact Lie group with semisimple connected component. Let μ be a probability measure on G such that $\text{supp}(\tilde{\mu} * \mu)$, $\tilde{\mu}$ defined by $\int f(x) d\tilde{\mu}(x) = \int f(x^{-1}) d\mu(x)$, generates a dense subgroup of G. Then there is a constant $c > 0$ depending only on μ such that the following holds.*

J. Bourgain (✉)
Institute for Advanced Study, Princeton, NJ 08540, USA
e-mail: bourgain@math.ias.edu

© Springer International Publishing AG 2017
B. Klartag, E. Milman (eds.), *Geometric Aspects of Functional Analysis*,
Lecture Notes in Mathematics 2169, DOI 10.1007/978-3-319-45282-1_4

Let $\varphi \in Lip\ (G)$, $\|\varphi\|_2 = 1$ and $\int_G \varphi = 0$. Then

$$\left\| \int \varphi(h^{-1}g)d\mu(h) \right\|_2 < 1 - c\ \log^{-A}(1 + \|\varphi\|_{Lip}) \tag{2}$$

with A depending on G.

Using (2) and decomposition of the regular representation of G in irreducibles (though this may be avoided), one deduces easily from (2) that it takes time at most $O(\log^A \frac{1}{\varepsilon})$ as $\varepsilon \to 0$ for the random walk governed by μ to produce an ε-approximation of uniform measure on G. Note that for $G = SU(d)$, this statement corresponds to the Solovay-Kitaev estimates on generation, cf. [7], which in fact turns out to be equivalent.

Let us focus on $G = SO(d)$ or $SU(d)$. While the exponent A in (2) is a constant, the prefactor c depends on μ, hence on G, and seems to have received little attention. Basically our aim is to prove a lower bound on c which is powerlike in $\frac{1}{d}$ and without the need for uniform spectral gaps (which may not be always available). We focus on the following model problem brought to the author's attention by T. Spencer (who was motivated by issues in random matrix theory that will not be pursued here). The general setting is as follows (we consider the $SU(d)$-version). Fix some probability measure η on $SU(2)$ such that its support generates a dense group, i.e. $\langle \text{supp } \eta \rangle = SU(2)$. This measure η may be Haar but could be taken discrete as well. Identify $\{0, 1, \ldots, d-1\}$ with the cyclic group $\mathbb{Z}/d\mathbb{Z}$ and denote ν_{ij} the measure η on $SU(2)$ acting on the space $[e_i, e_j]$. Consider the random walk on $SU(d)$ given by

$$Tf(x) = \frac{1}{d} \sum_{i=0}^{d-1} \int f(gx)\nu_{i,i+1}(dg). \tag{3}$$

How long does it take for this random walk to become an ε-approximation of uniform measure on G, with special emphasis on large d? Thus this is a particular instance of the more general issue formulated in the title. While we are unable to address the broader problem, specific cases such as (3) may be analyzed in a satisfactory way (based partly on arguments that are also relevant to the general setting).

We prove

Proposition 2 *In the above setting, ε-approximation of the uniform measure is achieved in time $C(d \log \frac{1}{\varepsilon})^C$, with C a constant independent of d.*

Comment

If η is taken to be a uniform measure on $SU(2)$, better results are available, exploiting Hurwitz' construction of Haar measure (see [8], Sect. 2). In this situation, the operator T displays in fact a uniform spectral gap and the power of $\log \frac{1}{\varepsilon}$ can be taken to be one (cf. [8], Theorem 1). Our interest in this presentation is a more robust approach however.

Basically, one could expect a more general phenomenon (though some additional assumptions are clearly needed). In some sense, it would give a continuous version of the conjecture of Babai and Seress [1] predicting poly-logarithmic diameter for the family of non-Abelian finite simple groups (independently of the choice of generators). Important progress in this direction for the symmetric group appears in [10].

Independently of Spencer's question, related spectral gap and mixing time issues for specific random walks in large (not necessarily compact) linear groups appear in the theory of Anderson localization for 'quasi-one-dimensional' methods in Math Phys.

Consider the strip $\mathbb{Z} \times \mathbb{Z}/d\mathbb{Z}$ and a random Schrödinger operator $\Delta + \lambda V$ with Δ the usual lattice Laplacian on $\mathbb{Z} \times \mathbb{Z}/d\mathbb{Z}$, V a random potential and $\lambda > 0$ the disorder. This model is well known to exhibit pure point spectrum with so-called Anderson localization for the eigenfunctions. The issue here is how the localization length (or equivalently, the Lyapounov exponents in the transfer matrix approach) depend on d when $d \to \infty$.

The classical approach based on Furstenberg's random matrix product theory (acting on exterior powers of $\mathbb{R}^d$), cf. [3], is not quantitative and sheds no light on the role of d. In fact, the first explicit lower bound on Lyapounov exponents seems to appear in [4] (using different techniques based on Green's function analysis), with, roughly speaking exponential dependence on d (while the 'true' behaviour is believed to be rather of the form d^{-C}). Clearly understanding the mixing time for the random walk in the symplectic group $Sp(2d)$ associated to the transfer matrix is crucial. Note that this group is non-compact, which is an added difficulty (for very small λ, depending on d, [11] provides the precise asymptotic of the exponents, based on a multi-dimensional extension of the Figotin-Pastur approach).

2 Some Preliminary Comments

The proof of Proposition 1 in [12] exploits the close relation between 'generation' and 'restricted spectral gaps'. This point of view is also the key idea here in establishing

Proposition 1′ *Let T be defined by (3). Then there is the following estimate*

$$\|Tf\|_2 < 1 - (Cd)^{-C}\big(\log(1 + \|f\|_{\mathrm{Lip}})^{-A}\big) \tag{4}$$

for $f \in \mathrm{Lip}(G)$. $\|f\|_2 = 1, \int_G f = 0$.

Here C and A are constants (denoted differently, because of their different appearance in the argument).

Unlike in [12], we tried to avoid the use of representation theory. The reason for this is the following. If one relies on decomposition of the regular representation of G in irreducibles and the Peter-Weyl theorem, one is faced in the absence of a uniform spectral gap with convergence issues of the generalized Fourier expansion of functions on G of given regularity. Conversely, we also need to understand the regularity of matrix coefficients of the representations of increasing dimension. While these are classical issues, understanding the role of the dimension d does not seem to have been addressed explicitly.

3   Proof of Proposition 1′

For simplicity, we take η to be a uniform measure on $SU(2)$ and indicate the required modifications for the general case in Sect. 5.

According to (3), denote

$$\nu = \frac{1}{d} \sum_{i=0}^{d-1} \nu_{i,i+1} \tag{5}$$

Thus $\nu = \tilde{\nu}$ and T is the corresponding averaging operator.

Let $f \in \mathrm{Lip}(G), \|f\|_2 = 1$ and $\int_G f = 0$. Assuming

$$\left\| \int \tau_g f \nu(dg) \right\|_2^2 = \|Tf\|_2^2 > 1 - \varepsilon \tag{6}$$

(denoting $\tau_g f(x) = f(gx)$) our aim is to obtain a lower bound on ε.

Clearly (6) implies that

$$\left\langle f, \int \tau_g f(\nu * \nu)(dg) \right\rangle > 1 - \varepsilon$$

and

$$\int \|f - \tau_g f\|_2^2 (\nu * \nu)(dg) < 2\varepsilon. \tag{7}$$

Fix $\varepsilon_1 > 0$ to be specified later and denote B_{ε_1} an ε_1-neighborhood (for the operator norm) of Id in $SU(d)$. It is clear from (5) that $\nu(B_{\varepsilon_1}) \gtrsim \varepsilon_1^3$ and hence (7) implies

$$\int \|f - \tau_{g'} f\|_2^2 \nu(dg) \lesssim \varepsilon_1^{-3}\varepsilon \tag{8}$$

for some $g' \in B_{\varepsilon_1}$. Next, partitioning $SU(2)$ in ε_1-cells Ω_α and denoting

$$\Omega_{\alpha,i} = \{g \in SU(d); g(e_j) = e_j \text{ for } j \notin \{i, i+1\} \text{ and } g|_{[e_i,e_{i+1}]} \in \Omega_\alpha\}$$

observe that $\nu(\Omega_{\alpha,i}) \geq \frac{1}{d}\varepsilon_1^3$ so that by (8)

$$\oint_{\Omega_{\alpha,i}} \|f - \tau_{g'g}f\|_2^2 \, \nu(dg) \lesssim d\varepsilon_1^{-6}\varepsilon \ll 1. \tag{9}$$

Exploiting (9), it is clear that we may introduce a collection $\mathcal{G} \subset SU(d)$ with the following properties

$$\|f - \tau_g f\|_2 \lesssim \sqrt{d}\,\varepsilon_1^{-3}\sqrt{\varepsilon} \text{ for } g \in \mathcal{G}. \tag{10}$$

and

Given an element $\gamma \in SU(2)$ and $1 \leq i < j \leq d$, denote γ_{ij} in $SU(d)$ the element defined by

$$\begin{cases} \gamma_{ij}(e_k) & = e_k \text{ for } k \notin \{i,j\} \\ \gamma_{ij}|_{[e_i,e_j]} & = \gamma. \end{cases} \tag{11}$$

Then, for each $\gamma \in SU(2)$ and $1 \leq i \leq d$, there is $g \in \mathcal{G}$ s.t.

$$\|g - \gamma_{i,i+1}\|_2 < \varepsilon_1. \tag{12}$$

At this point, we will invoke generation. Since $\int_G f = 0$,

$$\int_{SU(d)} \|f - \tau_g f\|_2^2 dg = 2$$

and we take some $h_0 \in SU(d)$ s.t.

$$\|f - \tau_{h_0} f\|_2 \geq \sqrt{2}.$$

If $\|h_0 - h_1\| < \delta \sim \frac{1}{\|f\|_{\text{Lip}}}$, then

$$\|\tau_{h_0}f - \tau_{h_1}f\|_2 \leq (\|f\|_{\text{Lip}}\delta)^{\frac{1}{2}} < \frac{1}{2}$$

and consequently

$$\|f - \tau_{h_1}f\|_2 > 1 \text{ if } \|h_0 - h_1\| < \delta. \tag{13}$$

In order to get a contradiction, we need to produce a word $h_1 = g_1 \cdots g_\ell; g_1, \ldots, g_\ell \in \mathcal{G}$ such that

$$\|h_0 - g_1 \cdots g_\ell\| < \delta \tag{14}$$

and

$$\ell < \frac{\varepsilon_1^3}{\sqrt{\varepsilon}\sqrt{d}}. \tag{15}$$

Indeed, (10) implies then that

$$\|f - \tau_{h_1}f\|_2 \leq \|f - \tau_{g_1}f\|_2 + \cdots + \|f - \tau_{g_\ell}f\|_2 < 1.$$

For $1 \leq i < d$, let $\sigma_{i,i+1} \in \mathrm{Sym}(d)$ be the transposition of i and $i+1$. Denote $\tilde{\sigma}_{i,i+1}$ the corresponding unitary operator. Since

$$\{\sigma_{i,i+1}; i = 1, \ldots, d-1\}$$

is a generating set for $\mathrm{Sym}(d)$ consisting of cycles of bounded length, it follows from a result in [9] that the corresponding Cayley graph on $\mathrm{Sym}(d)$ has diameter at most Cd^2. In particular, given $i, j \notin \mathbb{Z}/d\mathbb{Z}, i \neq j, \tilde{\sigma}_{i,j}$ may be realized as a composition of a string of elements $\tilde{\sigma}_{i,i+1}$ of length at most Cd^2. In view of (11), this implies that if $\gamma \in SU(2)$ and $1 \leq i < j \leq d$, then

$$\|\gamma_{ij} - g\| < cd^2\varepsilon_1 \tag{16}$$

for some $g \in \mathcal{G}_{\ell_1}, \ell_1 < cd^2$ ($\mathcal{G}_\ell = $ words of size ℓ written in g).

Let $\kappa > 0$,

$$\kappa^2 > cd^2\varepsilon_1. \tag{17}$$

Adopting the Lie-algebra point of view, the preceding implies that given $s \in \mathbb{R}$, $|s| < 1$ and $z \in \mathbb{C}, |z| < 1$, then

$$\mathrm{dist}\left(Id + \kappa\big(is(e_i \otimes e_i) - is(e_j \otimes e_j) + z(e_i \otimes e_j) - \bar{z}(e_j \otimes e_i)\big), \mathcal{G}_{\ell_1}\right) < \kappa^2 \tag{18}$$

and therefore

$$\mathrm{dist}\,(I + \kappa A, \mathcal{G}_{d^2\ell_1}) < d^2\kappa^2 \tag{19}$$

for skew-symmetric $A, \|A\| \leq 2\pi$.

Let $h \in SU(d), h = e^A$ with A as above. Taking $\kappa = \frac{1}{r}$, we have

$$e^A = (e^{\frac{1}{r}A})^r = \left(1 + \frac{1}{r}A\right)^r + O\left(\frac{1}{r}\right)$$

and therefore, by (19)

$$\mathrm{dist}\,(h_0, \mathcal{G}_{rd^2\ell_1}) \leq rd^2\kappa^2 = \frac{d^2}{r}. \tag{20}$$

Taking $\kappa = \frac{1}{r} = d^{-C}$ and $\varepsilon_1 = d^{-2C-2}$, (20) ensure that

$$\text{dist}\,(h, \mathcal{G}_d c_1) < d^{-C} \text{ for all } h \in SU(d). \tag{21}$$

Next, we rely on the Solovay-Kitaev commutator technique to produce approximations at smaller scale. This procedure is in fact dimensional free (see the comment in [7] following Lemma 2 in order to eliminate a polynomial prefactor in d—which actually would be harmless if we start from scales $\varepsilon_0 = d^{-C}$). The conclusion is that

$$\text{dist}\,(h, \mathcal{G}_\ell) < \tau \text{ for all } h \in SU(d)$$

may be achieved with

$$\ell < d^{C_1}\left(\log\frac{1}{\tau}\right)^A.$$

Returning to (14), (15), we obtain the condition

$$d^{C_0}\log^A(1 + \|f\|_{\text{Lip}}) < \frac{\varepsilon_1^3}{\sqrt{\varepsilon}\sqrt{d}} = d^{-C_2}\varepsilon^{-\frac{1}{2}} \tag{22}$$

and Proposition 1$'$ follows.

4   Proof of Proposition 2

The disadvantage of our approach is that T is not restricted to finite dimensional invariant subspaces of $L^2(G)$ so that strictly speaking, one can not rely on a spectral gap argument to control the norm of iterates of T.

But Proposition 1$'$ nevertheless permit to derive easily the following

Proposition 3 *Assume $f \in \text{Lip}(G)$, $\|f\|_2 = 1$, $\int_G f = 0$. Let $0 < \rho < \frac{1}{2}$. Then*

$$\|T^\ell f\|_2 < \rho \tag{23}$$

provided

$$\ell > Cd^C.\log^A(1 + \|f\|_{\text{Lip}}).\left(\log\frac{1}{\rho}\right)^{A+1}. \tag{24}$$

Proof Let $B = \|f\|_{\text{Lip}}$. Clearly $\|T^\ell f\|_{\text{Lip}} \leq B$ also.

Fix some ℓ and let $f_1 = \frac{T^\ell f}{\|T^\ell f\|_2}$. Hence $\|f_1\|_{\text{Lip}} \leq \frac{B}{\|T^\ell f\|_2}$.

Applying Proposition 1′, it follows that

$$\|T^{\ell+1}f\|_2 \le \|T^\ell f\|_2 (1 - \varepsilon_\ell)$$

with

$$\varepsilon_\ell = cd^{-C}\left(\log\left(1 + \frac{B}{\|T^\ell f\|_2}\right)\right)^{-A} > cd^{-C}\left(\log(1+B)\right)^{-A}\left(\log\left(1 + \frac{1}{\|T^\ell f\|_2}\right)\right)^{-A}.$$

Hence, assuming $\|T^\ell f\|_2 > \rho$, we obtain

$$\rho < \left(1 - cd^{-C}\left(\log(1+B)\right)^{-A}\left(\log\frac{1}{\rho}\right)^{-A}\right)^\ell$$

implying (24).   □

Proof of Proposition 2 Apply Proposition 3 with $\log B \sim \log\frac{1}{\varepsilon}$ and $\log\frac{1}{\rho} \sim d^2 \log\frac{1}{\varepsilon}$.

5 Variants

The previous argument is clearly very flexible and may be applied in other situations.

Returning to Sect. 3, assume more generally η a probability measure on $SU(2)$ satisfying $\overline{\langle \text{supp } \eta \rangle} = SU(2)$. Note that by Proposition 1, $\eta^{(\ell)}$ with $\ell \sim (\log\frac{1}{\varepsilon_1})^c \sim (\log d)^c$ provides an ε_1-approximation of Haar measure on $SU(2)$. It follows from (3), (7) that

$$\int \|f - \tau_g f\|_2^2 (\nu_{i,i+1} * \nu_{i,i+1})(dg) < 2d^2\varepsilon$$

and hence

$$\int \|f - \tau_g f\|_2^2 \nu_{i,i+1}^{(\ell)}(dg) < \ell d^2 \varepsilon$$

$$\fint_{\Omega_{a,i}} \|f - \tau_g f\|_2^2 \, \nu_{i,i+1}^{(\ell)}(dg) \lesssim \ell d^2 \varepsilon_1^{-3}\varepsilon.$$

The collection $\mathcal{G}$ may then be introduced similarly. Proposition 2 remains valid.

Let us point out that it is unknown if in general the density assumption $\overline{\langle \text{supp } \eta \rangle} = SU(2)$ implies a uniform spectral gap (see the discussion on Sect. 1).

Instead of (3), one may introduce at time $k = \mathbb{Z}_+$ the discrete average $T_k = \frac{1}{2}(\tau_g + \tau_{g-1})\cdots$ where we first pick some $i \in \mathbb{Z}/d\mathbb{Z}$ and then choose a random element $g \in SU(2)$ acting on $[e_i, e_{i+1}]$ according to η. In this situation, one obtains random walks on $SU(d)$ indexed by an additional probability space $\otimes(\mathbb{Z}/d\mathbb{Z} \otimes SU(2))$

$$T^\omega = \cdots T_k T_{k-1} \cdots T_1 \tag{25}$$

and may ask for the typical mixing time of a realization.

Rather straightforward adjustments of the arguments appearing in the proof of Proposition 1′ combined with some Markovian considerations permit us to establish the analogue of Proposition 2 for T^ω. Thus

Proposition 4 *Let T^ω be defined by (25). Then, with large probability in ω, ε-approximation of uniform measure on $SU(d)$ may be achieved in time $C(d \log \frac{1}{\varepsilon})^C$.*

Acknowledgements The author is grateful to P. Varju for his comments and also for pointing out various references.

P. Varju also reported the following somewhat related question of A. Lubotzky: Does $SU(d)$ admit a finite set of generators with a spectral gap that is uniform in d?

This work was partially supported by NSF grants DMS-1301619.

References

1. L. Babai, A. Seress, On the diameter of permutation groups. Eur. J. Combin. **13**, 231–243 (1992)
2. Y. Benoist, N. de Saxcé, A spectral gap theorem in simple Lie groups. Invent. Math. **205**(2), 337–361 (2016)
3. A. Bougerol, J. Lacroix, *Product of Random Matrices with Applications to Schrödinger Operators* (Birkhäuser, Basel, 1985)
4. J. Bourgain, A lower bound for the Lyapounov exponents of the random Schrödinger operator on a strip. J. Stat. Phys. **153**(1), 1–9 (2013)
5. J. Bourgain, A. Gamburd, On the spectral gap for finitely generated subgroups of $SU(2)$. Invent. Math. A1 **201**, 83–121 (2008)
6. J. Bourgain, A. Gamburd, A spectral gap theorem in $SU(d)$. J. Eur. Math. Soc. **14**(5), 1455–1511 (2012)
7. C. Dawson, M. Nielsen, The Solovay-Kitaev algorithm. Quantum Inf. Comput. **6**(1), 81–95 (2006)
8. P. Diaconis, L. Saloff-Coste, Bounds on the Kac's master equation. Commun. Math. Phys. **209**(3), 729–755 (2000)
9. J. Driscoll, M. Furst, Computing short generator sequences. Inf. Comput. **72**, 117–132 (1987)
10. H. Helfgott, A. Seress, On the diameter of permutation groups. Ann. Math. (2) **179** (2), 611–658 (2014)
11. H. Schulz-Baldes, Perturbation theory for the Lyapounov exponents of an Anderson model on a strip. Geom. Funct. Anal. **14**, 1029–1117 (2004)
12. P. Varju, Random walks in compact groups. Doc. Math. **18**, 1137–1175 (2013)

On a Problem of Farrell and Vershynin in Random Matrix Theory

Jean Bourgain

Abstract We settle a question of Farrell and Vershynin on the inverse of the perturbation of a given arbitrary symmetric matrix by a GOE element.

1 Introduction

In [1], the authors consider the invertibility of $d \times d$-matrices of the form $D + R$, with D an arbitrary symmetric deterministic matrix and R a symmetric random matrix whose independent entries have continuous distributions with bounded densities. In this setting, a uniform estimate

$$\|(D + R)^{-1}\| = O(d^2) \tag{1}$$

is shown to hold with high probability. The authors conjecture that (1) may be improved to $O(\sqrt{d})$. The purpose of this short Note is to prove this in the case R is Gaussian. Thus we have (stated in the ℓ_d^2-normalized setting).

Proposition *Let T be an arbitrary matrix in $Sym(d)$. Then, for A (normalized) in GOE, there is a uniform estimate*

$$\|(A + T)^{-1}\| = O(d) \tag{2}$$

with large probability.

J. Bourgain (✉)
Institute for Advanced Study, Princeton, NJ 08540, USA
e-mail: bourgain@math.ias.edu

© Springer International Publishing AG 2017
B. Klartag, E. Milman (eds.), *Geometric Aspects of Functional Analysis*,
Lecture Notes in Mathematics 2169, DOI 10.1007/978-3-319-45282-1_5

2 Proof of the Proposition

By invariance of GOE under orthogonal transformations, we may assume T diagonal. Let K be a suitable constant and partition

$$\{1,\ldots,d\} = \Omega_1 \cup \Omega_2$$

with

$$\Omega_1 = \{j = 1,\ldots,d; |T_{jj}| > K\}.$$

Denote $T^{(i)} = \pi_{\Omega_i} T \pi_{\Omega_i} (i = 1, 2)$ and $A^{(i,j)} = \pi_{\Omega_i} A \pi_{\Omega_j} (i,j = 1, 2)$. Since

$$(A^{(1,1)} + T^{(1)})^{-1} = (I + (T^{(1)})^{-1} A^{(1,1)})(T^{(1)})^{-1}$$

and

$$\|(T^{(1)})^{-1} A^{(1,1)}\| \leq \frac{1}{K} \|A^{(1,1)}\| < \frac{1}{2}$$

with large probability, we ensure that

$$\|(A^{(1,1)} + T^{(1)})^{-1}\| < 1. \tag{3}$$

Next, write by the Schur complement formula

$$(A + T)^{-1}$$
$$= \begin{pmatrix} (A^{(1,1)} + T^{(1)})^{-1} + (A^{(1,1)} + T^{(1)})^{-1} A^{(1,2)} S^{-1} A^{(2,1)} (A^{(1,1)} + T^{(1)})^{-1} & -(A^{(1,1)} + T^{(1)})^{-1} A^{(1,2)} S^{-1} \\ -S^{-1} A^{(2,1)} (A^{(1,1)} + T^{(1)})^{-1} & S^{-1} \end{pmatrix}$$

$$\tag{4}$$

defining

$$S = A^{(2,2)} + T^{(2)} - A^{(2,1)} (A^{(1,1)} + T^{(1)})^{-1} A^{(1,2)}. \tag{5}$$

Hence by (4)

$$\|(A + T)^{-1}\| \leq C(1 + \|(A^{(1,1)} + T^{(1)})^{-1}\|^2)(1 + \|A\|^2)\|S^{-1}\|$$
$$\leq C_1 \|S^{-1}\|. \tag{6}$$

Note that $A^{(2,2)}$ and $A^{(2,1)}(A^{(1,1)}+T^{(1)})^{-1}A^{(1,2)}$ are independent in the A randomness. Thus S may be written in the form

$$S = A^{(2,2)} + S_0 \tag{7}$$

with $S_0 \in \mathrm{Sym}(d)$, $\|S_0\| < O(1)$ (by construction, $\|T^{(2)}\| \leq K$) and $A^{(2,2)}$ and S_0 independent.

Fixing S_0, we may again exploit the invariance to put S_0 in diagonal form, obtaining

$$A^{(2,2)} + S_0' \text{ with } S_0' \text{ diagonal}. \tag{8}$$

Hence, we reduced the original problem to the case T is diagonal and $\|T\| < K + 1$.

Note however that (8) is a $(d_1 \times d_1)$-matrix and since d_1 may be significantly smaller than d, $A^{(2,2)}$ is not necessarily normalized anymore. Thus after renormalization of $A^{(2,2)}$, setting

$$A_1 = \left(\frac{d}{d_1}\right)^{\frac{1}{2}} A^{(2,2)} \tag{9}$$

and denoting

$$T_1 = \left(\frac{d}{d_1}\right)^{\frac{1}{2}} S_0' \tag{10}$$

we have

$$\|T_1\| < \left(\frac{d}{d_1}\right)^{\frac{1}{2}}(K + 1) \tag{11}$$

while the condition [cf. (6)]

$$\|(A^{(2,2)} + S_0')^{-1}\| = O(d) \tag{12}$$

becomes

$$\|(A_1 + T_1)^{-1}\| = O(\sqrt{dd_1}). \tag{13}$$

At this point, we invoke Theorem 1.2 from [2]. As Vershynin kindly pointed out to the author, the argument in [2] simplifies considerably in the Gaussian case. Examination of the proof shows that in fact the statement from [2], Theorem 1.2 can be improved in this case as follows.

Claim *Let A be a d × d normalized GOE matrix and T a deterministic, diagonal (d × d)-matrix. Then*

$$\mathbb{P}[\|(A+T)^{-1}\| > \lambda d] \le C(1+\|T\|)\lambda^{-\frac{1}{9}}. \tag{14}$$

We distinguish two cases. If $d_1 \ge \frac{1}{C_2}d, C_2 > C_1^3$, immediately apply the above claim with d replaced by d_1, A by A_1 and T by T_1. Thus by (11)

$$\mathbb{P}[\|(A_1+T_1)^{-1}\| > \lambda\sqrt{dd_1}] \le C(1+\|T_1\|)\left(\frac{d_1}{d}\right)^{-\frac{1}{18}}\lambda^{-\frac{1}{9}} < C\left(1+\sqrt{C_2}(K+1)\right)\lambda^{-\frac{1}{9}} \tag{15}$$

and (12) follows. If $d_1 < \frac{1}{C_2}d$, repeat the preceding replacing A by A_1, T by T_1. In the definition of Ω_1, replace K by $K_1 = 2K$, so that (3) will hold with probability at least

$$1 - e^{-cK_1^2} = 1 - e^{-4cK^2} \tag{16}$$

the point being of making the measure bounds $e^{-c4^sK^2}$, $s = 0, 1, 2, \ldots$ obtained in an iteration, sum up to $e^{-c_1K^2} = o(1)$.

Note that in (13), we only seek for an estimate

$$\|(A_1+T_1)^{-1}\| < O\left(\frac{\sqrt{C_2}}{C_1}d_1\right) \tag{17}$$

hence, cf. (12)

$$\|(A_1^{(2,2)} + S_{1,0}')^{-1}\| < O\left(\frac{\sqrt{d_2}}{C_1}d_1\right) \tag{18}$$

where $A_1^{(2,2)}$ and $S_{1,0}'$ are defined as before, considering now A_1 and T_1. Hence (13) gets replaced by

$$\|(A_2+T_2)^{-1}\| = O\left(\frac{\sqrt{C_2}}{C_1}\sqrt{d_1d_2}\right) \tag{19}$$

where A_2, T_2 are $(d_2 \times d_2)$-matrices,

$$\|T_2\| < \left(\frac{d_1}{d_2}\right)^{\frac{1}{2}}(2K + 1). \tag{20}$$

Assuming $d_2 \geq \frac{1}{C_2}d_1$, we obtain instead of (15)

$$\mathbb{P}[\|(A_2 + T_2)^{-1}\| > \lambda\frac{\sqrt{C_2}}{C_1}\sqrt{d_1d_2}] \leq C\big(1 + \sqrt{C_2}(K_1 + 1)\big)\Big(\frac{\sqrt{C_2}}{C_1}\lambda\Big)^{-\frac{1}{9}}$$

$$< C\big(1 + \sqrt{C_2}(K + 1)\big)\,(2C_1^{\frac{1}{9}}C_2^{-\frac{1}{18}})\lambda^{-\frac{1}{9}}$$

(21)

and we take C_2 to ensure that $2C_1^{\frac{1}{9}}C_2^{-\frac{1}{18}} < \frac{1}{2}$.

The continuation of the process is now clear and terminates in at most $^2\log d$ steps. At step s, we obtain if $d_{s+1} \geq \frac{1}{C_2}d_s$

$$\mathbb{P}\Big[\|(A_{s+1} + T_{s+1})^{-1}\| > \lambda\Big(\frac{\sqrt{C_2}}{C_1}\Big)^s\sqrt{d_sd_{s+1}}\Big] < C\big(1 + \sqrt{C_2}(K + 1)\big)2^{-s}\lambda^{-\frac{1}{9}}.$$

(22)

Summation over s gives a measure estimate $O(\lambda^{-\frac{1}{9}}) = o(1)$.

This concludes the proof of the Proposition. From quantitative point of view, previous argument shows

Proposition' *Let T and A be as in the Proposition. Then*

$$\mathbb{P}[\|(A + T)^{-1}\| > \lambda d] < O(\lambda^{-\frac{1}{10}}).$$

(23)

Acknowledgements The author is grateful to the referee for his comments on an earlier version. This work was partially funded by NSF grant DMS-1301619.

Note The author's interest in this issue came up in the study (joint with I. Goldsheid) of quantitative localization of eigenfunctions of random band matrices. The purpose of this Note is to justify some estimates in this forthcoming work.

References

1. B. Farrell, R. Vershynin, Smoothed analysis of symmetric random matrices with continuous distributions. Proc. AMS **144**(5), 2259–2261 (2016)
2. R. Vershynin, Invertibility of symmetric random matrices. Random Struct. Algoritm. **44**(2), 135–182 (2014)

Valuations on the Space of Quasi-Concave Functions

Andrea Colesanti and Nico Lombardi

Abstract We characterize the valuations on the space of quasi-concave functions on $\mathbb{R}^N$, that are rigid motion invariant and continuous with respect to a suitable topology. Among them we also provide a specific description of those which are additionally monotone.

1 Introduction

A *valuation* on a space of functions X is an application $\mu : X \to \mathbb{R}$ such that

$$\mu(f \vee g) + \mu(f \wedge g) = \mu(f) + \mu(g) \tag{1}$$

for every $f, g \in X$ s.t. $f \vee g, f \wedge g \in X$; here "$\vee$" and "$\wedge$" denote the point-wise maximum and minimum, respectively. The condition (1) can be interpreted as a finite additivity property (typically verified by integrals).

The study of valuations on spaces of functions stems principally from the theory of valuations on classes of sets, in which the main current concerns *convex bodies*. We recall that a convex body is simply a compact convex subset of $\mathbb{R}^N$, and the family of convex bodies is usually denoted by $\mathcal{K}^N$. An application $\sigma : \mathcal{K}^N \to \mathbb{R}$ is called a valuation if

$$\sigma(K \cup L) + \sigma(K \cap L) = \sigma(K) + \sigma(L) \tag{2}$$

for every $K, L \in \mathcal{K}^N$ such that $K \cup L \in \mathcal{K}^N$ (note that the intersection of convex bodies is a convex body). Hence, in passing from (2) to (1) union and intersection are replaced by maximum and minimum respectively. A motivation is that the characteristic function of the union (resp. the intersection) of two sets is the maximum (resp. the minimum) of their characteristic functions.

The theory of valuations is an important branch of modern convex geometry (the theory of convex bodies). The reader is referred to the monograph [16]

A. Colesanti (✉) • N. Lombardi
Dipartimento di Matematica e Informatica "U. Dini", Viale Morgagni 67/A, 50134 Firenze, Italy
e-mail: andrea.colesanti@unifi.it; colesant@math.unifi.it; nico.lombardi@unifi.it

© Springer International Publishing AG 2017
B. Klartag, E. Milman (eds.), *Geometric Aspects of Functional Analysis*,
Lecture Notes in Mathematics 2169, DOI 10.1007/978-3-319-45282-1_6

for an exhaustive description of the state of the art in this area, and for the corresponding bibliography. The valuations on $\mathcal{K}^N$, continuous with respect to the Hausdorff metric and rigid motion invariant, i.e. invariant with respect to composition with translations and proper rotations (elements of $O(N)$), have been completely classified in a celebrated result by Hadwiger (see [5–7]). Hadwiger's theorem asserts that any valuation σ with these properties can be written in the form

$$\sigma(K) = \sum_{i=0}^{N} c_i V_i(K) \quad \forall K \in \mathcal{K}^N, \tag{3}$$

where $c_1, \ldots, c_N$ are constants and $V_1, \ldots, V_N$ denote the *intrinsic volumes* (see Sect. 2, for the definition). This fact will be of great importance for the results presented here.

Let us give a brief account of the main known results in the area of valuations on function spaces. Wright, in his PhD thesis [21] and subsequently in collaboration with Baryshnikov and Ghrist [2], characterized rigid motion invariant and continuous valuations on the class of *definable functions* (we refer to the quoted papers for the definition). Their result is very similar to Hadwiger's theorem; roughly speaking it asserts that every valuation is the linear combination of integrals of intrinsic volumes of level sets. This type of valuations will be crucial in our results as well.

Rigid motion invariant and continuous valuations on $L^p(\mathbb{R}^N)$ and on $L^p(\mathbb{S}^{n-1})$ $(1 \leq p < \infty)$ have been studied and classified by Tsang in [17]. Basically, Tsang proved that every valuation μ with these properties is of the type

$$\mu(f) = \int \phi(f) dx \tag{4}$$

(here the integral is performed on $\mathbb{R}^N$ or $\mathbb{S}^{n-1}$) for some function ϕ defined on $\mathbb{R}$ verifying suitable growth conditions. Subsequently, the results of Tsang have been extended to Orlicz spaces by Kone in [8]. Also, the special case $p = \infty$ was studied by Cavallina in [3].

Valuations on the space of functions of bounded variations and on Sobolev spaces have been recently studied by Wang and Ma respectively, in [14, 19, 20] and [13].

In [4] the authors consider rigid motion invariant and continuous valuations (with respect to a certain topology that will be recalled later on) on the space of convex functions, and found some partial characterization results under the assumption of monotonicity and homogeneity.

Note that the results that we have mentioned so far concern *real-valued* valuations, but there are also studies regarding other types of valuations (e.g. matrix-valued valuations, or Minkowski and Blaschke valuations, etc.) that are interlaced with the results mentioned previously. A strong impulse to these studies have been given by Ludwig in the works [9–12]; the reader is referred also to [18] and [15].

Here we consider the space $\mathcal{C}^N$ of *quasi-concave* functions of N real variables. A function $f : \mathbb{R}^N \to \mathbb{R}$ is quasi-concave if it is non-negative and for every $t > 0$ the level set

$$L_t(f) = \{x \in \mathbb{R}^N : f(x) \geq t\}$$

is (either empty or) a compact convex set. $\mathcal{C}^N$ includes log-concave functions and characteristic functions of convex bodies as significant examples.

We consider valuations $\mu : \mathcal{C}^N \to \mathbb{R}$ which are rigid motion invariant (with the same notion as before for rigid motion transformations), i.e.

$$\mu(f) = \mu(f \circ T)$$

for every $f \in \mathcal{C}^N$ and for every rigid motion T of $\mathbb{R}^N$. We also impose a continuity condition on μ: if f_i, $i \in \mathbb{N}$, is a *monotone* (either increasing or decreasing) sequence in $\mathcal{C}^N$, converging to $f \in \mathcal{C}^N$ point-wise in $\mathbb{R}^N$, then we must have

$$\lim_{i \to \infty} \mu(f_i) = \mu(f).$$

In Sect. 4.1 we provide some motivation for this definition, comparing this notion of continuity with other possible choices.

There is a simple way to construct valuations on $\mathcal{C}^N$. To start with, note that if $f, g \in \mathcal{C}^N$ and $t > 0$

$$L_t(f \vee g) = L_t(f) \cup L_t(g), \quad L_t(f \wedge g) = L_t(f) \cap L_t(g). \tag{5}$$

Let ψ be a function defined on $(0, \infty)$ and fix $t_0 > 0$. Define, for every $f \in \mathcal{C}^N$,

$$\mu_0(f) = V_N(L_{t_0}(f))\psi(t_0).$$

Using (5) and the additivity of volume we easily deduce that μ_0 is a rigid motion invariant valuation. More generally, we can overlap valuations of this type at various levels t, and we can further replace V_N by any intrinsic volume V_k:

$$\mu(f) = \int_{(0,\infty)} V_k(L_t(f))\psi(t)\, dt = \int_{(0,\infty)} V_k(L_t(f))\, d\nu(t), \quad f \in \mathcal{C}^N, \tag{6}$$

where ν is the measure with density ψ. This is now a rather ample class of valuations; as we will see, basically every *monotone* valuation on $\mathcal{C}^N$ can be written in this form. To proceed, we observe that the function

$$t \to V_k(L_t(f))$$

is decreasing. In particular it admits a distributional derivative which is a non-positive measure. For ease of notation we write this measure in the form $-S_k(f; \cdot)$ where now $S_k(f; \cdot)$ is a (non-negative) Radon measure on $(0, \infty)$. Then, integrating by parts in (6) (boundary terms can be neglected, as it will be clear in the sequel) we obtain:

$$\mu(f) = \int_{(0,\infty)} \phi(t) \, dS_k(f;t) \tag{7}$$

where ϕ is a primitive of ψ. Our first result is the fact that functionals of this type exhaust, by linear combinations, all possible rigid motion invariant and continuous valuations on $\mathcal{C}^N$.

Theorem 1.1 *A map $\mu : \mathcal{C}^N \to \mathbb{R}$ is an invariant and continuous valuation on $\mathcal{C}^N$ if and only if there exist $(N + 1)$ continuous functions ϕ_k, $k = 0, \ldots, N$ defined on $[0, \infty)$, and $\delta > 0$ such that: $\phi_k \equiv 0$ in $[0, \delta]$ for every $k = 1, \ldots, N$, and*

$$\mu(f) = \sum_{k=0}^{N} \int_{[0,\infty)} \phi_k(t) dS_k(f;t) \quad \forall f \in \mathcal{C}^N.$$

The condition that each ϕ_k, except for ϕ_0, vanishes in a right neighborhood of the origin guarantees that the integral in (7) is finite for every $f \in \mathcal{C}^N$ (in fact, it is equivalent to this fact). As in the case of Hadwiger theorem, the proof of this result is based on a preliminary step in which valuations that are additionally *simple* are classified. A valuation μ on $\mathcal{C}^N$ is called simple if

$$f = 0 \text{ a.e. in } \mathbb{R}^N \quad \Rightarrow \quad \mu(f) = 0.$$

Note that for $f \in \mathcal{C}^N$, being zero a.e. is equivalent to say that the dimension of the support of f (which is a convex set) is strictly smaller than N. The following result is in a sense analogous to the so-called *volume theorem* for convex bodies.

Theorem 1.2 *A map $\mu : \mathcal{C}^N \to \mathbb{R}$ is an invariant, continuous and simple valuation on $\mathcal{C}^N$ if and only if there exists a continuous function ϕ defined on $[0, \infty)$, with $\phi \equiv 0$ in $[0, \delta]$ for some $\delta > 0$, such that*

$$\mu(f) = \int_{\mathbb{R}^n} \phi(f(x)) dx \quad \forall f \in \mathcal{C}^N,$$

or, equivalently,

$$\mu(f) = \int_{[0,\infty)} \phi(t) dS_N(f;t).$$

Here the equivalence of the two formulas follows from the layer cake principle. The representation formula of Theorem 1.1 becomes more legible in the case of monotone valuations. Here, each term of the sum is clearly a weighted mean of the intrinsic volumes of the level sets of f.

Theorem 1.3 *A map μ is an invariant, continuous and monotone increasing valuation on $\mathcal{C}^N$ if and only if there exists $(N + 1)$ Radon measures on $[0, \infty)$, v_k, $k = 0, \ldots, N$, such that each v_k is non-negative, non-atomic and, for $k \geq 1$, the support of v_k is contained in $[\delta, \infty)$ for a suitable $\delta > 0$, and*

$$\mu(f) = \sum_{k=0}^{N} \int_{[0,\infty)} V_k(L_t(f)) \, dv_k(t), \quad \forall f \in \mathcal{C}^N.$$

We remark that the non-negativity of v_k depends on the monotone increasing property of μ, as we will see in Sect. 8.

As we already mentioned, and it will be explained in details in Sect. 5.3, the passage

$$\int_{[0,\infty)} \phi_k(t) dS_k(f; t) \longrightarrow \int_{[0,\infty)} V_k(L_t(f)) \, dv_k(t)$$

is provided merely by an integration by parts, when this is permitted by the regularity of the function ϕ_k.

The paper is organized as follows. In the next section we provide some notions from convex geometry. Section 3 is devoted to the basic properties quasi-convex functions, while in Sect. 4 we define various types of valuations on the space $\mathcal{C}^N$. In Sect. 5 we introduce the integral valuations, which occur in Theorems 1.1 and 1.3. Theorem 1.2 is proved in Sect. 6, while Sects. 6 and 7 contain the proof of Theorems 1.1 and 1.3, respectively.

2 Notations and Preliminaries

We work in the N-dimensional Euclidean space $\mathbb{R}^N$, $N \geq 1$, endowed with the usual scalar product $(\cdot, \cdot)$ and norm $\| \cdot \|$. Given a subset A of $\mathbb{R}^N$, int(A), cl(A) and ∂A denote the interior, the closure and the topological boundary of A, respectively. For every $x \in \mathbb{R}^N$ and $r \geq 0$, $B_r(x)$ is the closed ball of radius r centered at x; in particular, for simplicity we will write B_r instead of $B_r(0)$. We recall that a *rigid motion* of $\mathbb{R}^N$ will be the composition of a translation and a rotation of $\mathbb{R}^N$ (i.e. an isometry). The Lebesgue measure in $\mathbb{R}^N$ will be denoted by V_N.

2.1 Convex Bodies

We recall some notions and results from convex geometry that will be used in the sequel. Our main reference on this subject is the monograph by Schneider [16]. As stated in the introduction the class of convex bodies is denoted by $\mathcal{K}^N$. For $K, L \in$

$\mathcal{K}^N$, we define the *Hausdorff distance* of K and L as

$$\delta(K,H) = \max\{\sup_{x\in K} \text{dist}(x,H), \sup_{y\in H} \text{dist}(K,y)\}.$$

Accordingly, a sequence of convex bodies $\{K_n\}_{n\in\mathbb{N}} \subseteq \mathcal{K}^N$ is said to converge to $K \in \mathcal{K}^N$ if

$$\delta(K_n, K) \to 0, \text{ as } n \to +\infty.$$

Remark 2.1 $\mathcal{K}^N$ with respect to Hausdorff distance is a complete metric space.

Remark 2.2 For every convex subset C of $\mathbb{R}^N$, and consequently for convex bodies, its dimension dim (C) can be defined as follows: dim(C) is the smallest integer such that there exists an affine sub-space of $\mathbb{R}^N$ containing C.

We are ready, now, to introduce some functionals operating on $\mathcal{K}^N$, the intrinsic volumes, which will be of fundamental importance in this paper. Among the various ways to define intrinsic volumes, we choose the one based on the Steiner formula. Given a convex body K and $\epsilon > 0$, the *parallel set* of K is

$$K_\epsilon = \{x \in \mathbb{R}^N \mid \text{dist}(x,K) \le \epsilon\}.$$

The following result asserts that the volume of the parallel body is a polynomial in ϵ, and contains the definition of intrinsic volumes.

Theorem 2.3 (Steiner Formula) *There exist N functions $V_0,\ldots,V_{N-1} : \mathcal{K}^N \to \mathbb{R}_+$ such that, for all $K \in \mathcal{K}^N$ and for all $\epsilon \ge 0$, we have*

$$V_N(K_\epsilon) = \sum_{i=0}^{N} V_i(K)\omega_{N-i}\epsilon^{N-i},$$

where ω_j denotes the volume of the unit ball in the space $\mathbb{R}^j$. $V_0(K),\ldots,V_N(K)$ are called the intrinsic volumes of K.

Hence one of the intrinsic volumes is the Lebesgue measure. Moreover V_0 is the Euler characteristic, so that for every K we have $V_0(K) = 1$. The name intrinsic volumes comes from the following fact: assume that K has dimension $j \in \{0,\ldots,N\}$, i.e. there exists a j-dimensional affine subspace of $\mathbb{R}^N$ containing K, and j is the lowest number with this property (we will write $\dim(K) = j$). Then K can be seen as a subset of $\mathbb{R}^j$ and $V_j(K)$ is the Lebesgue measure of K as a subset of $\mathbb{R}^j$. Intrinsic volumes have many other properties, listed in the following proposition.

Proposition 2.4 (Properties of Intrinsic Volumes) *For every $k \in \{0,\ldots,N\}$ the function V_k is:*

- *rigid motion invariant;*

- *continuous with respect to the Hausdorff metric;*
- *monotone increasing: $K \subset L$ implies $V_k(K) \le V_k(L)$;*
- *a valuation:*

$$V_k(K \cup L) + V_k(K \cap L) = V_k(K) + V_k(L) \quad \forall K, L \in \mathcal{K}^N \ s.t. \ K \cup L \in \mathcal{K}^N.$$

We also set conventionally

$$V_k(\varnothing) = 0, \quad \forall k = 0, \dots, N.$$

The previous properties essentially characterizes intrinsic volumes as stated by the following result proved by Hadwiger, already mentioned in the introduction.

Theorem 2.5 (Hadwiger) *If σ is a continuous and rigid motion invariant valuation, then there exist $(N + 1)$ real coefficients $c_0, \dots, c_N$ such that*

$$\sigma(K) = \sum_{i=0}^{N} c_i V_i(K),$$

for all $K \in \mathcal{K}^N \cup \{\varnothing\}$.

The previous theorem claims that $\{V_0, \dots, V_N\}$ spans the vector space of all continuous and invariant valuations on $\mathcal{K}^N \cup \{\varnothing\}$. It can be also proved that $V_0, \dots, V_N$ are linearly independent, so they form a basis of this vector space. In Hadwiger's Theorem continuity can be replaced by monotonicity hypothesis, obtaining the following result.

Theorem 2.6 *If σ is a monotone increasing (resp. decreasing) rigid motion invariant valuation, then there exist $(N + 1)$ coefficients $c_0, \dots, c_N$ such that $c_i \ge 0$ (resp. $c_i \le 0$) for every i and*

$$\sigma(K) = \sum_{i=0}^{N} c_i V_i(K),$$

for all $K \in \mathcal{K}^N \cup \{\varnothing\}$.

A special case of the preceding results concerns *simple* valuations. A valuation μ is said to be simple if

$$\mu(K) = 0 \quad \forall K \in \mathcal{K}^N \ s.t. \ \dim(K) < N.$$

Corollary 2.7 (Volume Theorem) *Let $\sigma : \mathcal{K}^N \cup \{\varnothing\} \to \mathbb{R}$ be a rigid motion invariant, simple and continuous valuation. Then there exists a constant c such that*

$$\mu = c V_N.$$

Remark 2.8 In the previous theorem continuity can be replaced by the following weaker assumption: for every *decreasing* sequence K_i, $i \in \mathbb{N}$, in $\mathcal{K}^N$, converging to $K \in \mathcal{K}^N$,

$$\lim_{i \to \infty} \sigma(K_i) = \sigma(K).$$

This follows, for instance, from the proof of the volume theorem given in [6].

3 Quasi-Concave Functions

3.1 The Space $\mathcal{C}^N$

Definition 3.1 A function $f : \mathbb{R}^N \to \mathbb{R}$ is said to be *quasi-concave* if

- $f(x) \geq 0$ for every $x \in \mathbb{R}^N$,
- for every $t > 0$, the set

$$L_t(f) = \{x \in \mathbb{R}^N : f(x) \geq t\}$$

is either a convex body or is empty.

We will denote with $\mathcal{C}^N$ the set of all quasi-concave functions.

Typical examples of quasi-convex functions are (positive multiples of) characteristic functions of convex bodies. For $A \subseteq \mathbb{R}^N$ we denote by I_A its characteristic function

$$I_A : \mathbb{R}^N \to \mathbb{R}, \quad I_A(x) = \begin{cases} 1 & \text{if } x \in A, \\ 0 & \text{if } \notin A. \end{cases}$$

Then we have that $s I_K \in \mathcal{C}^N$ for every $s > 0$ and $K \in \mathcal{K}^N$. We can also describe the sets $L_t(s I_K)$, indeed

$$L_t(s I_K) = \begin{cases} \varnothing & \text{if } t > s, \\ K & \text{if } 0 < t \leq s. \end{cases}$$

The following proposition gathers some of the basic properties of quasi-concave functions.

Proposition 3.2 *If* $f \in \mathcal{C}^N$ *then*

- $\lim\limits_{\|x\| \to +\infty} f(x) = 0$,
- *f is upper semi-continuous,*

- *f admits a maximum in* $\mathbb{R}^n$, *in particular*

$$\sup_{\mathbb{R}^N} f < +\infty.$$

Proof To prove the first property, let $\epsilon > 0$; as $L_\epsilon(f)$ is compact, there exists $R > 0$ such that $L_\epsilon(f) \subset B_R$. This is equivalent to say that

$$f(x) \le \epsilon \quad \forall x \text{ s.t. } \|x\| \ge R.$$

Upper semi-continuity follows immediately from compactness of super-level sets. Let $M = \sup_{\mathbb{R}^N} f$ and assume that $M > 0$. Let x_n, $n \in \mathbb{N}$, be a maximizing sequence:

$$\lim_{n \to \infty} f(x_n) = M.$$

As f decays to zero at infinity, the sequence x_n is compact; then we may assume that it converges to $\bar{x} \in \mathbb{R}^N$. Then, by upper semi-continuity

$$f(\bar{x}) \ge \lim_{n \to \infty} f(x_n) = M.$$

$\square$

For simplicity, given $f \in \mathcal{C}^N$, we will denote by $M(f)$ the maximum of f in $\mathbb{R}^N$.

Remark 3.3 Let $f \in \mathcal{C}^N$, we denote with $\text{supp}(f)$ the support of f, that is

$$\text{supp}(f) = \text{cl}(\{x \in \mathbb{R}^N : f(x) > 0\}).$$

This is a convex set; indeed

$$\text{supp}(f) = \text{cl}(\bigcup_{k=1}^{\infty} \{x \in \mathbb{R}^N : f(x) \ge 1/k\}).$$

The sets

$$\{x \in \mathbb{R}^N : f(x) \ge 1/k\} \quad k \in \mathbb{N},$$

forms an increasing sequence of convex bodies and their union is convex.

Remark 3.4 A special sub-class of quasi-concave functions is that formed by *log-concave* functions. Let u be a function defined on all $\mathbb{R}^N$, with values in $\mathbb{R} \cup \{+\infty\}$, convex and such that $\lim_{\|x\| \to +\infty} f(x) = +\infty$. Then the function $f = e^{-u}$ is quasi-concave (here we adopt the convention $e^{-\infty} = 0$). If f is of this form is said to be a *log-concave* function.

3.2 Operations with Quasi-Concave Functions

Let $f, g : \mathbb{R}^N \to \mathbb{R}$; we define the point-wise maximum and minimum function between f and g as

$$f \vee g(x) = \max\{f(x), g(x)\}, \quad f \wedge g(x) = \min\{f(x), g(x)\},$$

for all $x \in \mathbb{R}^N$. These operations, applied on $\mathcal{C}^N$, will replace the union and intersection in the definition of valuations on $\mathcal{K}^N \cup \{\varnothing\}$. The proof of the following equalities is straightforward.

Lemma 3.5 *If f and g belong to $\mathcal{C}^N$ and $t > 0$:*

$$L_t(f \wedge g) = L_t(f) \cap L_t(g), \quad L_t(f \vee g) = L_t(f) \cup L_t(g).$$

As the intersection of two convex bodies is still a convex body, we have the following consequence.

Corollary 3.6 *For all $f, g \in \mathcal{C}^N, f \wedge g \in \mathcal{C}^N$.*

On the other hand, in general $f, g \in \mathcal{C}^N$ does not imply that $f \vee g$ does, as it is shown by the example in which f and g are characteristic functions of two convex bodies with empty intersection.

The following lemma follows from the definition of quasi-concave function and the fact that if T is a rigid motion of $\mathbb{R}^N$ and $K \in \mathcal{K}^N$, then $T(K) \in \mathcal{K}^N$.

Lemma 3.7 *Let $f \in \mathcal{C}^N$ be a quasi concave function and $T : \mathbb{R}^N \to \mathbb{R}^N$ a rigid motion, then $f \circ T \in \mathcal{C}^N$.*

3.3 Three Technical Lemmas

We are going to prove some lemmas which will be useful for the study of continuity of valuations.

Lemma 3.8 *Let $f \in \mathcal{C}^N$. For all $t > 0$, except for at most countably many values, we have*

$$L_t(f) = \mathrm{cl}(\{x \in \mathbb{R}^N : f(x) > t\}).$$

Proof We fix $t > 0$ and we define

$$\Omega_t(f) = \{x \in \mathbb{R}^N : f(x) > t\}, \quad H_t(f) = \mathrm{cl}(\Omega_t(f)).$$

$\Omega_t(f)$ is a convex set for all $t > 0$, indeed

$$\Omega_t = \bigcup_{k \in \mathbb{N}} L_{t+1/k}(f).$$

Consequently H_t is a convex body and $H_t \subseteq L_t(f)$. We define $D_t = L_t(f) \setminus H_t$; our aim is now to prove that the set of all $t > 0$ such that $D_t \neq \varnothing$ is at most countable. We first note that if K and L are convex bodies with $K \subset L$, $\text{int}(L) \neq \varnothing$ and $L \setminus K \neq \varnothing$ then $\text{int}(L \setminus K) \neq \varnothing$, therefore

$$D_t \neq \varnothing \quad \Leftrightarrow \quad V_N(D_t) > 0. \tag{8}$$

It follows from

$$D_t = L_t(f) \setminus H_t \subseteq L_t(f) \setminus \Omega_t(f) = \{x \in \mathbb{R}^N : f(x) = t\},$$

that

$$t_1 \neq t_2 \ \Rightarrow \ D_{t_1}(f) \cap D_{t_2}(f) = \varnothing. \tag{9}$$

For the rest of the proof we proceed by induction on N. For $N = 1$, we observe that if f is identically zero, then the lemma is trivially true. If $\text{supp}(f) = \{x_0\}$ and $f(x_0) = t_0 > 0$, then we have

$$L_t(f) = \{x_0\} = \text{cl}(\Omega_t(f)) \quad \forall t > 0, \ t \neq t_0,$$

and in particular the lemma is true. We suppose next that $\text{int}(\text{supp}(f)) \neq \varnothing$; let $t_0 > 0$ be a number such that $\dim(L_t(f)) = 1$, for all $t \in (0, t_0)$ and $\dim(L_t(f)) = 0$, for all $t > t_0$. Moreover, let $t_1 = \max_{\mathbb{R}} f \geq t_0$. We observe that

$$L_t(f) = \text{cl}(\Omega_t(f)) = \varnothing \quad \forall t > t_1 \quad \text{and} \quad L_t(f) = \text{cl}(\Omega_t(f)) \quad \forall t \in (t_0, t_1).$$

Next we deal with values of $t \in (0, t_0)$. Let us fix $\epsilon > 0$ and let K be a compact set in $\mathbb{R}$ such that $K \supseteq L_t(f)$ for every $t \geq \epsilon$. We define, for $i \in \mathbb{N}$,

$$T_i^\epsilon = \left\{ t \in [\epsilon, t_0) : \ V_1(D_t) \geq \frac{1}{i} \right\}.$$

As $D_t \subseteq K$ for all $t \geq \epsilon$ and taking (9) into account we obtain that T_i^ϵ is finite. So

$$T^\epsilon = \bigcup_{i \in \mathbb{N}} T_i^\epsilon$$

is countable for every $\epsilon > 0$. By (8)

$$\{t \geq \epsilon : D_t \neq \varnothing\} \quad \text{is countable}$$

for every $\epsilon > 0$, so that

$$\{t > 0 : D_t \neq \varnothing\}$$

is also countable. The proof for $N = 1$ is complete.

Assume now that the claim of the lemma is true up to dimension $(N - 1)$, and let us prove in dimension N. If the dimension of supp(f) is strictly smaller than N, then (as supp(f) is convex) there exists an affine subspace H of $\mathbb{R}^N$, of dimension $(N - 1)$, containing supp(f). In this case the assert of the lemma follows applying the induction assumption to the restriction of f to H. Next, we suppose that there exists $t_0 > 0$ such that

$$\dim(L_t(f)) = N, \quad \forall\, t \in (0, t_0)$$

and

$$\dim(L_t(f)) < N, \quad \forall\, t > t_0.$$

By the same argument used in the one-dimensional case we can prove that

$$\{t \in (0, t_0) : D_t \neq \varnothing\}$$

is countable. For $t > t_0$, there exists a $(N - 1)$-dimensional affine sub-space of $\mathbb{R}^N$ containing $L_t(f)$ for every $t > t_0$. To conclude the proof we apply the inductive hypothesis to the restriction of f to this hyperplane. □

Lemma 3.9 *Let $\{f_i\}_{i\in\mathbb{N}} \subseteq C^N$ and $f \in C^N$. Assume that $f_i \nearrow f$ point-wise in $\mathbb{R}^N$ as $i \to +\infty$. Then, for all $t > 0$, except at most for countably many values,*

$$\lim_{i\to\infty} L_t(f_i) = L_t(f).$$

Proof For every $t > 0$, the sequence of convex bodies $L_t(f_i)$, $i \in \mathbb{N}$, is increasing and $L_t(f_i) \subset L_t(f)$ for every i. In particular this sequence admits a limit $L_t \subset L_t(f)$. We choose $t > 0$ such that

$$L_t(f) = \mathrm{cl}(\{x \in \mathbb{R}^N : f(x) > t\}).$$

By the previous lemma we know that this condition holds for every t except at most countably many values. It is clear that for every x s.t. $f(x) > t$ we have $x \in L_t$, hence $L_t \supset \{x \in \mathbb{R}^N : f(x) > t\}$; on the other hand, as L_t is closed, we have that $L_t \supset L_t(f)$. Hence $L_t = L_t(f)$ and the proof is complete. □

Lemma 3.10 *Let $\{f_i\}_{i\in\mathbb{N}} \subseteq C^N$ and $f \in C^N$. Assume that $f_i \searrow f$ point-wise in $\mathbb{R}^N$ as $i \to +\infty$. Then for all $t > 0$*

$$\lim_{i\to\infty} L_t(f_i) = L_t(f).$$

Proof The sequence $L_t(f_i)$ is decreasing and its limit, denoted by L_t, contains $L_t(f)$. On the other hand, as now

$$L_t = \bigcap_{k \in \mathbb{N}} L_t(f_k)$$

(see Lemma 1.8.1 of [16]), if $x \in L_t$ then $f_i(x) \geq t$ for every i, so that $f(x) \geq t$ i.e. $x \in L_t(f)$. □

4   Valuations

Definition 4.1 A functional $\mu : \mathcal{C}^N \to \mathbb{R}$ is said to be a valuation if

- $\mu(\underline{0}) = 0$, where $\underline{0} \in \mathcal{C}^N$ is the function identically equal to zero;
- for all f and $g \in \mathcal{C}^N$ such that $f \vee g \in \mathcal{C}^N$, we have

$$\mu(f) + \mu(g) = \mu(f \vee g) + \mu(f \wedge g).$$

A valuation μ is said to be rigid motion invariant, or simply invariant, if for every rigid motion $T : \mathbb{R}^N \to \mathbb{R}^N$ and for every $f \in \mathcal{C}^N$, we have

$$\mu(f) = \mu(f \circ T).$$

In this paper we will always consider invariant valuations. We will also need a notion of continuity which is expressed by the following definition.

Definition 4.2 A valuation μ is said to be continuous if for every sequence $\{f_i\}_{i \in \mathbb{N}} \subseteq \mathcal{C}^N$ and $f \in \mathcal{C}^N$ such that f_i converges point-wise to f in $\mathbb{R}^N$, and f_i is either monotone increasing or decreasing w.r.t. i, we have

$$\mu(f_i) \to \mu(f), \text{ for } i \to +\infty.$$

To conclude the list of properties that a valuation may have and that are relevant to our scope, we say that a valuation μ is monotone increasing (resp. decreasing) if, given $f, g \in \mathcal{C}^N$,

$$f \leq g \text{ point-wise in } \mathbb{R}^N \text{ implies } \mu(f) \leq \mu(g) \text{ (resp. } \mu(f) \geq \mu(g)).$$

4.1   A Brief Discussion on the Choice of the Topology in $\mathcal{C}^N$

A natural choice of a topology in $\mathcal{C}^N$ would be the one induced by point-wise convergence. Let us see that this choice would too restrictive, with respect to the theory of continuous and rigid motion invariant (but translations would be enough) valuations. Indeed, any translation invariant valuation μ on $\mathcal{C}^N$ such that

$$\lim_{i\to\infty} \mu(f_i) = \mu(f)$$

for every sequence f_i, $i \in \mathbb{N}$, in $\mathcal{C}^N$, converging to some $f \in \mathcal{C}^N$ point-wise, must be the valuation constantly equal to 0. To prove this claim, let $f \in \mathcal{C}^N$ have compact support, let e_1 be the first vector of the canonical basis of $\mathbb{R}^N$ and set

$$f_i(x) = f(x - i e_1) \quad \forall x \in \mathbb{R}^N, \quad \forall i \in \mathbb{N}.$$

The sequence f_i converges point-wise to the function $f_0 \equiv 0$ in $\mathbb{R}^N$, so that, by translation invariance, and as $\mu(f_0) = 0$, we have $\mu(f) = 0$. Hence μ vanishes on each function f with compact support. On the other hand every element of $\mathcal{C}^N$ is the point-wise limit of a sequence of functions in $\mathcal{C}^N$ with compact support. Hence $\mu \equiv 0$.

A different choice could be based on the following consideration: we have seen that $\mathcal{C}^N \subset L^\infty(\mathbb{R}^N)$, hence it inherits the topology of this space. In [3], Cavallina studied translation invariant and continuous valuations on $L^\infty(\mathbb{R}^N)$. In particular he proved that there exists non-trivial translation invariant and continuous valuations on this space, which vanishes on functions with compact support. In particular they cannot be written in integral form as those found in the present paper. Noting that in dimension $N = 1$ translation and rigid motion invariance provide basically the same condition, this suggest that the choice of the topology on $L^\infty(\mathbb{R}^N)$ on $\mathcal{C}^N$ would lead us to a completely different type of valuations.

5 Integral Valuations

A class of examples of invariant valuations which will be crucial for our characterization results is that of integral valuations.

5.1 Continuous Integral Valuations

Let $k \in \{0, \ldots, N\}$. For $f \in \mathcal{C}^N$, consider the function

$$t \to u(t) = V_k(L_t(f)) \quad t > 0.$$

This is a decreasing function, which vanishes for $t > M(f) = \max_{\mathbb{R}^N} f$. In particular u has bounded variation in $[\delta, M(f)]$ for every $\delta > 0$, hence there exists a Radon measure defined in $(0, \infty)$, that we will denote by $S_k(f; \cdot)$, such that

$$-S_k(f; \cdot) \text{ is the distributional derivative of } u$$

(see, for instance, [1]). Note that, as u is decreasing, we have put a minus sign in this definition to have a non-negative measure. The support of $S_k(f; \cdot)$ is contained in $[0, M(f)]$.

Let ϕ be a continuous function defined on $[0, \infty)$, such that $\phi(0) = 0$. We consider the functional on C^N defined by

$$\mu(f) = \int_{(0,\infty)} \phi(t) dS_k(f; t) \quad f \in C^N. \tag{10}$$

The aim of this section is to prove that this is a continuous and invariant valuation on C^N. As a first step, we need to find some condition on the function ϕ which guarantee that the above integral is well defined for every f.

Assume that

$$\exists \delta > 0 \text{ s.t. } \phi(t) = 0 \text{ for every } t \in [0, \delta]. \tag{11}$$

Then

$$\int_{(0,\infty)} \phi_+(t) dS_k(f; t) = \int_{[\delta, M(f)]} \phi_+(t) \, dS_k(f, t)$$

$$\leq M \left(V_k(L_\delta(f)) - V_k(M(f)) \right) < \infty,$$

where $M(f) = \max_{\mathbb{R}^N} f$, $M = \max_{[\delta, \max_{\mathbb{R}^N} f]} \phi_+$ and ϕ_+ is the positive part of ϕ. Analogously we can prove that the integral of the negative part of ϕ, denoted by ϕ_-, is finite, so that μ is well defined.

We will prove that, for $k \geq 1$, condition (11) is necessary as well. Clearly, if $\mu(f)$ is well defined (i.e. is a real number) for every $f \in C^N$, then

$$\int_{(0,\infty)} \phi_+(t) dS_k(f; t) < \infty \quad \text{and} \quad \int_{(0,\infty)} \phi_-(t) dS_k(f; t) < \infty \quad \forall f \in C^N.$$

Assume that ϕ_+ does not vanish identically in any right neighborhood of the origin. Then we have

$$\psi(t) := \int_0^t \phi_+(\tau) \, d\tau > 0 \quad \forall t > 0.$$

The function

$$t \to h(t) = \int_t^1 \frac{1}{\psi(s)} ds, \quad t \in (0, 1],$$

is strictly decreasing. As $k \geq 1$, we can construct a function $f \in C^N$ such that

$$V_k(L_t(f)) = h(t) \quad \text{for every } t > 0. \tag{12}$$

Indeed, consider a function of the form

$$f(x) = w(\|x\|), \quad x \in \mathbb{R}^N,$$

where $w \in C^1([0, +\infty))$ is positive and strictly decreasing. Then $f \in C^N$ and $L_t(f) = B_{r(t)}$, where

$$r(t) = w^{-1}(t)$$

for every $t \in (0, f(0)]$ (note that $f(0) = M(f)$). Hence

$$V_k(L_t(f)) = c\,(w^{-1}(t))^k$$

where c is a positive constant depending on k and N. Hence if we choose

$$w = \left[\left(\frac{1}{c}h\right)^{1/k}\right]^{-1},$$

(12) is verified. Hence

$$dS_k(f;t) = \frac{1}{\psi(t)}dt,$$

and

$$\int_{(0,\infty)} \phi_+(t)dS_k(f;t) = \int_{(0,M(f))} \frac{\psi'(t)}{\psi(t)}dt = \infty.$$

In the same way we can prove that ϕ_- must vanish in a right neighborhood of the origin. We have proved the following result.

Lemma 5.1 *Let $\phi \in C([0, \infty))$ and $k \in \{1, \ldots, N\}$. Then ϕ has finite integral with respect to the measure $S_k(f; \cdot)$ for every $f \in C^N$ if and only if ϕ verifies* (11).

In the special case $k = 0$, as the intrinsic volume V_0 is the Euler characteristic,

$$u(t) = \begin{cases} 1 & \text{if } 0 < t \leq M(f), \\ 0 & \text{if } t > M(f). \end{cases}$$

That is, S_0 is the Dirac point mass measure concentrated at $M(f)$ and μ can be written as

$$\mu(f) = \phi(M(f)) \quad \forall f \in C^N.$$

Next we show that (10) defines a continuous and invariant valuation.

Proposition 5.2 *Let $k \in \{0,\ldots,N\}$ and $\phi \in C([0,\infty))$ be such that $\phi(0) = 0$. If $k \geq 1$ assume that (11) is verified. Then (10) defines an invariant and continuous valuation on $\mathcal{C}^N$.*

Proof For every $f \in \mathcal{C}^N$ we define the function $u_f : [0, M(f)] \to \mathbb{R}$ as

$$u_f(t) = V_k(L_t(f)).$$

As already remarked, this is a decreasing function. In particular it has bounded variation in $[\delta, M(f)]$. Let ϕ_i, $i \in \mathbb{N}$, be a sequence of functions in $C^\infty([0,\infty))$, with compact support, converging uniformly to ϕ on compact sets. As $\phi \equiv 0$ in $[0, \delta]$, we may assume that the same holds for every ϕ_i. Then we have

$$\mu(f) = \lim_{i \to \infty} \mu_i(f),$$

where

$$\mu_i(f) = \int_{[0,\infty)} \phi_i(t) dS_k(f;t) \quad \forall f \in \mathcal{C}^N.$$

By the definition of distributional derivative of a measure, we have, for every f and for every i:

$$\int_{[0,\infty)} \phi_i(t) dS_k(f;t) = \int_{[0,\infty)} u_f(t)\phi_i'(t)dt = \int_{[0,M(f)]} V_k(L_t(f))\phi_i'(t)dt.$$

On the other hand, if $f, g \in \mathcal{C}^N$ are such that $f \vee g \in \mathcal{C}^N$, for every $t > 0$

$$L_t(f \vee g) = L_t(f) \cup L_t(g), \quad L_t(f \wedge g) = L_t(f) \cap L_t(g). \tag{13}$$

As intrinsic volumes are valuations

$$V_k(L_t(f \vee g)) + V_k(L_t(f \wedge g)) = V_k(L_t(f)) + V_k(L_t(g)).$$

Multiplying both sides times $\phi_i'(t)$ and integrating on $[0, \infty)$ we obtain

$$\mu_i(f \vee g) + \mu_i(f \wedge g) = \mu_i(f) + \mu_i(g).$$

Letting $i \to \infty$ we deduce the valuation property for μ.

In order to prove the continuity of μ, we first consider the case $k \geq 1$. Let $f_i, f \in \mathcal{C}^N$, $i \in \mathbb{N}$, and assume that the sequence f_i is either increasing or decreasing with respect to i, and it converges point-wise to f in $\mathbb{R}^N$. Note that in each case there exists a constant $M > 0$ such that $M(f_i), M(f) \leq M$ for every i. Consider now the sequence of functions u_{f_i}. By the monotonicity of the sequence f_i, and that of intrinsic volumes, this is a monotone sequence of decreasing functions, and it converges a.e. to u_f in $(0, \infty)$, by Lemmas 3.9 and 3.10. In particular the sequence u_{f_i} has uniformly bounded total variation in $[\delta, M]$. Consequently, the sequence of measures $S_k(f_i; \cdot)$, $i \in \mathbb{N}$, converges weakly to the measure $S_k(f; \cdot)$ as $i \to \infty$. Hence, as ϕ is continuous

$$\lim_{i \to \infty} \mu(f_i) = \lim_{i \to \infty} \int_{[\delta,M]} \phi(t) \, dS_k(f_i; t) = \int_{[0,M]} \phi(t) \, dS_k(f; t) = \mu(f).$$

If $k = 0$ then we have seen that

$$\mu(f) = \phi(M(f)) \quad \forall f \in \mathcal{C}^N.$$

Hence in this case continuity follows from the following fact: if f_i, $i \in \mathbb{N}$, is a monotone sequence in $\mathcal{C}^N$ converging point-wise to f, then

$$\lim_{i \to \infty} M(f_i) = M(f).$$

This is a simple exercise that we leave to the reader.

Finally, the invariance of μ follows directly from the invariance of intrinsic volumes with respect to rigid motions. $\square$

5.2 Monotone (and Continuous) Integral Valuations

In this section we introduce a slightly different type of integral valuations, which will be needed to characterize all possible continuous and monotone valuations on $\mathcal{C}^N$. Note that, as it will be clear in the sequel, when the involved functions are smooth enough, the two types can be reduced one to another by an integration by parts.

Let $k \in \{0, \dots, N\}$ and let ν be a Radon measure on $(0, +\infty)$; assume that

$$\int_0^{+\infty} V_k(L_t(f)) d\nu(t) < +\infty, \quad \forall f \in \mathcal{C}^N. \tag{14}$$

We will return later on explicit condition on ν such that (14) holds. Then define the functional $\mu : \mathcal{C}^N \to \mathbb{R}$ by

$$\mu(f) = \int_0^{+\infty} V_k(L_t(f)) d\nu(t) \quad \forall f \in \mathcal{C}^N. \tag{15}$$

Proposition 5.3 *Let ν be a Radon measure on $(0, \infty)$ which verifies (14); then the functional defined by (15) is a rigid motion invariant and monotone increasing valuation.*

Proof The proof that μ is a valuation follows from (13) and the valuation property for intrinsic volumes, as in the proof of Proposition 5.2. The same can be done for invariance. As for monotonicity, note that if $f, g \in \mathcal{C}^N$ and $f \le g$, then

$$L_t(f) \subset L_t(g) \quad \forall t > 0.$$

Therefore, as intrinsic volumes are monotone, $V_k(L_t(f)) \le V_K(L_t(g))$ for every $t > 0$. $\square$

If we do not impose any further assumption the valuation μ needs not to be continuous. Indeed, for example, if we fix $t = t_0 > 0$ and let $v = \delta_{t_0}$ be the delta Dirac measure at t_0; then the valuation

$$\mu(f) = V_N(L_{t_0}(f)), \ \forall f \in \mathcal{C}^N,$$

is not continuous. To see it, let $f = t_0 I_{B_1}$ (recall that B_1 is the unit ball of $\mathbb{R}^N$) and let

$$f_i = t_0 \left(1 - \frac{1}{i}\right) I_{B_1} \quad \forall i \in \mathbb{N}.$$

Then f_i is a monotone sequence of elements of $\mathcal{C}^N$ converging point wise to f in $\mathbb{R}^N$. On the other hand

$$\mu(f_i) = 0 \quad \forall i \in \mathbb{N},$$

while $\mu(f) = V_N(B_1) > 0$. The next results asserts that the presence of atoms is the only possible cause of discontinuity for μ. We recall that a measure v defined on $[0, \infty)$ is said non-atomic if $v(\{t\}) = 0$ for every $t \ge 0$.

Proposition 5.4 *Let v be a Radon measure on $(0, +\infty)$ such that (14) holds and let μ be the valuation defined by (14). Then the two following conditions are equivalent:*

i) v is non-atomic,
ii) μ is continuous.

Proof Suppose that *i)* does not hold, than there exists t_0 such that $v(\{t_0\}) = \alpha > 0$. Define $\varphi : \mathbb{R}_+ \to \mathbb{R}$ by

$$\varphi(t) = \int_{(0,t]} dv(s).$$

φ is an increasing function with a jump discontinuity at t_0 of amplitude α. Now let $f = t_0 I_{B_1}$ and $f_i = t_0(1 - \frac{1}{i})I_{B_1}$, for $i \in \mathbb{N}$. Then f_i is an increasing sequence in $\mathcal{C}^N$, converging point-wise to f in $\mathbb{R}^N$. On the other hand

$$\mu(f) = \int_0^{t_0} V_k(B)dv(s) = V_k(B)v((0, t_0]) = V_k(B_1)\, \varphi(t_0)$$

and similarly

$$\mu(f_i) = V_k(B_1)\, \varphi\left(t_0 - \frac{1}{i}\right).$$

Consequently

$$\lim_{i \to +\infty} \mu(f_i) < \mu(f).$$

Vice versa, suppose that *i)* holds. We observe that, as v is non-atomic, every countable subset has measure zero with respect to v. Let $f_i \in C^N$, $i \in \mathbb{N}$, be a sequence such that either $f_i \nearrow f$ or $f_i \searrow f$ as $i \to +\infty$, point-wise in $\mathbb{R}^N$, for some $f \in C^N$. Set

$$u_i(t) = V_k(L_t(f_i)), \quad u(t) = V_k(L_t(f)) \quad \forall t \geq 0, \quad \forall k \in \mathbb{N}.$$

The sequence u_i is monotone and, by Lemmas 3.9 and 3.10, converges to u v-a.e. Hence, by the continuity of intrinsic volumes and the monotone convergence theorem, we obtain

$$\lim_{i \to \infty} \mu(f_i) = \lim_{i \to \infty} \int_{(0,\infty)} u_i(t)\, dv = \int_{(0,\infty)} u(t)\, dv(t) = \mu(f).$$

$\square$

Now we are going to find a more explicit form of condition (14). We need the following lemma.

Lemma 5.5 *Let $\phi : [0, +\infty) \to \mathbb{R}$ be an increasing, non negative and continuous function with $\phi(0) = 0$ and $\phi(t) > 0$, for all $t > 0$. Let v be a Radon measure such that $\phi(t) = v([0, t])$, for all $t \geq 0$. Then*

$$\int_0^1 \frac{1}{\phi^k(t)}\, dv(t) = +\infty, \ \forall k \geq 1.$$

Proof Fix $\alpha \in [0, 1]$. The function $\psi : [\alpha, 1] \to \mathbb{R}$ defined by

$$\psi(t) = \begin{cases} \dfrac{1}{k-1}\phi^{1-k}(t) & \text{if } k > 1, \\ \ln(\phi(t)) & \text{if } k = 1, \end{cases}$$

is continuous and with bounded variation in $[\alpha, 1]$. Its distributional derivative is

$$\frac{1}{\phi^k(t)}\, v.$$

Hence, for $k > 1$,

$$\frac{1}{k-1}[\phi^{1-k}(\alpha) - \phi^{1-k}(1)] = \psi(1) - \psi(\alpha) = \int_{[\alpha,1]} \frac{dv}{\phi^k(t)}.$$

The claim of the lemma follows letting $\alpha \to 0^+$. A similar argument can be applied to the case $k = 1$. □

Proposition 5.6 *Let v be a non-atomic Radon measure on $[0, +\infty)$ and let $k \in \{1, \ldots, N\}$. Then (14) holds if and only if:*

$$\exists \delta > 0 \text{ such that } v([0, \delta]) = 0. \tag{16}$$

Proof We suppose that there exists $\delta > 0$ such that $[0, \delta] \cap \text{supp}(v) = \varnothing$. Then we have, for every $f \in C^N$,

$$\mu(f) = \int_\delta^{M(f)} V_i(L_t(f))dv(t) \leq V_i(L_\delta(f)) \int_\delta^{M(f)} dv(t) \tag{17}$$

$$= V_i(L_\delta(f))(v([0, M(f)]) - v([0, \delta])) < +\infty. \tag{18}$$

with $M(f) = \max_{\mathbb{R}^N} f$.

Vice versa, assume that (14) holds. By contradiction, we suppose that for all $\delta > 0$, we have $v([0, \delta]) > 0$. We define

$$\phi(t) = v([0, t]), \quad t \in [0, 1]$$

then ϕ is continuous (as v is non-atomic) and increasing; moreover $\phi(0) = 0$ and $\phi(t) > 0$, for all $t > 0$. The function

$$\psi(t) = \frac{1}{t\phi(t)}, \quad t \in (0, 1],$$

is continuous and strictly decreasing. Its inverse ψ^{-1} is defined in $[\psi(1), \infty)$; we extend it to $[0, \psi(1))$ setting

$$\psi^{-1}(r) = 1 \quad \forall r \in [0, \psi(1)).$$

Then

$$V_1(\{r \in [0, +\infty) : \psi^{-1}(r) \geq t\}) = \begin{cases} \psi(t), \forall t \in (0, 1] \\ \\ 0 \quad \forall t > 1. \end{cases}$$

We define now the function $f : \mathbb{R}^N \to \mathbb{R}$ as

$$f(x) = \psi^{-1}(||x||), \quad \forall x \in \mathbb{R}^N.$$

Then

$$L_t(f) = \{x \in \mathbb{R}^N : \psi(||x||) \geq t\} = B_{\frac{1}{t\phi(t)}}(0),$$

and

$$V_k(L_t(f)) = c \frac{1}{t^k \phi^k(t)} \quad \forall t \in (0, 1],$$

where $c > 0$ depends on N and k. Hence, by Lemma 5.5

$$\int_0^{+\infty} V_k(L_t(f))dv(t) = \int_0^1 V_k(L_t(f))dv(t) \geq c \int_0^{+\infty} \frac{dv(t)}{\phi^k(t)} = +\infty.$$

□

The following proposition summarizes some of the results we have found so far.

Proposition 5.7 *Let $k \in \{0, \ldots, N\}$ and let v be a Radon measure on $[0, \infty)$ which is non-atomic and, if $k \geq 1$, verifies condition (16). Then the map $\mu : C^N \to \mathbb{R}$ defined by (15) is an invariant, continuous and increasing valuations.*

5.3 The Connection Between the Two Types of Integral Valuations

When the regularity of the involved functions permits, the two types of integral valuations that we have seen can be obtained one from each other by a simple integration by parts.

Let $k \in \{0, \ldots, N\}$ and $\phi \in C^1([0, \infty))$ be such that $\phi(0) = 0$. For simplicity, we may assume also that ϕ has compact support. Let $f \in C^N$. By the definition of distributional derivative of an increasing function we have:

$$\int_{[0,\infty)} \phi(t) \, dS_k(f; t) = \int_{[0,\infty)} \phi'(t) V_k(L_t(f))dt.$$

If we further decompose $-\phi'$ as the difference of two non-negative functions, and we denote by v_1 and v_2 the Radon measures having those functions as densities, we get

$$\int_{[0,\infty)} \phi(t) \, dS_k(f; t) = \int_{[0,\infty)} V_k(L_t(f))dv_1(t) - \int_{[0,\infty)} V_k(L_t(f))dv_2(t).$$

The assumption that ϕ has compact support can be removed by a standard approximation argument. In his way we have seen that each valuation of the form (10), if ϕ is regular, is the difference of two monotone integral valuations of type (15).

Vice versa, let v be a Radon measure (with support contained in $[\delta, \infty)$, for some $\delta > 0$), and assume that it has a smooth density with respect to the Lebesgue measure:

$$dv(t) = \phi'(t)dt$$

where $\phi \in C^1([0, \infty))$, and it has compact support. Then

$$\int_{[0,\infty)} V_k(L_t(f)) \, dv(t) = \int_{[0,\infty)} \phi(t) \, dS_k(f; t).$$

Also in this case the assumption that the support of v is compact can be removed. In other words each integral monotone valuation, with sufficiently smooth density, can be written in the form (10).

5.4 The Case $k = N$

If μ is a valuation of the form (10) and $k = N$, the Layer Cake principle provides and alternative simple representation.

Proposition 5.8 Let ϕ be a continuous function on $[0, \infty)$ verifying (16). Then for every $f \in C^N$ we have

$$\int_{[0,\infty)} \phi(t) \, dS_N(f; t) = \int_{\mathbb{R}^N} \phi(f(x))dx. \tag{19}$$

Proof As ϕ can be written as the difference of two non-negative continuous function, and (19) is linear with respect to ϕ, there is no restriction if we assume that $\phi \geq 0$. In addition we suppose initially that $\phi \in C^1([0, \infty))$ and it has compact support. Fix $f \in C^N$; by the definition of distributional derivative, we have

$$\int_{[0,\infty)} \phi(t) \, dS_N(f; t) = \int_{[0,\infty)} V_N(L_t(f))\phi'(t)dt.$$

There exists $\phi_1, \phi_2 \in C^1([0, \infty))$, strictly increasing, such that $\phi = \phi_1 - \phi_2$. Now:

$$\int_{[0,\infty)} V_N(L_t(f))\phi_1'(t)dt = \int_{[0,\infty)} V_N(\{x \in \mathbb{R}^N : \phi_1(f(x)) \geq s\})ds = \int_{\mathbb{R}^N} \phi_1(f(x))dx,$$

where in the last equality we have used the Layer Cake principle. Applying the same argument to ϕ_2 we obtain (19) when ϕ is smooth and compactly supported.

For the general case, we apply the result obtained in the previous part of the proof to a sequence ϕ_i, $i \in \mathbb{N}$, of functions in $C^1([0, \infty))$, with compact support, which converges uniformly to ϕ on compact subsets of $(0, \infty)$. The conclusion follows from a direct application of the dominated convergence theorem.   □

6 Simple Valuations

Throughout this section μ will be an invariant and continuous valuation on C^N. We will also assume that μ is *simple*.

Definition 6.1 A valuation μ on C^N is said to be simple if, for every $f \in C^N$ with $\dim(\operatorname{supp}(f)) < N$, we have $\mu(f) = 0$.

Note that $\dim(\operatorname{supp}(f)) < N$ implies that $f = 0$ a.e. in $\mathbb{R}^N$, hence each valuation of the form (19) is simple. We are going to prove that in fact the converse of this statement is true.

Fix $t \geq 0$ and define a real-valued function σ_t on $\mathcal{K}^N \cup \{\varnothing\}$ as

$$\sigma_t(K) = \mu(tI_K) \quad \forall K \in \mathcal{K}^N, \quad \sigma_t(\varnothing) = 0.$$

Let $K, L \in \mathcal{K}^N$ be such that $K \cup L \in \mathcal{K}^N$. As, trivially,

$$tI_K \vee tI_L = tI_{K \cup L} \quad \text{and} \quad tI_K \wedge tI_L = tI_{K \cap L},$$

using the valuation property of μ we infer

$$\sigma_t(K \cup L) + \sigma_t(K \cap L) = \sigma_t(K) + \sigma_t(L),$$

i.e. σ_t is a valuation on $\mathcal{K}^N$. It also inherits directly two properties of μ: it is invariant and simple. Then, by the continuity of μ, Corollary 2.7 and the subsequent remark, there exists a constant c such that

$$\sigma_t(K) = cV_N(K) \tag{20}$$

for every $K \in \mathcal{K}^N$. The constant c will in general depend on t, i.e. it is a real-valued function defined in $[0, \infty)$. We denote this function by ϕ_N. Note that, as $\mu(f) = 0$ for $f \equiv 0$, $\phi_N(0) = 0$. Moreover, the continuity of μ implies that for every $t_0 \geq 0$ and for every monotone sequence t_i, $i \in \mathbb{N}$, converging to t_0, we have

$$\phi_N(t_0) = \lim_{i \to \infty} \phi_N(t_i).$$

From this it follows that ϕ_N is continuous in $[0, \infty)$.

Proposition 6.2 *Let μ be an invariant, continuous and simple valuation on $\mathcal{C}^N$. Then there exists a continuous function ϕ_N on $[0, \infty)$, such that*

$$\mu(tI_K) = \phi_N(t) V_N(K)$$

for every $t \geq 0$ and for every $K \in \mathcal{K}^N$.

6.1 Simple Functions

Definition 6.3 A function $f : \mathbb{R}^N \to \mathbb{R}$ is called simple if it can be written in the form

$$f = t_1 I_{K_1} \vee \cdots \vee t_m I_{K_m} \tag{21}$$

where $0 < t_1 < \cdots < t_m$ and $K_1, \ldots, K_m$ are convex bodies such that

$$K_1 \supset K_2 \supset \cdots \supset K_m.$$

The proof of the following fact is straightforward.

Proposition 6.4 *Let f be a simple function of the form (21) and let $t > 0$. Then*

$$L_t(f) = \{x \in \mathbb{R}^N : f(x) \geq t\} = \begin{cases} K_i & \text{if } t \in (t_{i-1}, t_i] \text{ for some } i = 1, \ldots m, \\ \\ \varnothing & \text{if } t > t_m, \end{cases}$$

$$\tag{22}$$

where we have set $t_0 = 0$.

In particular simple functions are quasi-concave. Let $k \in \{0, \ldots, N\}$, and let f be of the form (21). Consider the function

$$t \to u(t) := V_k(L_t(f)), \quad t > 0.$$

By Proposition 6.4, this is a decreasing function that is constant on each interval of the form $(t_{i-1}, t_i]$, on which it has the value $V_k(K_i)$. Hence its distributional derivative is $-S_k(f; \cdot)$, where

$$S_k(f; \cdot) = \sum_{i=1}^{m-1} (V_k(K_i) - V_k(K_{i+1})) \, \delta_{t_i}(\cdot) + V_k(K_m) \delta_{t_m}(\cdot). \tag{23}$$

6.2 Characterization of Simple Valuations

In this section we are going to prove Theorem 1.2. We will first prove it for simple functions and then pass to the general case by approximation.

Lemma 6.5 *Let μ be an invariant, continuous and simple valuation on $\mathcal{C}^N$, and let $\phi = \phi_N$ be the function whose existence is established in Proposition 6.2. Then, for every simple function $f \in \mathcal{C}^N$ we have*

$$\mu(f) = \int_{[0,\infty)} \phi(t)\, dS_N(f;t).$$

Proof Let f be of the form (21). We prove the following formula

$$\mu(f) = \sum_{i=1}^{m-1} \phi(t_i)(V_N(K_i) - V_N(K_{i+1})) + \phi(t_m)V_N(K_m); \tag{24}$$

by (23), this is equivalent to the statement of the lemma. Equality (24) will be proved by induction on m. For $m = 1$ its validity follows from Proposition 6.2. Assume that it has been proved up to $(m - 1)$. Set

$$g = t_1 I_{K_1} \vee \cdots \vee t_{m-1} I_{K_{m-1}}, \quad h = t_m I_{K_m}.$$

We have that $g, h \in \mathcal{C}^N$ and

$$g \vee h = f \in \mathcal{C}^N, \quad g \wedge h = t_{m-1} I_{K_m}.$$

Using the valuation property of μ and Proposition 6.2 we get

$$\mu(f) = \mu(g \vee h) = \mu(g) + \mu(h) - \mu(g \wedge h)$$
$$= \mu(g) + \phi(t_m)V_N(K_m) - \phi(t_{m-1})V_N(K_m).$$

On the other hand, by induction

$$\mu(g) = \sum_{i=1}^{m-2} \phi(t_i)(V_N(K_i) - V_N(K_{i+1})) + \phi(t_{m-1})V_N(K_{m-1}).$$

The last two equalities complete the proof. $\square$

Proof of Theorem 1.2 As before, $\phi = \phi_N$ is the function coming from Proposition 6.2. We want to prove that

$$\mu(f) = \int_{[0,\infty)} \phi(t)\, dS_N(f;t) \tag{25}$$

for every $f \in \mathcal{C}^N$. This, together with Proposition 5.8, provides the proof.

Step 1.   Our first step is to establish the validity of this formula when the support of f bounded, i.e. there exists some convex body K such that

$$L_t(f) \subset K \quad \forall t > 0. \tag{26}$$

Given $f \in \mathcal{C}^N$ with this property, we build a monotone sequence of simple functions, f_i, $i \in \mathbb{N}$, converging point-wise to f in $\mathbb{R}^N$. Let $M = M(f)$ be the maximum of f on $\mathbb{R}^N$. Fix $i \in \mathbb{N}$. We consider the dyadic partition $\mathcal{P}_i$ of $[0, M]$:

$$\mathcal{P}_i = \left\{ t_j = j\frac{M}{2^i} : j = 0, \dots, 2^i \right\}.$$

Set

$$K_j = L_{t_j}(f), \quad f_i = \bigvee_{j=1}^{2^i} t_j I_{K_j}.$$

f_i is a simple function; as $t_j I_{K_j} \leq f$ for every j we have that $f_i \leq f$ in $\mathbb{R}^N$. The sequence of function f_i is increasing, since $\mathcal{P}_i \subset \mathcal{P}_{i+1}$. The inequality $f_i \leq f$ implies that

$$\lim_{i \to \infty} f_i(x) \leq f(x) \quad \forall x \in \mathbb{R}^N$$

(in particular the support of f_i is contained in K, for every $i \in \mathbb{N}$). We want to establish the reverse inequality. Let $x \in \mathbb{R}^N$; if $f(x) = 0$ then trivially

$$f_i(x) = 0 \quad \forall i \quad \text{hence} \quad \lim_{i \to \infty} f_i(x) = f(x).$$

Assume that $f(x) > 0$ and fix $\epsilon > 0$. Let $i_0 \in \mathbb{N}$ be such that $2^{-i_0} M < \epsilon$. Let $j \in \{1, \dots, 2^{i_0} - 1\}$ be such that

$$f(x) \in \left(j\frac{M}{2^{i_0}}, (j+1)\frac{M}{2^{i_0}} \right].$$

Then

$$f(x) \leq j\frac{M}{2^{i_0}} + \frac{M}{2^{i_0}} \leq f_{i_0}(x) + \epsilon \leq \lim_{i \to \infty} f_i(x) + \epsilon.$$

Hence the sequence f_i converges point-wise to f in $\mathbb{R}^N$. In particular, by the continuity of μ we have that

$$\mu(f) = \lim_{i \to \infty} \mu(f_i) = \lim_{i \to \infty} \int_{[0,\infty)} \phi(t) \, dS_N(f_i; t).$$

By Lemma 3.9, a further consequence is that

$$\lim_{i\to\infty} u_i(t) = u(t) \quad \text{for a.e. } t \in (0,\infty),$$

where

$$u_i(t) = V_N(L_t(f_i)), \quad i \in \mathbb{N}, \quad u(t) = V_N(L_t(f))$$

for $t > 0$. We consider now the sequence of measures $S_N(f_i; \cdot)$, $i \in \mathbb{N}$; the total variation of these measures in $(0,\infty)$ is uniformly bounded by $V_N(K)$, moreover they are all supported in $(0, M)$. As they are the distributional derivatives of the functions u_i, which converges a.e. to u, we have that (see for instance [1, Proposition 3.13]) the sequence $S_N(f_i; \cdot)$ converges weakly in the sense of measures to $S_N(f; \cdot)$. This implies that

$$\lim_{i\to\infty} \int_{(0,\infty)} \bar\phi(t)\, dS_N(f_i; t) = \int_{(0,\infty)} \bar\phi(t)\, dS_N(f; t) \tag{27}$$

for every function $\bar\phi$ continuous in $(0,\infty)$, such that $\bar\phi(0) = 0$ and $\bar\phi(t)$ is identically zero for t sufficiently large. In particular (recalling that $\phi(0) = 0$), we can take $\bar\phi$ such that it equals ϕ in $[0, M]$. Hence, as the support of the measures $S_N(f_i; \cdot)$ is contained in this interval, we have that (27) holds for ϕ as well. This proves the validity of (25) for functions with bounded support.

Step 2. This is the most technical part of the proof. The main scope here is to prove that ϕ is identically zero in some right neighborhood of the origin. Let $f \in C^N$. For $i \in \mathbb{N}$, let

$$f_i = f \wedge (M(f)I_{B_i})$$

where B_i is the closed ball centered at the origin, with radius i. The function f_i coincides with f in B_i and vanishes in $\mathbb{R}^N \backslash B_i$; in particular it has bounded support. Moreover, the sequence f_i, $i \in \mathbb{N}$, is increasing and converges point-wise to f in $\mathbb{R}^N$. Hence

$$\mu(f) = \lim_{i\to\infty} \mu(f_i) = \lim_{i\to\infty} \int_{(0,\infty)} \phi(t)\, dS_N(f_i; t).$$

Let ϕ_+ and ϕ_- be the positive and negative parts of ϕ, respectively. We have that

$$\lim_{i\to\infty} \left[\int_{(0,\infty)} \phi_+(t)\, dS_N(f_i; t) + \int_{(0,\infty)} \phi_-(t)\, dS_N(f_i; t) \right]$$

exists and it is finite. We want to prove that this implies that ϕ_+ and ϕ_- vanishes identically in $[0, \delta]$ for some $\delta > 0$.

By contradiction, assume that this is not true for ϕ_+. Then there exists three sequences t_i, r_i and ϵ_i, $i \in \mathbb{N}$, with the following properties: t_i tends decreasing to zero; $r_i > 0$ is such that the intervals $C_i = [t_i - r_i, t_i + r_i]$ are contained in $(0, 1]$ and pairwise disjoint; $\phi_+(t) \geq \epsilon_i > 0$ for $t \in C_i$. Let

$$C = \bigcup_{i \in \mathbb{N}} C_i, \quad \Omega = (0, 1] \setminus C.$$

Next we define a function $\gamma : (0, 1] \to [0, \infty)$ as follows. $\gamma(t) = 0$ for every $t \in \Omega$ while, for every $i \in \mathbb{N}$, γ is continuous in C_i and

$$\gamma(t_i \pm r_i) = 0, \quad \int_{C_i} \gamma(t)dt = \frac{1}{\epsilon_i}.$$

Note in particular that γ vanishes on the support of ϕ_- intersected with $(0, 1]$. We also set

$$g(t) = \gamma(t) + 1 \quad \forall t > 0.$$

Observe that

$$\int_0^1 \phi_-(t)g(t)dt = \int_0^1 \phi_-(t)dt < \infty.$$

On the other hand

$$\int_0^1 \phi_+(t)g(t)dt \geq \int_0^1 \phi(t)\gamma(t)dt = \sum_{i=1}^{\infty} \int_{C_i} \phi_+(t)\gamma(t)dt$$

$$\geq \sum_{i=1}^{\infty} \epsilon_i \int_{C_i} \gamma(t)dt = +\infty.$$

Let

$$G(t) = \int_t^1 g(s)ds \quad \text{and} \quad \rho(t) = [G(t)]^{1/N}, \quad 0 < t \leq 1.$$

As γ is non-negative, g is strictly positive, and continuous in $(0, 1)$. Hence G is strictly decreasing and continuous, and the same holds for ρ. Let

$$S = \sup_{(0,1]} \rho = \lim_{t \to 0^+} \rho(t),$$

and let $\rho^{-1} : [0, S) \to \mathbb{R}$ be the inverse function of ρ. If $S < \infty$, we extend ρ^{-1} to be zero in $[S, \infty)$. In this way, ρ^{-1} is continuous in $[0, \infty)$, and $C^1([0, S))$. Let

$$f(x) = \rho^{-1}(\|x\|), \quad \forall x \in \mathbb{R}^N.$$

For $t > 0$ we have

$$L_t(f) = \begin{cases} \{x \in \mathbb{R}^N \ : \ \|x\| \le \rho(t)\} & \text{if } t \le 1, \\ \varnothing & \text{if } t > 1. \end{cases}$$

In particular $f \in C^N$. Consequently,

$$V_N(L_t(f)) = c \rho^N(t) = c\, G(t) \quad \forall\, t \in (0, 1],$$

where $c > 0$ is a dimensional constant, and then

$$dS_N(f; t) = c\, g(t)dt.$$

By the previous considerations

$$\int_{[0,\infty)} \phi_+(t)dS_N(f, t) = c \int_{[0,\infty)} \phi_+(t)g(t)dt = \infty, \quad \int_{[0,\infty)} \phi_+(t)dS_N(f, t) < \infty.$$

Clearly we also have that

$$\int_{[0,\infty)} \phi_+(t)dS_N(f, t) = \lim_{i \to \infty} \int_{[0,\infty)} \phi_+(t)dS_N(f_i, t),$$

and the same holds for ϕ_-; here f_i is the sequence approximating f defined before. We reached a contradiction.

Step 3. The conclusion of the proof proceeds as follows. Let $\bar{\mu} \ : \ C^N \to \mathbb{R}$ be defined by

$$\bar{\mu}(f) = \int_{(0,\infty)} \phi(t)\, dS_N(f; t).$$

By the previous step, and by the results of Sect. 5.1, this is well defined, and is an invariant and continuous valuation. Hence the same properties are shared by $\mu - \bar{\mu}$; on the other hand, by Step 1 and the definition of $\bar{\mu}$, this vanishes on functions with bounded support. As for any element f of C^N there is a monotone sequence of functions in C^N, with bounded support and converging point-wise to f in $\mathbb{R}^N$, and as $\mu - \bar{\mu}$ is continuous, it must be identically zero on C^N.

$\square$

7   Proof of Theorem 1.1

We proceed by induction on N. For the first step of induction, let μ be an invariant and continuous valuation on C^1. For $t > 0$ let

$$\phi_0(t) = \mu(tI_{\{0\}}).$$

This is a continuous function in $\mathbb{R}$, with $\phi_0(0) = 0$. We consider the application $\mu_0 : \mathcal{C}^1 \to \mathbb{R}$:

$$\mu_0(f) = \phi_0(M(f))$$

where as usual $M(f) = \max_{\mathbb{R}} f$. By what we have seen in Sect. 5.1, this is an invariant and continuous valuation. Note that it can be written in the form

$$\mu_0(f) = \int_{(0,\infty)} \phi_0(t)\, dS_0(f;t).$$

Next we set $\bar{\mu} = \mu - \mu_0$; this is still an invariant and continuous valuation, and it is also simple. Indeed, if $f \in \mathcal{C}^1$ is such that $\dim(\text{supp}(f)) = 0$, this is equivalent to say that

$$f = t I_{\{x_0\}}$$

for some $t \geq 0$ and $x_0 \in \mathbb{R}$. Hence

$$\mu(f) = \mu(t I_{\{0\}}) = \phi_0(t) = \mu_0(f).$$

Therefore we may apply Theorem 1.2 to μ_1 and deduce that there exists a function $\phi_1 \in C([0,\infty))$, which vanishes identically in $[0,\delta]$ for some $\delta > 0$, and such that

$$\bar{\mu}(f) = \int_{(0,\infty)} \phi_1(t)\, dS_1(f;t) \quad \forall f \in \mathcal{C}^1.$$

The proof in the one-dimensional case is complete.

We suppose that the Theorem holds up to dimension $(N - 1)$. Let H be an hyperplane of $\mathbb{R}^N$ and define $\mathcal{C}_H^N = \{f \in \mathcal{C}^N : \text{supp}(f) \subseteq H\}$. $\mathcal{C}_H^N$ can be identified with $\mathcal{C}^{N-1}$; moreover μ restricted to $\mathcal{C}_H^N$ is trivially still an invariant and continuous valuation. By the induction assumption, there exists $\phi_k \in C([0,\infty))$, $k = 0, \ldots, N - 1$, such that

$$\mu(f) = \sum_{k=0}^{N-1} \int_{(0,\infty)} \phi_k(t)\, dS_k(f;t) \quad \forall f \in \mathcal{C}_H^N.$$

In addition, there exists $\delta > 0$ such that $\phi_1, \ldots, \phi_{N-1}$ vanish in $[0,\delta]$. Let $\bar{\mu} : \mathcal{C}^N \to \mathbb{R}$ as

$$\bar{\mu}(f) = \sum_{k=0}^{N-1} \int_{(0,\infty)} \phi_k(t)\, dS_k(f;t).$$

This is well defined for $f \in C^N$ and it is an invariant and continuous valuation. The difference $\mu - \bar{\mu}$ is simple; applying Theorem 1.2 to it, as in the one-dimensional case, we complete the proof. $\square$

8 Monotone Valuations

In this section we will prove Theorem 1.3; in particular we will assume that μ is an invariant, continuous and increasing valuation on C^N throughout. Note that, as $\mu(f_0) = 0$, where f_0 is the function identically zero in $\mathbb{R}^N$, we have that $\mu(f) \geq 0$ for every $f \in C^N$.

The proof is divided into three parts.

8.1 Identification of the Measures v_k, $k = 0, \ldots, N$

We proceed as in the proof of Proposition 6.2. Fix $t > 0$ and consider the application $\sigma_t : \mathcal{K}^N \to \mathbb{R}$:

$$\sigma_t(K) = \mu(tI_K), \quad K \in \mathcal{K}^N.$$

This is a rigid motion invariant valuation on $\mathcal{K}^N$ and, as μ is increasing, σ_t has the same property. Hence there exists $(N + 1)$ coefficients, depending on t, that we denote by $\psi_k(t)$, $k = 0, \ldots, N$, such that

$$\sigma_t(K) = \sum_{k=0}^{N} \psi_k(t) V_k(K) \quad \forall K \in \mathcal{K}^N. \tag{28}$$

We prove that each ψ_k is continuous and monotone in $(0, \infty)$. Let us fix the index $k \in \{0, \ldots, N\}$, and let Δ_k be a closed k-dimensional ball in $\mathbb{R}^N$, of radius 1. We have

$$V_j(\Delta_k) = 0 \quad \forall j = k + 1, \ldots, N,$$

and

$$V_k(\Delta_k) =: c(k) > 0.$$

Fix $r \geq 0$; for every j, V_j is positively homogeneous of order j, hence, for $t > 0$,

$$\mu(tI_{r\Delta_k}) = \sum_{j=0}^{k} r^j V_j(\Delta_k) \psi_j(t).$$

Consequently

$$\psi_k(t) = V_k(\Delta_k) \cdot \lim_{r \to \infty} \frac{\mu(tI_{r\Delta_k})}{r^k}.$$

By the properties of μ, the function $t \to \mu(tI_{r\Delta_k})$ is non-negative, increasing and vanishes for $t = 0$, for every $r \geq 0$; these properties are inherited by ψ_k.

As for continuity, we proceed in a similar way. To prove that ψ_0 is continuous we observe that the function

$$t \to \mu(t\Delta_0) = \psi_0(t)$$

is continuous, by the continuity of μ. Assume that we have proved that $\psi_0, \dots, \psi_{k-1}$ are continuous. Then by the equality

$$\mu(tI_{\Delta_k}) = \sum_{j=1}^{k} V_j(\Delta_k)\psi_j(t),$$

it follows that ψ_k is continuous.

Proposition 8.1 *Let μ be an invariant, continuous and increasing valuation on $\mathcal{C}^N$. Then there exists $(N + 1)$ functions $\psi_0, \dots, \psi_N$ defined in $[0, \infty)$, such that (28) holds for every $t \geq 0$ and for every K. In particular each ψ_k is continuous, increasing, and vanishes at $t = 0$.*

For every $k \in \{0, \dots, N\}$ we denote by ν_k the distributional derivative of ψ_k. In particular as ψ_k is continuous, ν_k is non-atomic and

$$\psi_k(t) = \nu_k([0, t)), \quad \forall t \geq 0.$$

Since ψ_k are non-negative functions, by Theorem 2.6, then ν_k are non-negative measures.

8.2   The Case of Simple Functions

Let f be a simple function:

$$f = t_1 I_{K_1} \vee \cdots \vee t_m I_{K_m}$$

with $0 < t_1 < \cdots < t_m$, $K_1 \supset \cdots \supset K_m$ and $K_i \in \mathcal{K}^N$ for every i. The following formula can be proved with the same method used for (24)

$$\mu(f) = \sum_{k=0}^{N} \sum_{i=1}^{m} (\psi_k(t_i) - \psi_k(t_{i-1}))V_k(L_{t_i}(f)), \tag{29}$$

where we have set $t_0 = 0$. As

$$\psi_k(t_i) - \psi_k(t_{i-1}) = v_k((t_{i-1}, t_i])$$

and $L_t(f) = K_i$ for every $t \in (t_{i-1}, t_i]$, we have

$$\mu(f) = \sum_{k=0}^{N} \int_{[0,\infty)} V_k(L_t(f)) \, dv_k(t). \tag{30}$$

In other words, we have proved the theorem for simple functions.

8.3 Proof of Theorem 1.3

Let $f \in C^N$ and let f_i, $i \in \mathbb{N}$, be the sequence of functions built in the proof of Theorem 1.2, Step 2. We have seen that f_i is increasing and converges point-wise to f in $\mathbb{R}^N$. In particular, for every $k = 0, \ldots, N$, the sequence of functions $V_k(L_t(f_i))$, $t \geq 0$, $i \in \mathbb{N}$, is monotone increasing and it converges a.e. to $V_k(L_t(f))$ in $[0, \infty)$. By the B. Levi theorem, we have that

$$\lim_{i \to \infty} \int_{[0,\infty)} V_k(L_t(f_i)) \, dv_k(t) = \int_{[0,\infty)} V_k(L_t(f)) \, dv_k(t)$$

for every k. Using (30) and the continuity of μ we have that the representation formula (30) can be extended to every $f \in C^N$.

Note that in (21) each term of the sum in the right hand-side is non-negative, hence we have that

$$\int_{[0,\infty)} V_k(L_t(f)) \, dv_k(t) < \infty \quad \forall f \in C^N.$$

Applying Proposition 5.6 we obtain that, if $k \geq 1$, there exists $\delta > 0$ such that the support of v_k is contained in $[\delta, \infty)$. The proof is complete. $\square$

References

1. L. Ambrosio, N. Fusco, D. Pallara, *Functions of Bounded Variation and Free Discontinuity Problems* (Oxford University Press, Oxford, 2000)
2. Y. Baryshnikov, R. Ghrist, M. Wright, Hadwiger's theorem for definable functions. Adv. Math. **245**, 573–586 (2013)
3. L. Cavallina, Non-trivial translation-invariant valuations on L^∞. Preprint, 2015 (arXiv:1505.00089)

4. L. Cavallina, A. Colesanti, Monotone valuations on the space of convex functions. Anal. Geom. Metr. Spaces **1**(3) (2015) (electronic version)
5. H. Hadwiger, *Vorlesungen über Inhalt, Oberfläche und Isoperimetrie* (Springer, Berlin-Göttingen-Heidelberg, 1957)
6. D. Klain, A short proof of Hadwiger's characterization theorem. Mathematika **42**, 329–339 (1995)
7. D. Klain, G. Rota, *Introduction to Geometric Probability* (Cambridge University Press, New York, 1997)
8. H. Kone, Valuations on Orlicz spaces and L^ϕ-star sets. Adv. Appl. Math. **52**, 82–98 (2014)
9. M. Ludwig, Fisher information and matrix-valued valuations. Adv. Math. **226**, 2700–2711 (2011)
10. M. Ludwig, Valuations on function spaces. Adv. Geom. **11**, 745–756 (2011)
11. M. Ludwig, Valuations on Sobolev spaces. Am. J. Math. **134**, 824–842 (2012)
12. M. Ludwig, Covariance matrices and valuations. Adv. Appl. Math. **51**, 359–366 (2013)
13. D. Ma, Analysis of Sobolev Spaces in the Context of Convex Geometry and Investigations of the Busemann-Petty Problem. PhD thesis, Technische Universität, Vienna, 2015
14. D. Ma, Real-valued valuations on Sobolev spaces. Preprint, 2015 (arXiv: 1505.02004)
15. M. Ober, L^p-Minkowski valuations on L^q-spaces. J. Math. Anal. Appl. **414**, 68–87 (2014)
16. R. Schneider, *Convex Bodies: The Brunn-Minkowski Theory*, 2nd expanded edn. (Cambridge University Press, Cambridge, 2014)
17. A. Tsang, Valuations on L^p-spaces. Int. Math. Res. Not. **20**, 3993–4023 (2010)
18. A. Tsang, Minkowski valuations on L^p-spaces. Trans. Am. Math. Soc. **364**(12), 6159–6186 (2012)
19. T. Wang, Affine Sobolev inequalities. PhD thesis, Technische Universität, Vienna, 2013
20. T. Wang, Semi-valuations on $BV(\mathbb{R}^n)$. Indiana Univ. Math. J. **63**, 1447–1465 (2014)
21. M. Wright, Hadwiger integration on definable functions, PhD thesis, University of Pennsylvania, 2011

An Inequality for Moments of Log-Concave Functions on Gaussian Random Vectors

Nikos Dafnis and Grigoris Paouris

Abstract We prove sharp moment inequalities for log-concave and log-convex functions, on Gaussian random vectors. As an application we take a reverse form of the classical logarithmic Sobolev inequality, in the case where the function is log-concave.

1 Introduction and Main Results

A function $f : \mathbb{R}^k \to [0, +\infty)$ is called *log-concave* (*on its support*), if and only if

$$f\big((1 - \lambda)x + \lambda y\big) \geq f(x)^{(1-\lambda)} f(y)^{\lambda},$$

for every $\lambda \in [0, 1]$ and $x, y \in \mathrm{supp}(f)$. Respectively, f is called log-*convex* (*on its support*), if and only if

$$f\big((1 - \lambda)x + \lambda y\big) \leq f(x)^{(1-\lambda)} f(y)^{\lambda},$$

for every $\lambda \in [0, 1]$ and $x, y \in \mathrm{supp}(f)$. The aim of this note is to present a sharp inequality for Gaussian moments of log-concave and log-convex functions, stated below as Theorem 1.1.

We work on $\mathbb{R}^k$, equipped with the standard scalar product $\langle \cdot, \cdot \rangle$. We denote by $|\cdot|$ the corresponding Euclidean norm and the absolute value of a real number. We use the notation $X \sim N(\xi, T)$, if X is a Gaussian random vector in $\mathbb{R}^k$, with expectation $\xi \in \mathbb{R}^k$ and covariance the $k \times k$ positive semi-definite matrix T. We say that X is a *standard Gaussian* random vector if it is centered (i.e. $\mathbb{E}X = 0$) with covariance matrix the identity in $\mathbb{R}^k$, where in that case γ_k stands for its distribution law. Finally,

N. Dafnis (✉)
Department of Mathematics, Technion - Israel Institute of Technology, Haifa 32000, Israel
e-mail: nikdafnis@gmail.com

G. Paouris
Department of Mathematics, Texas A&M University, College Station, TX 77843, USA
e-mail: grigorios.paouris@gmail.com

© Springer International Publishing AG 2017
B. Klartag, E. Milman (eds.), *Geometric Aspects of Functional Analysis*,
Lecture Notes in Mathematics 2169, DOI 10.1007/978-3-319-45282-1_7

$\mathcal{L}^{p,s}(\gamma_k)$ stand for the class of all functions $f \in L^p(\gamma_k)$ whose partial derivatives up to order s, are also in $L^p(\gamma_k)$.

Theorem 1.1 *Let $k \in \mathbb{N}$ and X be a Gaussian random vector in $\mathbb{R}^k$. Let $f : \mathbb{R}^k \to [0, +\infty)$ be a log-concave and $g : \mathbb{R}^k \to [0, +\infty)$ be a log-convex function. Then,*

(i) for every $r \in [0, 1]$

$$\mathbb{E}f(\sqrt{r}X) \geq (\mathbb{E}f(X)^r)^{\frac{1}{r}} \quad \text{and} \quad \mathbb{E}g(\sqrt{r}X) \leq (\mathbb{E}g(X)^r)^{\frac{1}{r}}, \tag{1}$$

(ii) for every $q \in [1, +\infty)$

$$\mathbb{E}f(\sqrt{q}X) \leq (\mathbb{E}f(X)^q)^{\frac{1}{q}} \quad \text{and} \quad \mathbb{E}g(\sqrt{q}X) \geq (\mathbb{E}g(X)^q)^{\frac{1}{q}}. \tag{2}$$

In any case, equality holds if $r = 1 = q$ or if $f(x) = g(x) = e^{-\langle a,x \rangle + c}$, where $a \in \mathbb{R}^k$ and $c \in \mathbb{R}$.

We prove Theorem 1.1 in Sect. 2, where we combine techniques from [7] along with Barthe's inequality [2].

The *entropy* of a function $f : \mathbb{R}^k \to \mathbb{R}$, with respect to a random vector X in $\mathbb{R}^k$, is defined to be

$$\text{Ent}_X(f) := \mathbb{E}|f(X)| \log |f(X)| - \mathbb{E}|f(X)| \log \mathbb{E}|f(X)|,$$

provided all the expectations exist. Note that (for $f \geq 0$)

$$\text{Ent}_X(f) = \frac{d}{dq} \left[(\mathbb{E}f(X)^q)^{\frac{1}{q}} \right]_{q=1}$$

and so, Theorem 1.1 implies the following entropy inequality:

Corollary 1.2 *Let $f : \mathbb{R}^k \to [0, +\infty)$ and X be a Gaussian random vector in $\mathbb{R}^k$.*

(i) If f is log-concave, then

$$\text{Ent}_X(f) \geq \frac{1}{2}\mathbb{E}\langle X, \nabla f(X) \rangle. \tag{3}$$

(ii) If f is log-convex, then

$$\text{Ent}_X(f) \leq \frac{1}{2}\mathbb{E}\langle X, \nabla f(X) \rangle. \tag{4}$$

In any case, equality holds if $f(x) = \exp(\langle a, x \rangle + c)$, $a \in \mathbb{R}^k$, $c \in \mathbb{R}$.

Proof Let $m(q) := (\mathbb{E}f(X)^q)^{\frac{1}{q}}$ and $h(q) := \mathbb{E}f(\sqrt{q}X)$. Then we have

$$m(1) = \mathbb{E}f(X) = h(1), \quad m'(1) = \text{Ent}_X(f) \quad \text{and} \quad h'(1) = \frac{1}{2}\mathbb{E}\langle X, \nabla f(X) \rangle,$$

and Theorem 1.1 implies the desired result.   □

The logarithmic Sobolev inequality, proved by Gross in [10], states that if $X \sim N(0, I_k)$, then

$$\text{Ent}_X(f^2) \leq 2 \, \mathbb{E}|\nabla f(X)|^2, \tag{5}$$

for every function $f \in L^2(\gamma_k)$. Moreover, Carlen showed in [6], that equality holds if and only if f is an exponential function. For more details about the logarithmic Sobolev inequality we refer the reader to [4, 14, 19, 20] and to the references therein.

In Sect. 3, we show that Corollary 1.2, after an application of the Gaussian integration by parts formula (see Lemma 3.1), leads to the following reverse form of Gross' inequality, when the function is log concave:

Theorem 1.3 *Let X be a standard Gaussian random vector in $\mathbb{R}^k$ and $f = e^{-v} \in \mathcal{L}^{2,1}(\gamma_k)$, be a positive log-concave function (on its support). Then*

$$2 \, \mathbb{E}|\nabla f(X)|^2 - \mathbb{E}f(X)^2 \Delta v(X) \leq \text{Ent}_X(f^2). \tag{6}$$

Theorem 1.3, ensures that if a log-concave function $f = e^{-v}$ is close to be an exponential, in the sense that $\mathbb{E}f(X)^2 \Delta v(X)$ is small, then the logarithmic Sobolev inequality for f is close to be sharp.

For more properties and stability results on the logarithmic-Sobolev inequalities we refer to the papers [8, 9, 11] and the references therein.

2 Proof of the Main Result

The first ingredient of the proof of Theorem 1.1, is the following inequality for Gaussian random vectors, proved in [7]. We recall that for two square matrices A and B, we say that $A \leq B$ if and only if $B - A$ is positive semi-definite.

Theorem 2.1 *Let $m, n_1, \ldots, n_m \in \mathbb{N}$ and set $N = \sum_{i=1}^{m} n_i$. For every $i = 1, \ldots, m$, let X_i be a Gaussian random vector in $\mathbb{R}^{n_i}$, such that $\mathbf{X} := (X_1, \ldots, X_m)$ is a Gaussian random vector in $\mathbb{R}^N$ with covariance the $N \times N$ matrix $T = (T_{ij})_{1 \leq i,j \leq m}$, where T_{ij} is the covariance $n_i \times n_j$ matrix between X_i and X_j, $1 \leq i,j \leq m$. Let $p_1, \ldots, p_m \in \mathbb{R}$ and consider the $N \times N$ block diagonal matrix $P = \text{diag}(p_1 T_{11}, \ldots, p_m T_{mm})$. Then, for any set of nonnegative measurable functions f_i on $\mathbb{R}^{n_i}$, $i = 1, \ldots, m$,*

(i) *if $T \leq P$, then*

$$\mathbb{E} \prod_{i=1}^{m} f_i(X_i) \leq \prod_{i=1}^{m} \left(\mathbb{E}f_i(X_i)^{p_i} \right)^{\frac{1}{p_i}}, \tag{7}$$

(ii) if $T \geq P$, then

$$\mathbb{E}\prod_{i=1}^{m} f_i(X_i) \geq \prod_{i=1}^{m} \left(\mathbb{E}f_i(X_i)^{p_i}\right)^{\frac{1}{p_i}}. \tag{8}$$

Theorem 2.1 generalizes many fundamental results in analysis, such as Hölder inequality and its reverse, Young inequality with the best constant and its reverse [3] and [5], and Nelson's Gaussian Hypercontractivity and its reverse [17] and [15]. Actually, the first part of Theorem 2.1 is another formulation of the Brascamp-Lieb inequality [5, 13], while the second part provides a reverse form.

Moreover, (8) implies (see [7]) F. Barthe's reverse Brascamp-Lieb inequality [2], which the second main tool in our the proof of Theorem 1.1. For more extensions of Brascamp-Lieb inequality and similar results see [12] and [16].

For our purposes, we need the so-called *geometric* form (see [1]) of Barthe's theorem.

Theorem 2.2 *Let* $n, m, n_1, \ldots, n_m \in \mathbb{N}$ *with* $n_i \leq n$ *for every* $i = 1, \ldots, m$. *Let* U_i *be a* $n_i \times n$ *matrix with* $U_i U_i^* = I_{n_i}$ *for* $i = 1, \ldots, m$ *and* $c_1, \ldots, c_m$ *be positive real numbers such that*

$$\sum_{i=1}^{m} c_i U_i^* U_i = I_n.$$

Let $h : \mathbb{R}^n \to [0, +\infty)$ *and* $f_i : \mathbb{R}^{n_i} \to [0, +\infty)$, $i = 1, \ldots, m$, *be measurable functions such that*

$$h\left(\sum_{i=1}^{N} c_i U_i^* \xi_i\right) \geq \prod_{i=1}^{m} f_i(\xi_i)^{c_i} \quad \forall \, \xi_i \in \mathbb{R}^{n_i}, \tag{9}$$

$i = 1, \ldots, m$. *Then*

$$\int_{\mathbb{R}^n} h(x) \, d\gamma_n(x) \geq \prod_{i=1}^{m} \left(\int_{\mathbb{R}^{n_i}} f_i(x) \, d\gamma_{n_i}(x)\right)^{c_i}. \tag{10}$$

2.1 Decomposing the Identity

We will apply Theorem 2.1 in the special case where the covariance is the $kn \times kn$ matrix $T = \left([T_{ij}]\right)_{i,j \leq n}$, with $T_{ii} = I_k$ and $T_{ij} = t I_k$ if $i \neq j$, for some $t \in [-\frac{1}{n-1}, 1]$. Equivalently, in that case $\mathbf{X} := (X_1, \ldots, X_n) \sim N(0, T)$, where $X_1, \cdots, X_n$

are standard Gaussian random vectors in $\mathbb{R}^k$, such that

$$\mathbb{E}(X_i X_j^*) = \begin{cases} I_k, & i = j \\ t I_k, & i \neq j \end{cases}. \tag{11}$$

For any $t \in [0, 1]$, a natural way to construct such random vectors is to consider n independent copies $Z_1, \ldots, Z_n$, of a $Z \sim N(0, I_k)$ and set

$$X_i := \sqrt{t} Z + \sqrt{1 - t} Z_i, \quad i = 1, \ldots, n.$$

However, we are going to use a more geometric approach. First we will deal with the 1-dimensional case and then, by using a tensorization argument, we will pass to the general k-dimensional case, for any $k \in \mathbb{N}$. We begin with the definition of the SR-simplex.

Definition 2.3 We say that $S = \text{conv}\{v_1, \ldots, v_n\} \subseteq \mathbb{R}^{n-1}$ is the *spherico-regular simplex* (in short SR-simplex) in $\mathbb{R}^{n-1}$, if $v_1, \ldots, v_n$ are unit vectors in $\mathbb{R}^{n-1}$ with the following two properties:

(SR1) $\langle v_i, v_j \rangle = -\frac{1}{n-1}$, for any $i \neq j$,
(SR2) $\sum_{i=1}^n v_i = 0$.

Using the vertices of the SR-simplex in $\mathbb{R}^{n-1}$, we create n vectors in $\mathbb{R}^n$ with the same angle between them. This is done in the next lemma.

Lemma 2.4 *Let $n \geq 2$ and $v_1, \ldots, v_n$ be the vertices of any RS-Simplex in $\mathbb{R}^{n-1}$. For every $t \in [-\frac{1}{n-1}, 1]$, let $u_1, \ldots, u_n$ be the unit vectors in $\mathbb{R}^n$ with*

$$u_i = u_i(t) = \sqrt{\frac{t(n-1)+1}{n}} \, e_n + \sqrt{\frac{n-1}{n}(1-t)} \, v_i, \tag{12}$$

$i = 1, \ldots, n$. *Then we have that*

$$\langle u_i, u_j \rangle = t, \quad \forall i \neq j. \tag{13}$$

Moreover,

(i) if $t \in [0, 1]$, then

$$\frac{1}{t(n-1)+1} \sum_{i=1}^n u_i u_i^* + \frac{nt}{t(n-1)+1} \sum_{j=1}^{n-1} e_j e_j^* = I_n, \tag{14}$$

(ii) if $t \in [-\frac{1}{n-1}, 0]$, then

$$\frac{1}{1-t} \sum_{i=1}^n u_i u_i^* + \frac{-nt}{1-t} e_n e_n^* = I_n. \tag{15}$$

Proof A direct computation, using the properties (*SR*1), (*SR*2) and the fact that

$$\frac{n-1}{n} \sum_{i=1}^{n} v_i v_i^* = I_{n-1},$$

shows that (13)–(15) holds true.   □

Remark 2.5 If $Z \sim N(0, I_n)$, then $X_i := \langle u_i, Z \rangle$, $i = 1, \ldots, n$, are standard Gaussian random variables, satisfying the condition (11) in the 1-dimensional case.

For the general case we first recall the definition of the *tensor product* of two matrices:

Definition 2.6 For any matrices $A \in \mathbb{R}^{m \times n}$ and $B \in \mathbb{R}^{k \times \ell}$, their tensor product is defined to be the $km \times \ell n$ matrix

$$A \otimes B = \begin{pmatrix} a_{11}B & \cdots & a_{1n}B \\ \vdots & \ddots & \vdots \\ a_{m1}B & \cdots & a_{mn}B \end{pmatrix}.$$

Every vector $a \in \mathbb{R}^n$ is considered to be a $n \times 1$ column matrix and with this notation, we state some basic properties for the tensor product, that we will use.

Lemma 2.7 *1. Let $a = (a_1, \ldots, a_m)^* \in \mathbb{R}^m$ and $b = (b_1, \ldots, b_n)^* \in \mathbb{R}^n$. Then*

$$a \otimes b^* = ab^* = \begin{pmatrix} a_1 b_1 & \cdots & a_1 b_n \\ \vdots & \ddots & \vdots \\ a_m b_1 & \cdots & a_m b_n \end{pmatrix} \in \mathbb{R}^{m \times n},$$

and as a linear transformation, $a \otimes b^* = ab^* : \mathbb{R}^n \to \mathbb{R}^m$ with

$$(a \otimes b^*)(x) = (ab^*)(x) = \langle x, b \rangle a, \quad x \in \mathbb{R}^n.$$

2. Let $A_i \in \mathbb{R}^{m \times n}$ and $B \in \mathbb{R}^{k \times \ell}$. Then $\left(\sum_i A_i \right) \otimes B = \sum_i A_i \otimes B$.
3. Let $A_1 \in \mathbb{R}^{m \times n}$, $B_1 \in \mathbb{R}^{k \times \ell}$, and $A_2 \in \mathbb{R}^{n \times r}$, $B_2 \in \mathbb{R}^{\ell \times s}$. Then

$$(A_1 \otimes B_1)(A_2 \otimes B_2) = (A_1 A_2) \otimes (B_1 B_2) \in \mathbb{R}^{km \times rs}.$$

4. For any matrices A and B,

$$(A \otimes B)^* = A^* \otimes B^*.$$

For our k-dimensional construction, we consider the $k \times kn$ matrices

$$U_i := u_i^* \otimes I_k = \left(\begin{bmatrix} u_{i1}I_k \end{bmatrix} \cdots \begin{bmatrix} u_{in}I_k \end{bmatrix} \right), \tag{16}$$

$$E_j := e_j^* \otimes I_k = \left(\begin{bmatrix} e_{j1}I_k \end{bmatrix} \cdots \begin{bmatrix} e_{jn}I_k \end{bmatrix} \right), \tag{17}$$

for $i = 1 \ldots, n$. Note that

$$U_i^* U_i = (u_i^* \otimes I_k)^* (u_i^* \otimes I_k) = u_i u_i^* \otimes I_k$$

and

$$E_j^* E_j = (e_j^* \otimes I_k)^* (e_j^* \otimes I_k) = e_j e_j^* \otimes I_k,$$

for every $i, j = 1, \ldots, n$. Thus by taking the tensor product with I_k, in both sides of (14), we get that

$$\frac{1}{p} \sum_{i=1}^n U_i^* U_i + \frac{nt}{p} \sum_{j=1}^{n-1} E_j^* E_j = I_{kn}, \tag{18}$$

for every $t \in [0, 1]$, where $p := (n-1)t + 1$. Moreover, we can now construct the general case describing in (11). We summarize in the next lemma.

Lemma 2.8 *Suppose that $Z_1, \ldots, Z_n$ are iid standard Gaussian random vectors in $\mathbb{R}^k$ and set $\mathbf{Z} := (Z_1, \ldots, Z_n) \sim N(0, I_{kn})$. Consider the random vectors*

$$X_i := U_i \mathbf{Z} = \sum_{a=1}^n u_{ia} Z_a, \qquad i = 1, \ldots, n, \tag{19}$$

where U_i, $i = 1, \ldots, n$, are the matrices defined in (16). Then X_i is a standard Gaussian random vector in $\mathbb{R}^k$, for every $i = 1, \ldots, n$ and

$$\mathbb{E}\left[X_i \otimes X_j^* \right] = \left(\mathbb{E}[X_{ir}X_{j\ell}] \right)_{r,\ell \leq k} = \left(t\delta_{r\ell} \right)_{r,\ell \leq k} = tI_k, \tag{20}$$

for every $i \neq j$.

Proof Clearly, $\mathbb{E}X_i = 0$, for every $i, j = 1, \ldots, n$, and since

$$\mathbb{E}\left[Z_a \otimes Z_b^* \right] = \left(\mathbb{E}[Z_{ar}Z_{b\ell}] \right)_{r,\ell \leq k} = \delta_{\alpha\beta} I_k$$

we have that

$$
\begin{aligned}
\mathbb{E}\left[X_{ir}X_{j\ell}\right] &= \mathbb{E}\left[\left(\sum_{a=1}^{n} u_{ia}Z_{ar}\right)\left(\sum_{b=1}^{n} u_{jb}Z_{b\ell}\right)\right] \\
&= \sum_{a=1}^{n}\sum_{b=1}^{n} u_{ia}u_{jb}\,\mathbb{E}\left[Z_{ar}Z_{b\ell}\right] \\
&= \sum_{a=1}^{n} u_{ia}u_{ja}\,\mathbb{E}\left[Z_{ar}Z_{a\ell}\right] \\
&= \sum_{a=1}^{n} u_{ia}u_{ja}\,\delta_{r\ell} \\
&= \langle u_i, u_j\rangle\,\delta_{r\ell}.
\end{aligned}
$$

The proof is complete, since $|u_i| = 1$ for all i's and by (13) $\langle u_i, u_j\rangle = t$ for all $i \neq j$.
□

2.2 Proof of Theorem 1.1

The next proposition is the main ingredient for the proof of Theorem 1.1.

Proposition 2.9 *Let $t \in [0, 1]$, $k, n \in \mathbb{N}$, $p = t(n-1)+1$, X be a standard Gaussian random vector in $\mathbb{R}^k$ and $X_1, \cdots, X_n$ be copies of X such that*

$$
\mathbb{E}[X_i \otimes X_j^*] = \left(\mathbb{E}[X_{ir}X_{j\ell}]\right)_{r,\ell \leq k} = tI_k, \quad \forall\, i \neq j.
$$

Then, for any log-concave *(on its support) function $f : \mathbb{R}^k \to [0, +\infty)$, we have that*

$$
\mathbb{E}\left(\prod_{i=1}^{n} f(X_i)\right)^{\frac{1}{n}} \leq \left(\mathbb{E}f(X)^{\frac{\ell}{n}}\right)^{\frac{n}{p}} \leq \mathbb{E}f\left(\frac{1}{n}\sum_{i=1}^{n} X_i\right) \tag{21}
$$

Note that, the log-concavity of f implies that

$$
\left(\prod_{i=1}^{n} f(X_i)\right)^{\frac{1}{n}} \leq f\left(\frac{1}{n}\sum_{i=1}^{n} X_i\right),
$$

where equality is achieved for the exponential function $f(x) = e^{\langle a,x\rangle+c}$, $a \in \mathbb{R}^k$ and $c \in \mathbb{R}$.

Proof of Proposition 2.9 In order to prove the left-hand side inequality in (21), we will apply Theorem 2.1. Note that the assumption of log-concavity will not be used. The left-hand side inequality in (21) holds true for any non-negative measurable function f.

To be more precise, let $X_1, \ldots, X_n$ be standard Gaussian random vectors in $\mathbb{R}^k$ satisfying condition (20) and $t \in [-\frac{1}{n-1}, 1]$. Then, $\mathbf{X} := (X_1, \ldots, X_n)$, is a centered Gaussian vector in $\mathbb{R}^{kn}$ with covariance the $kn \times kn$ matrix $T = (T_{ij})_{i,j \leq n}$, with block entries the $k \times k$ matrices $T_{ii} = I_k$ and $T_{ij} = tI_k$, for $i \neq j$. Setting

$$p := (n-1)t + 1 \quad \text{and} \quad q := 1 - t,$$

it's not hard to check that, for any $t \in [0.1]$, p is the biggest and q is the smallest singular value of T, while for any $t \in [-\frac{1}{n-1}, 0]$, q is the biggest and p is the smallest singular value of T. Thus,

(i) if $t \geq 0$, then

$$qI_{kn} \leq T \leq pI_{kn},$$

(ii) if $t \leq 0$, then

$$pI_{kn} \leq T \leq qI_{kn}$$

In the above situation, Theorem 2.1 reads as follows:

Theorem 2.10 *Let $k, n \in \mathbb{N}$, $t \in [-\frac{1}{n-1}, 1]$ and let $X_1, \ldots, X_n$ be standard Gaussian random vectors in $\mathbb{R}^k$, with $\mathbb{E}[X_i \otimes X_j^*] = tI_k$, for all $i \neq j$. Set $p := (n-1)t+1, q := 1 - t$, and then for every measurable functions $f_i : \mathbb{R}^k \to [0, +\infty)$, $i = 1, \ldots, n$,*

(i) if $t \in [0, 1]$, then

$$\prod_{i=1}^{n} \left(\mathbb{E} f_i(X_i)^q \right)^{1/q} \leq \mathbb{E} \prod_{i=1}^{n} f_i(X_i) \leq \prod_{i=1}^{n} \left(\mathbb{E} f_i(X_i)^p \right)^{1/p}, \tag{22}$$

(ii) if $t \in [-\frac{1}{n-1}, 0]$, then

$$\prod_{i=1}^{n} \left(\mathbb{E} f_i(X_i)^p \right)^{1/p} \leq \mathbb{E} \prod_{i=1}^{n} f_i(X_i) \leq \prod_{i=1}^{n} \left(\mathbb{E} f_i(X_i)^q \right)^{1/q}. \tag{23}$$

Now, the left-hand side inequality of (21) follows immediately from (22), by taking $f_i = f^{1/n}$ for every $i = 1, \ldots, n$.

In order to prove the right-hand side inequality of (21) we apply Barthe's theorem, using the decomposition of the identity in (18). In the following lemma we gather some technical facts.

Lemma 2.11 *Let U_i and E_i, $i = 1, \ldots, n$ the matrices defined in (16) and (17), and set $p = (n-1)t + 1$, $q = 1 - t$. Then*

$$U_i^* = \sqrt{\frac{p}{n}}\, e_n \otimes I_k + \sqrt{\frac{n-1}{n}}\, q\, v_i \otimes I_k \, \in \mathbb{R}^{kn \times k}.$$

$$U_i U_j^* = \langle u_i, u_j \rangle I_k$$

$$U_i E_j^* = \sqrt{\frac{n-1}{n}}\, q\, \langle v_i, e_j \rangle I_k$$

for every $i \le n$ and $j \le n - 1$.

Proof The first and the second assertion can be verified, just by using the definitions. For the third one, we have

$$U_i E_j^* = (u_i^* \otimes I_k)(e_j^* \otimes I_k)^*$$

$$= \left(\sqrt{\frac{p}{n}}\, e_n^* \otimes I_k + \sqrt{\frac{n-1}{n}}\, q\, v_i^* \otimes I_k \right) (e_j \otimes I_k)$$

$$= \sqrt{\frac{p}{n}}\, (e_n^* \otimes I_k)(e_j \otimes I_k) + \sqrt{\frac{n-1}{n}}\, q\, (v_i^* \otimes I_k)(e_j \otimes I_k)$$

$$= \sqrt{\frac{p}{n}}\, e_n^* e_j \otimes I_k + \sqrt{\frac{n-1}{n}}\, q\, v_i^* e_j \otimes I_k$$

$$= \sqrt{\frac{p}{n}}\, \langle e_n, e_j \rangle I_k + \sqrt{\frac{n-1}{n}}\, q\, \langle v_i, e_j \rangle I_k$$

$$= \mathbb{0} + \sqrt{\frac{n-1}{n}}\, q\, \langle v_i, e_j \rangle I_k.$$

$\square$

To finish the proof of Proposition 2.9, we apply Barthe's Theorem 2.2, using the decomposition of the identity appearing in (18). We choose the parameters: $n \leftrightarrow kn$, $m := 2n - 1$, $n_i := k$ for all $i = 1, \ldots, 2n - 1$, and

$$c_i := \begin{cases} \frac{1}{p} & , i = 1, \ldots, n \\ \frac{nt}{p} & , i = n + 1, \ldots, 2n - 1 \end{cases}.$$

Then, we apply Theorem 2.2 to the functions

$$\tilde{f}_i(x) := \begin{cases} f(x)^{\frac{p}{n}} & , i = 1, \ldots, n \\ 1 & , i = n + 1, \ldots, 2n - 1 \end{cases}, \quad x \in \mathbb{R}^k$$

and

$$h(x) := f\left(\frac{1}{n}\sum_{i=1}^{n} U_i x\right), \quad x \in \mathbb{R}^{kn}.$$

For any $\xi_1, \ldots, \xi_n \in \mathbb{R}^k$, by Lemma 2.11, we get that

$$h\left(\sum_{j=1}^{n}\frac{1}{p}U_j^*\xi_j + \sum_{a=1}^{n-1}\frac{nt}{p}E_a^*\xi_{n+a}\right)$$

$$= f\left(\frac{1}{n}\sum_{i=1}^{n}\sum_{j=1}^{n}\frac{1}{p}U_iU_j^*\xi_j + \frac{1}{n}\sum_{i=1}^{n}\sum_{a=1}^{n-1}\frac{nt}{p}U_iE_a^*\xi_{n+a}\right)$$

$$= f\left(\frac{1}{n}\sum_{i=1}^{n}\sum_{j=1}^{n}\frac{1}{p}U_iU_j^*\xi_j + \frac{1}{n}\sum_{i=1}^{n}\sum_{a=1}^{n-1}\frac{nt}{p}\sqrt{\frac{n-1}{n}}q\langle v_i, e_a\rangle\xi_{n+a}\right)$$

$$= f\left(\frac{1}{n}\sum_{i=1}^{n}\sum_{j=1}^{n}\frac{1}{p}U_iU_j^*\xi_j\right) \qquad \left(\text{since } \sum v_i = 0\right)$$

$$= f\left(\frac{1}{n}\sum_{i=1}^{n}\sum_{j=1}^{n}\frac{1}{p}\langle u_i, u_j\rangle\xi_j\right)$$

$$= f\left(\frac{1}{n}\sum_{i=1}^{n}\left(\frac{1}{p}\xi_i + \sum_{j\neq i}\frac{t}{p}\xi_j\right)\right)$$

$$= f\left(\frac{1}{n}\sum_{i=1}^{n}\left(\frac{1}{p} + (n-1)\frac{t}{p}\right)\xi_i\right)$$

$$= f\left(\frac{1}{n}\sum_{i=1}^{n}\xi_i\right)$$

$$\geq \prod_{i=1}^{n}f(\xi_i)^{\frac{1}{n}} = \prod_{i=1}^{n}\left(f(\xi_i)^{\frac{p}{n}}\right)^{\frac{1}{p}} = \prod_{i=1}^{n}\tilde{f}(\xi_i)^{c_i}.$$

Thus, Theorem 2.2 implies

$$\mathbb{E}f\left(\frac{1}{n}\sum_{i=1}^{n}X_i\right) = \mathbb{E}f\left(\frac{1}{n}\sum_{i=1}^{n}U_iZ\right) \geq \prod_{i=1}^{n}\left(\mathbb{E}f(X_i)^{\frac{p}{n}}\right)^{\frac{1}{p}} = \left(\mathbb{E}f(X)^{\frac{p}{n}}\right)^{\frac{n}{p}}$$

and the proof is complete. $\square$

We close this section with the proof of our primary result.

Proof of Theorem 1.1 Suppose first that $X \sim N(0, I_k)$. Then, under the notation of Lemma 2.8 we have that

$$
\frac{1}{n} \sum_{i=1}^{n} U_i \mathbf{Z} = \frac{1}{n} \sum_{i=1}^{n} \sqrt{\frac{p}{n}} (e_n^* \otimes I_k) \mathbf{Z} + \frac{1}{n} \sum_{i=1}^{n} \sqrt{\frac{n-1}{n}} q (v_i^* \otimes I_k) \mathbf{Z}
$$

$$
= \sqrt{\frac{p}{n}} (e_n^* \otimes I_k) \mathbf{Z} + \frac{1}{n} \sqrt{\frac{n-1}{n}} q \left(\sum_{i=1}^{n} v_i^* \right) \otimes I_k \mathbf{Z}
$$

$$
= \sqrt{\frac{p}{n}} E_n \mathbf{Z} + \frac{1}{n} \sqrt{\frac{n-1}{n}} q \left(\sum_{i=1}^{n} v_i \right)^* \otimes I_k \mathbf{Z}
$$

$$
= \sqrt{\frac{p}{n}} Z_n.
$$

Thus, the right hand side of (21) can be written as

$$
\mathbb{E} f \left(\sqrt{\frac{p}{n}} X \right) \geq \left(f(X)^{\frac{p}{n}} \right)^{\frac{n}{p}}. \tag{24}
$$

where $p = (n-1)t + 1$, $n \in \mathbb{N}$, and $t \in [0, 1]$.

Consequently, if $f : \mathbb{R}^k \to [0, +\infty)$ is a log-concave function and $r \in (0, 1]$, then there exist $t \in [0, 1]$ and $n \in \mathbb{N}$, such that $r = \frac{p}{n} = \frac{(n-1)t+1}{n}$ and so by (24) we get that

$$
\mathbb{E} f(\sqrt{r} X) \geq (\mathbb{E} f(X)^r)^{\frac{1}{r}} \tag{25}
$$

for every $r \in (0, 1]$. We consider now the case where $r = 0$. Since f is *log*-concave, there exists a convex function $v : \mathbb{R}^k \to \mathbb{R}$ such that $f = e^{-v}$. Then, for $r = 0$, inequality (1) is equivalent to Jensen's inequality

$$
v(0) = v(\mathbb{E} X) \leq \mathbb{E} v(X), \tag{26}
$$

and the proof of (1) is complete.

For every $q \geq 1$ consider $r = \frac{1}{q} \in (0, 1]$. Let $F(x) = f(x/\sqrt{r})^{1/r}$ which is also log-concave and so (25) for F and r implies

$$
\mathbb{E} f(X)^q \geq \left(\mathbb{E} f(\sqrt{q} X) \right)^q, \tag{27}
$$

and (2) follows.

Assume now that $g : \mathbb{R}^n \to [0, +\infty)$ is log-convex and $r \in (0, 1]$. By the log-convexity of g and Theorem 2.10(i), we have that

$$\mathbb{E}g\left(\frac{1}{n}\sum_{i=1}^n X_i\right) \le \mathbb{E}\prod_{i=1}^n g(X_i)^{\frac{1}{n}} \le \left(\mathbb{E}g(X)^{\frac{p}{n}}\right)^{\frac{n}{p}}. \tag{28}$$

As we have seen at the beginning of the proof $\frac{1}{n}\sum_{i=1}^n X_i \stackrel{d}{=} \sqrt{\frac{p}{n}} X$. So, using (28) for $t \in [0, 1]$ and $n \in \mathbb{N}$ such that $\frac{p}{n} = \frac{(n-1)t+1}{n} = r$, we derive that

$$\mathbb{E}g\left(\sqrt{r}X\right) \le \left(\mathbb{E}g(X)^r\right)^{\frac{1}{r}},$$

for every $r \in (0, 1]$. The rest of the proof for a log-convex function g is identical to the log-concave case.

For the equality case, a straightforward computation shows that for $f(x) = e^{\langle a,x\rangle + c}$, we have that

$$\mathbb{E}f\left(\sqrt{q}X\right) = C\exp\left(\frac{q}{2}|a|^2\right) = \left(\mathbb{E}f(X)^q\right)^{\frac{1}{q}}.$$

for every $q \ge 0$.

Finally, suppose that X is a general Gaussian random vector in $\mathbb{R}^k$ with expectation $\xi \in \mathbb{R}^k$ and covariance matrix $T = UU^*$ where $U \in \mathbb{R}^{k \times k}$. Note, that if f is log-concave (or log-convex) and positive function on $\mathbb{R}^k$, then so is $F(x) := f(Ux - \xi)$. Moreover, if $Z \sim N(0, I_k)$ then $UZ - \xi \stackrel{d}{=} X \sim N(0, T)$. The general case follows then, by applying the previous case on function F. $\square$

3 Reverse Logarithmic Sobolev Inequality

In the next lemma, we state the *Gaussian Integration by Parts* formula (see [18, Appendix 4] for a simple proof).

Lemma 3.1 *Let $X, Y_1, \ldots, Y_n$ be centered jointly Gaussian random variables, and F be a real valued function on $\mathbb{R}^n$, that satisfy the growth condition*

$$\lim_{|x|\to\infty} |F(x)| \exp\left(-a|x|^2\right) = 0 \qquad \forall\, a > 0. \tag{29}$$

Then

$$\mathbb{E}\left[XF(Y_1, \ldots, Y_n)\right] = \sum_{i=1}^n \mathbb{E}\left[XY_i\right] \mathbb{E}\left[\partial_i F(Y_1, \ldots, Y_n)\right]. \tag{30}$$

Involving this formula, we can further elaborate Corollary 1.2.

Let $\mathscr{G}_k$, be the class of all positive functions in $\mathbb{R}^k$, such that their first derivatives satisfy the growth condition (29). Then for any $f \in \mathscr{G}_k$, by Lemma 3.1, we get that

$$\mathbb{E}\big[\langle X, \nabla f(X)\rangle\big] = \sum_{i=1}^{k}\mathbb{E}\big[X_i \partial_i f(X)\big]$$

$$= \sum_{i=1}^{k}\sum_{j=1}^{k}\mathbb{E}\big[X_i X_j\big]\mathbb{E}\big[\partial_{ij}f(X)\big] = \mathbb{E}\big[\mathrm{tr}\big(T\,H_f(X)\big)\big],$$

where T is the covariance matrix of X and $H_f(x)$ stands for the Hessian matrix of f at $x \in \mathbb{R}^k$. In the special case where $X \sim N(0, I_k)$, Corollary 1.2 implies the following:

Corollary 3.2 *Let $k \in \mathbb{N}$, and X be a standard Gaussian vector in $\mathbb{R}^k$. Then*

(i) for every log-concave function $f \in \mathscr{G}_k$, we have

$$\mathrm{Ent}_X(f) \geq \frac{1}{2}\mathbb{E}\Delta f(X), \tag{31}$$

(ii) for every log-convex function $f \in \mathscr{G}_k$, we have

$$\mathrm{Ent}_X(f) \leq \frac{1}{2}\mathbb{E}\Delta f(X). \tag{32}$$

Proof of Theorem 1.3 Let $f \in \mathcal{L}^{2,1}(\gamma_k)$. Without loss of generality we may also assume that $\mathbb{E}f^2(X) = 1$. Suppose first that f has a bounded support. Then $f^2 \in \mathscr{G}_k$ and Corollary 3.2, after an application of the chain rule $\frac{1}{2}\Delta f^2 = |\nabla f|^2 + f\Delta f$, gives that

$$\mathbb{E}|\nabla f(X)|^2 + \mathbb{E}f(X)\Delta f(X) \leq \mathrm{Ent}_X(f^2) \leq 2\,\mathbb{E}|\nabla f(X)|^2. \tag{33}$$

Let $f = e^{-v}$, where $v : supp(f) \to \mathbb{R}$ is a convex function. Again by the chain rule we have $f\Delta f = |\nabla f|^2 - f^2\Delta v$, and so

$$\mathbb{E}f(X)\Delta f(X) = \mathbb{E}|\nabla f(X)|^2 - \mathbb{E}f(X)^2\Delta v(X). \tag{34}$$

Equations (33) and (34), prove Theorem 1.3 in this case.

To drop the assumption of the bounded support, we consider the functions $f_n := f\,\mathbf{1}_{nB_2^k}$, where $\mathbf{1}_{nB_2^k}$ is the indicator function of the Euclidean Ball in $\mathbb{R}^k$ with radius $n \in \mathbb{N}$. Every f_n has bounded support and so by the previous case,

$$2\,\mathbb{E}|\nabla f_n(X)|^2 - \mathbb{E}f_n(X)^2\Delta v_n(X) \leq \mathrm{Ent}_X(f_n^2). \tag{35}$$

In order to avoid any possible problem of infiniteness of the derivatives of f_n, $n \in \mathbb{N}$, we define the functions

$$F_n = |\nabla f|^2 \cdot \mathbf{1}_{nB_2^k}, \qquad H_n = f^2 \Delta v \cdot \mathbf{1}_{nB_2^k}.$$

Notice that $F_n = |\nabla f_n|^2$ and $H_n = f_n^2 \Delta v_n$ almost everywhere, since they could only differ on the zero-measure set $\{x \in \mathbb{R}^k : |x| = n\}$. Thus,

$$0 \le f_n \nearrow f, \qquad 0 \le F_n \nearrow |\nabla f|^2, \qquad 0 \le H_n \nearrow f^2 \Delta v,$$

and by the monotone convergence theorem

$$\mathbb{E}|\nabla f_n(X)|^2 = \mathbb{E} F_n(X) \longrightarrow \mathbb{E}|\nabla f(X)|^2 \tag{36}$$

and

$$\mathbb{E} f_n(X)^2 \Delta v_n(X) = \mathbb{E} H_n(X) \longrightarrow \mathbb{E} f(X)^2 \Delta v(X). \tag{37}$$

Moreover, $f_n^2 \log f_n^2 \to f^2 \log f^2$ and $|f_n^2 \log f_n^2| \le |f^2 \log f^2|$, for every $n \in \mathbb{N}$ (where we have taken that $0 \log 0 = 0$). Since, by Gross' inequality, $f^2 \log f^2 \in L^1(\gamma_k)$, the Lebesgue's dominated convergence theorem implies that

$$\mathrm{Ent}_X(f_n^2) \longrightarrow \mathrm{Ent}_X(f^2). \tag{38}$$

Under the light of (36)–(38), the desired result follows by taking the limit in (35), as $n \to \infty$. □

Acknowledgements The research leading to these results is part of a project that has received funding from the European Research Council (ERC) under the European Union's Horizon 2020 research and innovation programme (grant agreement No 637851).

Part of this work was done while the first named author was a postdoctoral research fellow at the University of Crete, and he was supported by the Action Supporting Postdoctoral Researchers of the Operational Program Education and Lifelong Learning (Actions Beneficiary: General Secretariat for Research and Technology), co-financed by the European Social Fund (ESF) and the Greece State.

The second named author is supported by the US NSF grant CAREER-1151711 and BSF grant 2010288.

Finally, we would like to thank the anonymous referee whose valuable remarks improved the presentation of the paper.

References

1. K.M. Ball, *Volumes of Sections of Cubes and Related Problems*. Lecture Notes in Mathematics, vol. 1376 (Springer, Berlin, 1989), pp. 251–260
2. F. Barthe. On a reverse form of the Brascamp-Lieb inequality. Invent. Math. **134**, 335–361 (1998)

3. W. Beckner, Inequalities in Fourier analysis. Ann. Math. **102**, 159–182 (1975)
4. V. Bogachev, *Gaussian Measures*. Mathematical Surveys and Monographs, vol. 62 (American Mathematical Society, Providence, 1998)
5. H.J. Brascamp, E.H. Lieb, Best constants in Young's inequality, its converse, and its generalization to more than three functions. Adv. Math. **20**, 151–173 (1976)
6. E.A. Carlen, Superadditivity of Fisher's information and logarithmic Sobolev inequalities. J. Funct. Anal. **101**, 194–211 (1991)
7. W.-K. Chen, N. Dafnis, G. Paouris, Improved Hölder and reverse Hölder inequalities for Gaussian random vectors. Adv. Math. **280**, 643–689 (2015)
8. M. Fathi, E. Indrei, M. Ledoux, Quantitative logarithmic Sobolev inequalities and stability estimates. Discrete Contin. Dyn. Syst. 36(12), 6835–6853 (2016)
9. A. Figalli, F. Maggi, A. Pratelli. Sharp stability theorems for the anisotropic Sobolev and log-Sobolev inequalities on functions of bounded variation. Adv. Math. **242**, 80–101 (2013)
10. L. Gross. Logarithmic Sobolev inequalities. Am. J. Math. **97**, 1061–1083 (1975)
11. E. Indrei, D. Marcon, Quantitative log-Sobolev inequality for a two parameter family of functions. Int. Math. Res. Not. **20**, 5563–5580 (2014)
12. M. Ledoux, Remarks on Gaussian noise stability, Brascamp-Lieb and Slepian inequalities, in *Geometric Aspects of Functional Analysis*. Lecture Notes in Mathematics, vol. 2116 (Springer, Cham, 2014), pp. 309–333
13. E.H. Lieb, Gaussian kernels have only Gaussian maximizers. Inv. Math. **102**, 179–208 (1990)
14. E.H. Lieb, M. Loss, *Analysis*, 2nd edn. Graduate Studies in Mathematics, vol. 14 (American Mathematical Society, Providence, 2001)
15. E. Mossel, K. Oleszkiewicz, A. Sen, On reverse hypercontractivity. Geom. Funct. Anal. **23**(3), 1062–1097 (2013)
16. J. Neeman, A multi-dimensional version of noise stability. Electron. Commun. Probab. **19**(72), 1–10 (2014)
17. E. Nelson, The free Markov field. J. Funct. Anal. **12**, 211–227 (1973)
18. M. Talagrand, *Mean Field Models for Spin Glasses. Volume I* (Ergebnisse der Mathematik und ihrer Grenzgebiete. 3. Folge). A Series of Modern Surveys in Mathematics, vol. **54** (Springer, Berlin, 2010)
19. C. Villani, *Topics in Optimal Transportation*. Graduate Studies in Mathematics, vol. 58 (American Mathematical Society, Providence, 2003)
20. C. Villani, *Optimal Transport. Old and New* (Grundlehren der mathematischen Wissenschaften), vol. 338 (Springer, Berlin, 2009)

(s,p)-Valent Functions

Omer Friedland and Yosef Yomdin

Abstract We introduce the notion of $(\mathcal{F},p)$-valent functions. We concentrate in our investigation on the case, where $\mathcal{F}$ is the class of polynomials of degree at most s. These functions, which we call (s,p)-valent functions, provide a natural generalization of p-valent functions (see Hayman, Multivalent Functions, 2nd ed, Cambridge Tracts in Mathematics, vol 110, 1994). We provide a rather accurate characterizing of (s,p)-valent functions in terms of their Taylor coefficients, through "Taylor domination", and through linear non-stationary recurrences with uniformly bounded coefficients. We prove a "distortion theorem" for such functions, comparing them with polynomials sharing their zeroes, and obtain an essentially sharp Remez-type inequality in the spirit of Yomdin (Isr J Math 186:45–60, 2011) for complex polynomials of one variable. Finally, based on these results, we present a Remez-type inequality for (s,p)-valent functions.

1 Introduction

Let us introduce the notion of "$(\mathcal{F},p)$-valent functions". Let $\mathcal{F}$ be a class of functions to be specified later. A function f regular in a domain $\Omega \subset \mathbb{C}$ is called $(\mathcal{F},p)$-valent in Ω if for any $g \in \mathcal{F}$ the number of solutions of the equation $f(z) = g(z)$ in Ω does not exceed p.

For example, the classic p-valent functions are obtained for $\mathcal{F}$ being the class of constants, these are functions f for which the equation $f = c$ has at most p solutions in Ω for any c. There are many other natural classes $\mathcal{F}$ of interest, like rational functions, exponential polynomials, quasi-polynomials, etc. In particular, for the class $\mathcal{R}_s$ consisting of rational functions $R(z)$ of a fixed degree s, the number of zeroes of $f(z) - R(z)$ can be explicitly bounded for f solving linear ODEs with

---

O. Friedland (✉)
Institut de Mathématiques de Jussieu, Université Pierre et Marie Curie (Paris 6), 4 Place Jussieu, 75005 Paris, France
e-mail: omer.friedland@imj-prg.fr

Y. Yomdin
Department of Mathematics, The Weizmann Institute of Science, Rehovot 76100, Israel
e-mail: yosef.yomdin@weizmann.ac.il

© Springer International Publishing AG 2017
B. Klartag, E. Milman (eds.), *Geometric Aspects of Functional Analysis*,
Lecture Notes in Mathematics 2169, DOI 10.1007/978-3-319-45282-1_8

polynomial coefficients (see, e.g. [4]). Presumably, the collection of $(\mathcal{R}_s, p)$-valent functions with explicit bounds on p (as a function of s) is much wider, including, in particular, "monogenic" functions (or "Wolff-Denjoy series") of the form $f(z) = \sum_{j=1}^{\infty} \frac{\gamma_j}{z - z_j}$ (see, e.g. [13, 16] and references therein).

However, in this note we shall concentrate on another class of functions, for which $\mathcal{F}$ is the class of polynomials of degree at most s. We denote it in short as (s, p)-valent functions. For an (s, p)-valent function f the equation $f = P$ has at most p solutions in Ω for any polynomial P of degree s. We shall always assume that $p \geq s + 1$, as subtracting from f its Taylor polynomial of degree s we get zero of order at least $s + 1$. Note that this is indeed a generalization of p-valent functions, simply take $s = 0$, and every $(0, p)$-valent function is p-valent.

As we shall see this class of (s, p)-valent functions is indeed rich and appears naturally in many examples: algebraic functions, solutions of algebraic differential equations, monogenic functions, etc. In fact, it is fairly wide (see Sect. 2). It possesses many important properties: Distortion theorem, Bernstein-Markov-Remez type inequalities, etc. Moreover, this notion is applicable to any analytic function, under an appropriate choice of the domain Ω and the parameters s and p. In addition, it may provide a useful information in very general situations.

The following example shows that an (s, p)-valent function may not be $(s+1, p)$-valent:

Example 1.1 Let $f(z) = z^p + z^N$ for $N \geq 10p + 1$. Then, for $s = 0, \ldots, p - 1$, the function f is (s, p)-valent in the disk $D_{1/3}$, but only (p, N)-valent there.

Indeed, taking $P(z) = z^p + c$ we see that the equation $f(z) = P(z)$ takes the form $z^N = c$. So for c small enough, it has exactly N solutions in the $D_{1/3}$. Now, for $s = 0, \ldots, p - 1$, take a polynomial $P(z)$ of degree $s \leq p - 1$. Then, the equation $f(z) = P(z)$ takes the form $z^p - P(z) + z^N = 0$. Applying Chebyshev theorem (for more details see for example [17, Lemma 3.3]) to the polynomial $Q(z) = z^p - P(z)$ of degree p (with leading coefficient 1) we find a circle $S_\rho = \{|z| = \rho\}$ with $1/3 \leq \rho \leq 1/2$ such that $|Q(z)| \geq (1/2)^{10p}$ on S_ρ. On the other hand $z^N \leq (1/2)^{10p+1} < (1/2)^{10p}$ on S_ρ. Therefore, by the Rouché principle the number of zeroes of $Q(z) + z^N$ in the disk D_ρ is the same as for $Q(z)$, which is at most p. Thus, f is (s, p)-valent in the disk $D_{1/3}$, for $s = 0, \ldots, p - 1$.

This paper is organized as follows: in Sect. 2 we characterize (s, p)-valent functions in terms of their Taylor domination and linear recurrences for their coefficients. In Sect. 3 we prove a Distortion theorem for (s, p)-valent functions. In Sect. 4 we make a detour and investigate Remez-type inequalities for complex polynomials, which is interesting in its own right. Finally, in Sect. 5, we extend the Remez-type inequality to (s, p)-valent functions, via the Distortion theorem.

2   Taylor Domination, Bounded Recurrences

In this section we provide a rather accurate characterization of (s, p)-valent functions in a disk D_R in terms of their Taylor coefficients. "Taylor domination" for an analytic function $f(z) = \sum_{k=0}^{\infty} a_k z^k$ is an explicit bound of all its Taylor coefficients a_k through the first few of them. This property was classically studied, in particular, in relation with the Bieberbach conjecture: for univalent f we always have $|a_k| \leq k|a_1|$ (see [2, 3, 12] and references therein). To give an accurate definition, let us assume that the radius of convergence of the Taylor series for f is $\hat{R}$, for $0 < \hat{R} \leq +\infty$.

Definition 2.1 (Taylor Domination) Let $0 < R < \hat{R}$, $N \in \mathbb{N}$, and $S(k)$ be a positive sequence of a subexponential growth. The function f is said to possess an $(N, R, S(k))$-Taylor domination property if

$$|a_k| R^k \leq S(k) \max_{i=0,\dots,N} |a_i| R^i, \quad k \geq N + 1.$$

The following theorem shows that f is an (s, p)-valent function in D_R, essentially, if and only if its lower s-truncated Taylor series possesses a $(p - s, R, S(k))$-Taylor domination.

Theorem 2.2 *Let $f(z) = \sum_{k=0}^{\infty} a_k z^k$ be an (s, p)-valent function in D_R, and let $\hat{f}(z) = \sum_{k=1}^{\infty} a_{s+k} z^k$ be the lower s-truncation of f. Put $m = p - s$. Then, $\hat{f}$ possesses an $(m, R, S(k))$-Taylor domination, with $S(k) = \left(\frac{A_m k}{m}\right)^{2m}$, and A_m being a constant depending only on m.*

Conversely, if $\hat{f}$ possesses an $(m, R, S(k))$-Taylor domination, for a certain sequence $S(k)$ of a subexponential growth, then for $R' < R$ the function f is (s, p)-valent in $D_{R'}$, where $p = p(s + m, S(k), R'/R)$ depends only on $m + s$, the sequence $S(k)$, and the ratio R'/R. Moreover, p tends to ∞ for $R'/R \to 1$, and it is equal to $m + s$ for R'/R sufficiently small.

Proof First observe that if f is (s, p)-valent in D_R, then $\hat{f}$ is m-valent there, with $m = p - s$. Indeed, put $P(z) = \sum_{k=0}^{s} a_k z^k + cz^s$, with any $c \in \mathbb{C}$. Then, $f(z) - P(z) = z^s(\hat{f}(z) - c)$ may have at most p zeroes. Consequently, $\hat{f}(z) - c$ may have at most m zeroes in D_R, and thus $\hat{f}$ is m-valent there. Now we apply the following classic theorem:

Theorem 2.3 (Biernacki [3]) *If f is m-valent in the disk D_R of radius R centered at $0 \in \mathbb{C}$ then*

$$|a_k| R^k \leq \left(\frac{A_m k}{m}\right)^{2m} \max_{i=1,\dots,m} |a_i| R^i, \quad k \geq m + 1,$$

where A_m is a constant depending only on m.

In our situation, Theorem 2.3 claims that the function $\hat{f}$ which is m-valent in D_R, possesses an $(m, R, \left(\frac{A_m k}{m}\right)^{2m})$-Taylor domination property. This completes the proof in one direction.

In the opposite direction, for polynomial $P(z)$ of degree s the function $f - P$ has the same Taylor coefficients as $\hat{f}$, starting with the index $k = s+1$. Consequently, if $\hat{f}$ possesses an $(m, R, S(k))$-Taylor domination, then $f - P$ possesses an $(s+m, R, S(k))$-Taylor domination. An explicit bound for the number of zeroes of a function possessing Taylor domination can be obtained by using the following result [15, Proposition 2.2.2] (which is announced here as appears in [1]):

Theorem 2.4 ([1, Theorem 2.3]) *Let the function f possess an $(N, R, S(k))$-Taylor domination property. Then for each $R' < R$, f has at most $M = M(N, \frac{R'}{R}, S(k))$ zeros in $D_{R'}$, where M depends only on N, $\frac{R'}{R}$ and on the sequence $S(k)$, satisfying $\lim_{\frac{R'}{R} \to 1} M = \infty$ and $M = N$ for $\frac{R'}{R}$ sufficiently small.*

Now a straightforward application of the above theorem provides the required bound on the number of zeroes of $f - P$ in the disk D_R. □

A typical situation for natural classes of (s, p)-valent functions is that they are (s, p)-valent for any s with a certain $p = p(s)$ which depends on s. However, it is important to notice that essentially *any* analytic function possesses this property, with some $p(s)$.

Proposition 2.5 *Let $f(z)$ be an analytic function in an open neighbourhood U of the closed disk D_R. Assume that f is not a polynomial. Then, the function f is $(s, p(s))$-valent for any s with a certain sequence $p(s)$.*

Proof Let f be given by its Taylor series $f(z) = \sum_{k=0}^{\infty} a_k z^k$. By assumptions, the radius of convergence $\hat{R}$ of this series satisfies $\hat{R} > R$. Since f is not a polynomial, for any given s there is the index $k(s) > s$ such that $a_{k(s)} \neq 0$. Now, we need the following result of [1]:

Proposition 2.6 ([1, Proposition 1.1]) *If $0 < \hat{R} \leq +\infty$ is the radius of convergence of $f(z) = \sum_{k=0}^{\infty} a_k z^k$, with $f \not\equiv 0$, then for each finite and positive $0 < R \leq \hat{R}$, f satisfies the $(N, R, S(k))$-Taylor domination property with N being the index of its first nonzero Taylor coefficient, and $S(k) = R^k |a_k| (|a_N| R^N)^{-1}$, for $k > N$.*

Applying the above proposition to the lower truncated series $\hat{f}(z) = \sum_{k=1}^{\infty} a_{s+k} z^k$. Thus, we obtain, an $(m, \hat{R}, S(k))$-Taylor domination for $\hat{f}$, for certain m and $S(k)$. Now, the second part of Theorem 2.2 provides the required $(s, p(s))$-valency for f in the smaller disk D_R, with $p(s) = p(s + m, S(k), R/\hat{R})$. □

More accurate estimates of $p(s)$ can be provided via the lacunary structure of the Taylor coefficients of f. Consequently, (s, p)-valency becomes really interesting only for those *classes* of analytic functions f where we can specify the parameters in an explicit and uniform way. The following theorem provides still very general, but important such class. We remark that the second part is known, see [15, Lemma 2.2.3] and [1, Theorem 4.1].

Theorem 2.7 *Let* $f(z) = \sum_{k=0}^{\infty} a_k z^k$ *be* $(s, s + m)$-valent in D_R for any s. Then, the Taylor coefficients a_k of f satisfy a linear homogeneous non-stationary recurrence relation

$$a_k = \sum_{j=1}^{m} c_j(k) a_{k-j} \qquad (1)$$

with uniformly bounded (in k) coefficients $c_j(k)$ satisfying $|c_j(k)| \leq C\rho^j$, with $C = e^2 A_m^{2m}, \rho = R^{-1}$, where A_m is the constant in the Biernacki's Theorem 2.3.

Conversely, if the Taylor coefficients a_k of f satisfy recurrence relation (1), with the coefficients $c_j(k)$, bounded for certain $K, \rho > 0$ and for any k as $|c_j(k)| \leq K\rho^j$, $j = 1, \ldots, m$, then for any s, f is $(s, s + m)$-valent in a disk D_R, with $R = \frac{1}{2^{3m+1}(2K+2)\rho}$.

Proof We need to prove only the first part. Let us fix $s \geq 0$. As in the proof of Theorem 2.2, we notice that if f is $(s, s+m)$-valent in D_R, then its lower s-truncated series $\hat{f}$ is m-valent there. By Biernacki's Theorem 2.3 we conclude that

$$|a_{s+m+1}|R^{m+1} \leq \left(\frac{A_m(m+1)}{m}\right)^{2m} \max_{i=1,\ldots,m} |a_{s+i}|R^i \leq C \max_{i=1,\ldots,m} |a_{s+i}|R^i,$$

with $C = e^2 A_m^{2m}$. Putting $k = s + m + 1$, and $\rho = R^{-1}$ we can rewrite this as

$$|a_k| \leq C \max_{j=1,\ldots,m} |a_{k-j}|\rho^j.$$

Hence we can chose the coefficients $c_j(k)$, $k = s + m + 1$, in such a way that $a_k = \sum_{j=1}^{m} c_j(k) a_{k-j}$, and $|c_j(k)| \leq C\rho^j$, which completes the proof. $\qquad \square$

Notice that the bound on the recursion coefficients is sharp, e.g. take $f(z) = [1 - (\frac{z}{R})^m]^{-1}$, in this case, as well as for other lacunary series with the gap m, the coefficients $c_j(k)$ are defined uniquely.

3   Distortion Theorem

In this section we prove a distortion-type theorem for (s, p)-valent functions which shows that the behavior of these functions is controlled by the behavior of a polynomial with the same zeroes.

First, let us recall the following theorem for p-valent functions, which is our main tool in proof.

Theorem 3.1 ([12, Theorem 5.1]) *Let $g(z) = a_0 + a_1 z + \ldots$ be a regular non-vanishing p-valent function in D_1. Then, for any $z \in D_1$*

$$\left(\frac{1-|z|}{1+|z|}\right)^{2p} \leq |g(z)/a_0| \leq \left(\frac{1+|z|}{1-|z|}\right)^{2p}.$$

Now, we are ready to formulate a distortion-type theorem for (s,p)-valent functions.

Theorem 3.2 (Distortion Theorem) *Let f be an (s,p)-valent function in D_1 having there exactly s zeroes $z_1, \ldots, z_s$ (always assumed to be counted according to multiplicity). Define a polynomial*

$$P(z) = A \prod_{j=1}^{s} (z - z_j),$$

where the coefficient A is chosen such that the constant term in the Taylor series for $f(z)/P(z)$ is equal to 1. Then, for any $x \in D_1$

$$\left(\frac{1-|z|}{1+|z|}\right)^{2p} \leq |f(z)/P(z)| \leq \left(\frac{1+|z|}{1-|z|}\right)^{2p}.$$

Proof The function $g(z) = f(z)/P(z)$ is regular in D_1 and does not vanish there. Moreover, g is p-valent in D_1. Indeed, the equation $g(z) = c$ is equivalent to $f(z) = cP(z)$ so it has at most p solutions by the definition of (s,p)-valent functions. Now, apply Theorem 3.1 to the function g.   □

It is not clear whether the requirement for f to be (s,p)-valent is really necessary in this theorem. The ratio $g(z) = \frac{f(z)}{P(z)}$ certainly may not be p-valent for f being just p-valent, but not (s,p)-valent. Indeed, take $f(z) = z^p + z^N$ as in Example 1.1. By this example f is p-valent in $D_{1/3}$ and it has a root of multiplicity p at zero. So $g(z) = f(z)/z^p = 1 + z^{N-p}$ and the equation $g(z) = c$ has $N - p$ solutions in $D_{1/3}$ for c sufficiently close to 1. So g is not p-valent there.

4   Complex Polynomials

The distortion Theorem 3.2, proved in the previous section, allows us easily to extend deep properties from polynomials to (s,p)-valent functions, just by comparing them with polynomials having the same zeros. In this section we make a detour and investigate one specific problem for complex polynomials, which is interesting in its own right: a Remez-type inequality for complex polynomial (compare [14, 18]). Denote by

$$V_\rho(g) = \{z : |g(z)| \leq \rho\}$$

the ρ sub-level set of a function g. For polynomials in one complex variable a result similar to the Remez inequality is provided by the classic Cartan (or Cartan-Boutroux) lemma (see, for example, [11] and references therein):

Lemma 4.1 (Cartan's Lemma [7], as Appears in [11]) *Let $\alpha, \varepsilon > 0$, and let $P(z)$ be a monic polynomial of degree d. Then*

$$V_{\varepsilon^d}(P) \subset \cup_{j=1}^{p} D_{r_j},$$

where $p \leq d$, and $D_{r_1}, \ldots, D_{r_p}$ are balls with radii $r_j > 0$ satisfying $\sum_{j=1}^{p} r_j^{\alpha} \leq e(2\varepsilon)^{\alpha}$.

In [5, 6, 19, 20] some generalizations of the Cartan-Boutroux lemma to plurisubharmonic functions have been obtained, which lead, in particular, to the bounds on the size of sub-level sets. In [5] some bounds for the covering number of sublevel sets of complex analytic functions have been obtained, similar to the results of [18] in the real case. Now, we shall derive from the Cartan lemma both the definition of the invariant $c_{d,\alpha}$ and the corresponding Remez inequality.

Definition 4.2 Let $Z \subset D_1$. The (d, α)-Cartan measure of Z is defined as

$$c_{d,\alpha}(Z) = \min \left(\sum_{j=1}^{p} r_j^{\alpha} \right)^{1/\alpha}$$

where the minimum is taken over all covers of Z by $p \leq d$ balls with radii $r_j > 0$.

Clearly, the invariant $c_{d,\alpha}(Z)$ satisfies the following basic properties. It is monotone in Z, that is, for $Z_1 \subset Z_2$ we have $c_{d,\alpha}(Z_1) \leq c_{d,\alpha}(Z_2)$. And, also monotone in d, that is, for $d_1 \leq d$ we have $c_{d,\alpha}(Z) \leq c_{d_1,\alpha}(Z)$. Finally, for any $Z \subset D_1$ we have $c_{d,\alpha}(Z) \leq 1$. Note also that the α-dimensional Hausdorff content of Z is defined in a similar way

$$H_{\alpha}(Z) = \inf \left\{ \sum_{j} r_j^{\alpha} : \text{there is a cover of } Z \text{ by balls with radii } r_j > 0 \right\}.$$

Thus, by the above definitions, we have $H_{\alpha}^{\frac{1}{\alpha}}(Z) \leq c_{d,\alpha}(Z)$.

For $\alpha = 1$ the $(d, 1)$-Cartan measure $c_{d,\alpha}(Z)$ was introduced and used, under the name "d-th diameter", in [8, 9]. In particular, Lemma 3.3 of [8] is, essentially, equivalent to the case $\alpha = 1$ of our Theorem 4.3. In Sect. 4.1 below we provide some initial geometric properties of $c_{d,\alpha}(Z)$ and show that a proper choice of α may improve the geometric sensitivity of this invariant.

Now we can state and proof our generalized Remez inequality for complex polynomials:

Theorem 4.3 *Let $P(z)$ be a polynomial of degree d. Let $Z \subset D_1$. Then, for any*
$\alpha > 0$

$$\max_{D_1} |P(z)| \leq \left(\frac{6e^{1/\alpha}}{c_{d,\alpha}(Z)} \right)^d \max_{Z} |P(z)| \leq \left(\frac{6e}{H_\alpha(Z)} \right)^{\frac{d}{\alpha}} \max_{Z} |P(z)|.$$

Proof Assume that $|P(z)| \leq 1$ on Z. First, we prove that the absolute value A of the
leading coefficient of P satisfies

$$A \leq \left(\frac{2e^{1/\alpha}}{c_{d,\alpha}(Z)} \right)^d.$$

Indeed, we have $Z \subset V_1(P)$. By the definition of $c_{d,\alpha}(Z)$ for every covering of
$V_1(P)$ by p disks $D_{r_1}, \ldots, D_{r_p}$ of the radii $r_1, \ldots, r_d$ (which is also a covering of Z)
we have $\sum_{i=1}^d r_i^\alpha \geq c_{d,\alpha}(Z)^\alpha$. Denoting, as above, the absolute value of the leading
coefficient of $P(z)$ by A we have by the Cartan lemma that for a certain covering as
above

$$c_{d,\alpha}(Z)^\alpha \leq \sum_{i=1}^d r_i^\alpha \leq e \left(\frac{2}{A^{1/d}} \right)^\alpha.$$

Now, we write $P(z) = A \prod_{j=1}^d (z - z_j)$, and consider separately two cases:

(1) All $|z_j| \leq 2$. Thus, $\max_{D_1} |P(z)| \leq A3^d \leq \left(\frac{2e^{1/\alpha}}{c_{d,\alpha}(Z)} \right)^d 3^d$, as required.
(2) For $j = 1, \ldots, d_1 < d$, $|z_j| \leq 2$, while $|z_j| > 2$ for $j = d_1 + 1, \ldots, d$. Denote

$$P_1(z) = A \prod_{j=1}^{d_1} (z - z_j), \quad P_2(z) = \prod_{j=d_1+1}^d (z - z_j),$$

and notice that for any two points $v_1, v_2 \in D_1$ we have $|P_2(v_1)/P_2(v_2)| < 3^{d-d_1}$.
Consequently we get

$$\frac{\max_{D_1} |P(z)|}{\max_Z |P(z)|} < 3^{d-d_1} \frac{\max_{D_1} |P_1(z)|}{\max_Z |P_1(z)|}.$$

All the roots of P_1 are bounded in absolute value by 2, so by first part we have

$$\frac{\max_{D_1} |P_1(z)|}{\max_Z |P_1(z)|} \leq \left(\frac{2e^{1/\alpha}}{c_{d_1,\alpha}(Z)} \right)^{d_1} 3^{d_1} \leq \left(\frac{2e^{1/\alpha}}{c_{d,\alpha}(Z)} \right)^d 3^{d_1}$$

where the last inequality follows from the basic properties of the invariant $c_{d,\alpha}(Z)$
described after Definition 4.2. Finally, application of the inequality $H_\alpha(Z) \leq$
$c_{d,\alpha}(Z)^\alpha$ completes the proof.   □

Let us stress a possibility to chose an optimal α in the bound of Theorem 4.3. Let

$$K_d(Z) = \inf_{\alpha>0} \left(\frac{6e^{1/\alpha}}{c_{d,\alpha}(Z)} \right)^d, \quad K_d^H(Z) = \inf_{\alpha>0} \left(\frac{6e}{H_\alpha(Z)} \right)^{\frac{d}{\alpha}}.$$

Corollary 4.4 *Let $P(z)$ be a polynomial of degree d. Let $Z \subset D_1$. Then,*

$$\max_{D_1} |P(z)| \leq K_d(Z) \max_Z |P(z)| \leq K_d^H(Z) \max_Z |P(z)|.$$

4.1 Geometric and Analytic Properties of the Invariant $c_{d,\alpha}$

In addition to the basic properties of $c_{d,\alpha}$ we also have

Proposition 4.5 *Let $\alpha > 0$. Then, $c_{d,\alpha}(Z) > 0$ if and only if Z contains more than d points. In the latter case, $c_{d,\alpha}(Z)$ is greater than or equal to one half of the minimal distance between the points of Z.*

Proof Any d points can be covered by d disks with arbitrarily small radii. But, the radius of at least one disk among d disks covering more than $d + 1$ different points is greater than or equal to the one half of a minimal distance between these points. $\square$

The lower bound of Proposition 4.5 does not depend on α. However, in general, this dependence is quite prominent.

Example 4.6 Let $Z = [a, b]$. Then, for $\alpha \geq 1$ we have $c_{d,\alpha}(Z) = (b - a)/2$, while for $\alpha \leq 1$ we have $c_{d,\alpha}(Z) = d^{\frac{1}{\alpha}-1}(b - a)/2$.

Indeed, in the first case the minimum is achieved for $r_1 = (b - a)/2, r_2 = \cdots = r_d = 0$, while in the second case for $r_1 = r_2 = \cdots = r_d = (b - a)/2d$.

Proposition 4.7 *Let $\alpha > \beta > 0$. Then, for any Z*

$$c_{d,\alpha}(Z) \leq c_{d,\beta}(Z) \leq d^{(\frac{1}{\beta}-\frac{1}{\alpha})} c_{d,\alpha}(Z). \tag{2}$$

Proof Let $r = (r_1, \ldots, r_d)$ and $\gamma > 0$. Consider $||r||_\gamma = (\sum_{j=1}^d r_j^\gamma)^{\frac{1}{\gamma}}$. Then, by the definition, $c_{d,\gamma}(Z)$ is the minimum of $||r||_\gamma$ over all $r = (r_1, \ldots, r_d)$ being the radii of d balls covering Z. Now we use the standard comparison of the norms $||r||_\gamma$, that is, for any $x = (z_1, \ldots, z_d)$ and for $\alpha > \beta > 0$,

$$||z||_\alpha \leq ||z||_\beta \leq d^{(\frac{1}{\beta}-\frac{1}{\alpha})} ||z||_\alpha.$$

Take $r = (r_1, \ldots, r_d)$ for which the minimum of $||r||_\beta$ is achieved, and we get

$$c_{d,\alpha}(Z) \leq ||r||_\alpha \leq ||r||_\beta = c_{d,\beta}(Z).$$

Now taking r for which the minimum of $||r||_\alpha$ is achieved, exactly in the same way we get the second inequality.   □

Now, we compare $c_{d,\alpha}(Z)$ with some other metric invariants which may be sometimes easier to compute. In each case we do it for the most convenient value of α. Then, using the comparison inequalities of Proposition 4.7, we get corresponding bounds on $c_{d,\alpha}(Z)$ for any $\alpha > 0$. In particular, we can easily produce a simple lower bound for $c_{d,2}(Z)$ through the measure of Z:

Proposition 4.8 *For any measurable $Z \subset D_1$ we have*

$$c_{d,2}(Z) \geq (\mu_2(Z)/\pi)^{1/2}.$$

Proof For any covering of Z by d disks $D_1, \ldots, D_d$ of the radii $r_1, \ldots, r_d$ we have $\pi(\sum_{i=1}^{d} r_i^2) \geq \mu_2(Z)$.   □

However, in order to deal with discrete or finite subsets $Z \subset D_1$ we have to compare $c_{d,\alpha}(Z)$ with the covering number $M(\varepsilon, Z)$ (which is, by definition, the minimal number of ε-disks covering Z).

Definition 4.9 Let $Z \subset D_1$. Define

$$\omega_{cd}(Z) = \sup_\varepsilon \varepsilon(M(\varepsilon, Z) - d)^{1/2},$$

if $|Z| \geq d$, and $\omega_{cd}(Z) = 0$ otherwise. Put $\rho_d(Z) = d\varepsilon_0$, where ε_0 is the minimal ε for which there is a covering of Z with d ε-disks. Note that, writing $y = M(\varepsilon, Z) = \Psi(\varepsilon)$, and taking the inverse $\varepsilon = \Psi^{-1}(y)$, we have $\varepsilon_0 = \Psi^{-1}(d)$.

As it was mentioned above, a very similar invariant

$$\omega_d(Z) = \sup_\varepsilon \varepsilon(M(\varepsilon, Z) - d),$$

if $|Z| \geq d$, and $\omega_{cd}(Z) = 0$ otherwise, was introduced and used in [18] in the real case. We compare ω_{cd} and ω_d below.

Proposition 4.10 *Let $Z \subset D_1$. Then, $\omega_{cd}(Z)/2 \leq c_{d,2}(Z) \leq c_{d,1}(Z) \leq \rho_d(Z)$.*

Proof To prove the upper bound for $c_{d,1}(Z)$ we notice that it is the infimum of the sum of the radii in all the coverings of Z with d disks, while $\rho_d(Z)$ is such a sum for one specific covering.

To prove the lower bound, let us fix a covering of Z by d disks D_i of the radii r_i with $c_{d,2}(Z) = (\sum_{i=1}^{d} r_i^2)^{1/2}$. Let $\varepsilon > 0$. Now, for any disk D_j with $r_j \geq \varepsilon$ we need at most $4r_j^2/\varepsilon^2$ ε-disks to cover it. For any disk D_j with $r_j \leq \varepsilon$ we need exactly one ε-disk to cover it, and the number of such D_j does not exceed d. So, we conclude that $M(\varepsilon, Z)$ is at most $d + (4/\varepsilon^2)\sum_{i=1}^{d} r_i^2$. Thus, we get $c_{d,2}(Z) = (\sum_{i=1}^{d} r_i^2)^{1/2} \geq \varepsilon/2(M(\varepsilon, Z) - d)^{1/2}$. Taking supremum with respect to $\varepsilon > 0$ we get $c_{d,2}(Z) \geq \omega_{cd}(Z)/2$.   □

Since $M(\varepsilon, Z)$ is always an integer, we have

$$\omega_d(Z) \geq \omega_{cd}(Z).$$

For $Z \subset D_1$ of positive plane measure, $\omega_d(Z) = \infty$ while $\omega_{cd}(Z)$ remains bounded (in particular, by $\rho_d(Z)$).

Some examples of computing (or bounding) $\omega_d(Z)$ for "fractal" sets Z can be found in [18]. Computations for $\omega_{cd}(Z)$ are essentially the same. In particular, in an example given in [18] in connection to [10] we have that for $Z = Z_r = \{1, 1/2^r, 1/3^r, \ldots, 1/k^r, \ldots\}$

$$\omega_d(Z_r) \asymp \frac{r^r}{(r+1)^{r+1}d^r}, \quad \omega_{cd}(Z_r) \asymp \frac{(2r+1)^r}{(2r+2)^{r+1}d^{r+1/2}}.$$

The asymptotic behavior here is for $d \to \infty$, as in [10].

4.2 An Example

We conclude this section with one very specific example. Let

$$Z = Z(d, h) = \{z_1, z_2, \ldots, z_{2d-1}, z_{2d}\}, \quad z_i \in \mathbb{C}, \ d \geq 2.$$

We assume that Z consists of d, 2η-separated couples of points, with points in each couple being in a distance $2h$. Let $2D(Z)$ be the diameter of the smallest disk containing Z. Assume $h \ll 1, 2\eta \gg h$.

Proposition 4.11 *Let Z be as above. Then,*

(1) $\omega_d(Z) = dh$.
(2) $\omega_{cd}(Z) = \sqrt{d}h$.
(3) *For $\alpha > 0$, we have $c_{d,\alpha}(Z) \leq d^{\frac{1}{\alpha}}h$.*
(4) *For $\alpha \gg 1$, we have $c_{d,\alpha}(Z) = d^{\frac{1}{\alpha}}h$.*
(5) *For $\kappa = [\log_d(\frac{D(Z)}{h})]^{-1}$, we have $c_{d,\kappa}(Z) \geq \eta$.*

Proof For $\varepsilon > h$, we have $M(\varepsilon, Z) \leq d$, and hence $M(\varepsilon, Z) - d$ is non-positive. For $\varepsilon < h$, we have $M(\varepsilon, Z) = 2d$, and $M(\varepsilon, Z) - d = d$. Thus the supremum of $\varepsilon(M(\varepsilon, Z) - d)$, or the supremum of $\varepsilon(M(\varepsilon, Z) - d)^{\frac{1}{2}}$, is achieved as $\varepsilon < h$ tends to h. Therefore, $\omega_d(Z) = dh$, and $\omega_{cd}(Z) = \sqrt{d}h$.

Covering each couple with a separate ball of radius h, we get for any $\alpha > 0$ that $c_{d,\alpha}(Z) \leq d^{\frac{1}{\alpha}}h$. For $\alpha \gg 1$ it is easy to see that this uniform covering is minimal. Thus, for such α we have the equality $c_{d,\alpha}(Z) = d^{\frac{1}{\alpha}}h$.

Now let us consider the case of a "small" $\alpha = \kappa$. Take a covering of Z with certain disks $D_j, j \leq d$. If there is at least one disk D_j containing three points of Z or more, the radius of this disk is at least η. Thus, for this covering $(\sum_{j=1}^{d} r_j^\kappa)^{\frac{1}{\kappa}} \geq \eta$. If each disk in the covering contains at most two points, it must contain exactly two, otherwise these disks could not cover all the $2d$ points of Z. Hence, the radius of each disk D_j in such covering is at least h, and their number is exactly d. We have, by the choice of κ, that $(\sum_{j=1}^{d} r_j^\kappa)^{\frac{1}{\kappa}} \geq d^{\frac{1}{\kappa}} h = D(Z) \geq \eta$.   □

Let us use two choices of α in the Remez-type inequality of Theorem 4.3: $\alpha = 1$ and $\alpha = \kappa$. We get two bounds for the constant $K_d(Z)$:

$$ K_d(Z) \leq \left(\frac{6e}{c_{d,1}(Z)} \right)^d \quad \text{or} \quad K_d(Z) \leq \left(\frac{6e^{1/\kappa}}{c_{d,\kappa}(Z)} \right)^d . $$

By Proposition 4.11 we have $c_{d,1}(Z) \leq dh$, while $c_{d,\kappa}(Z) \geq \eta$. Therefore we get

$$ \left(\frac{6e}{c_{d,1}(Z)} \right)^d \geq \left(\frac{6e}{dh} \right)^d , \quad \text{while} \quad \left(\frac{6e^{1/\kappa}}{c_{d,\kappa}(Z)} \right)^d \leq \left(\frac{6e^{1/\kappa}}{\eta} \right)^d . \qquad (3) $$

But $e^{1/\kappa} = e^{\log_d(\frac{D(Z)}{h})} = (\frac{D(Z)}{h})^{\frac{1}{\ln d}}$. So the second bound of (3) takes a form

$$ K_d(Z) \leq \left(\frac{6D(Z)}{\eta^{\ln d} h} \right)^{\frac{d}{\ln d}} . $$

We see that for $d \geq 3$ and for $h \to 0$ the asymptotic behavior of this last bound, corresponding to $\alpha = \kappa$, is much better than of the first bound in (3), corresponding to $\alpha = 1$. Notice, that κ depends on h and $D(Z)$, i.e. on the specific geometry of the set Z.

5 Remez Inequality

Now, we present a Remez-type inequality for (s, p)-valent functions. We recall that by Proposition 2.5 above, any analytic function in an open neighborhood U of the closed disk D_R is $(s, p(s))$-valent in D_R for any s with a certain sequence $p(s)$. Consequently, the following theorem provides a non-trivial information for any analytic function in an open neighborhood of the unit disk D_1. Of course, this results becomes really interesting only in cases where we can estimate $p(s)$ explicitly.

Theorem 5.1 *Let f be an analytic function in an open neighborhood U of the closed disk D_1. Assume that f has in D_1 exactly s zeroes, and that it is (s, p)-valent in D_1. Let Z be a subset in the interior of D_1, and put $\rho = \rho(Z) = \min\{\eta : Z \subset D_\eta\}$. Then, for any $R < 1$ function f satisfies*

$$ \max_{D_R} |f(z)| \leq \sigma_p(R, \rho) K_s(Z) \max_Z |f(z)|, $$

where $\sigma_p(R, \rho) = \left(\frac{1+R}{1-R} \cdot \frac{1+\rho}{1-\rho}\right)^{2p}$.

Proof Assume that $|f(z)|$ is bounded by 1 on Z. Let $z_1, \ldots, z_s$ be zeroes of f in D_1. Consider, as in Theorem 3.2, the polynomial

$$P(z) = A \prod_{j=1}^{l}(z - z_j),$$

where the coefficient A is chosen in such a way that the constant term in the Taylor series for $g(z) = f(z)/P(z)$ is equal to 1. Then by Theorem 3.2 for g we have

$$\left(\frac{1 - |z|}{1 + |z|}\right)^{2p} \le |g(z)| \le \left(\frac{1 + |z|}{1 - |z|}\right)^{2p}.$$

We conclude that $P(z) \le (\frac{1+\rho}{1-\rho})^{2p}$ on Z. Hence by the polynomial Remez inequality provided by Theorem 4.3 we obtain

$$|P(z)| \le K_s(Z)\left(\frac{1+\rho}{1-\rho}\right)^{2p}$$

on D_1. Finally, we apply once more the bound of Theorem 3.2 to conclude that

$$|f(z)| \le K_s(Z)\left(\frac{1+R}{1-R}\right)^{2p}\left(\frac{1+\rho}{1-\rho}\right)^{2p}$$

on D_R. □

References

1. D. Batenkov, Y. Yomdin, Taylor domination, Turán lemma, and Poincaré-Perron sequences, in *Nonlinear Analysis and Optimization*. Contemporary Mathematics, vol. 659 (American Mathematical Society, Providence, 2016), pp. 1–15
2. L. Bieberbach, *Analytische Fortsetzung*. Ergebnisse der Mathematik und ihrer Grenzgebiete (N.F.), Heft 3 (Springer, Berlin/Göttingen/Heidelberg, 1955) (Russian)
3. M. Biernacki, Sur les fonctions multivalentes d'ordre p. C. R. Acad. Sci. Paris **203**, 449–451 (1936)
4. G. Binyamini, D. Novikov, S. Yakovenko, *Quasialgebraic Functions*. Algebraic Methods in Dynamical Systems, vol. 94. Polish Academy of Sciences, Institute of Mathematics, Warsaw (Banach Center Publications, 2011), pp. 61–81
5. A. Brudnyi, On covering numbers of sublevel sets of analytic functions. J. Approx. Theory **162**(1), 72–93 (2010)
6. A. Brudnyi, Y. Brudnyi, Remez type inequalities and Morrey-Campanato spaces on Ahlfors regular sets, in *Interpolation Theory and Applications*. Contemporary Mathematics, vol. 445 (American Mathematical Society, Providence, RI, 2007), pp. 19–44

7. H. Cartan, Sur les systèmes de fonctions holomorphes à variétés linéaires lacunaires et leurs applications. Ann. Sci. École Norm. Sup. (3) **45**, 255–346 (1928) (French)
8. D. Coman, E.A. Poletsky, Measures of transcendency for entire functions. Mich. Math. J. **51**(3), 575–591 (2003)
9. D. Coman, E.A. Poletsky, Transcendence measures and algebraic growth of entire functions. Invent. Math. **170**(1), 103–145 (2007)
10. J. Favard, Sur l'interpolation. Bull. Soc. Math. Fr. **67**, 102–113 (1939) (French)
11. E.A. Gorin, A. Cartan's lemma according to B. Ya. Levin with various applications. Zh. Mat. Fiz. Anal. Geom. **3**(1), 13–38 (2007) (Russian, with English and Russian summaries)
12. W.K. Hayman, *Multivalent Functions*, 2nd edn., Cambridge Tracts in Mathematics, vol. 110 (Cambridge University Press, Cambridge, 1994)
13. S. Marmi, D. Sauzin, A quasianalyticity property for monogenic solutions of small divisor problems. Bull. Braz. Math. Soc. (N.S.) **42**(1), 45–74 (2011)
14. E.J. Remez, Sur une propriété des polynômes de Tchebycheff. Commun. Inst. Sci. Kharkov **13**, 93–95 (1936)
15. N. Roytwarf, Y. Yomdin, Bernstein classes. Ann. Inst. Fourier (Grenoble) **47**(3), 825–858 (1997) (English, with English and French summaries)
16. R.V. Sibilev, A uniqueness theorem for Wolff-Denjoy series. Algebra i Analiz **7**(1), 170–199 (1995) (Russian, with Russian summary); English transl., St. Petersburg Math. J. **7**(1), 145–168 (1996)
17. Y. Yomdin, emphAnalytic reparametrization of semi-algebraic sets. J. Complex. **24**(1), 54–76 (2008)
18. Y. Yomdin, Remez-type inequality for discrete sets. Isr. J. Math. **186**, 45–60 (2011)
19. A. Zeriahi, Volume and capacity of sublevel sets of a Lelong class of plurisubharmonic functions. Indiana Univ. Math. J. **50**(1), 671–703 (2001)
20. A. Zeriahi, A minimum principle for plurisubharmonic functions. Indiana Univ. Math. J. **56**(6), 2671–2696 (2007)

A Remark on Projections of the Rotated Cube to Complex Lines

Efim D. Gluskin and Yaron Ostrover

Abstract Motivated by relations with a symplectic invariant known as the "cylindrical symplectic capacity", in this note we study the expectation of the area of a minimal projection to a complex line for a randomly rotated cube.

1 Introduction and Result

Consider the complex vector space $\mathbb{C}^n$ with coordinates $z = (z_1, \ldots, z_n)$, and equipped with its standard Hermitian structure $\langle z, w \rangle_{\mathbb{C}} = \sum_{j=1}^{n} z_j \overline{w}_j$. By writing $z_j = x_j + iy_j$, we can look at $\mathbb{C}^n$ as a real $2n$-dimensional vector space $\mathbb{C}^n \simeq \mathbb{R}^{2n} = \mathbb{R}^n \oplus \mathbb{R}^n$ equipped with the usual complex structure J, i.e., J is the linear map $J : \mathbb{R}^{2n} \to \mathbb{R}^{2n}$ given by $J(x_j, y_j) = (-y_j, x_j)$. Moreover, note that the real part of the Hermitian inner product $\langle \cdot, \cdot \rangle_{\mathbb{C}}$ is just the standard inner product on $\mathbb{R}^{2n}$, and the imaginary part is the standard symplectic structure on $\mathbb{R}^{2n}$. As usual, we denote the orthogonal and symplectic groups associated with these two structures by $O(2n)$ and $\mathrm{Sp}(2n)$, respectively. It is well known that $O(2n) \cap \mathrm{Sp}(2n) = U(n)$, where the unitary group $U(n)$ is the subgroup of $\mathrm{GL}(n, \mathbb{C})$ that preserves the above Hermitian inner product.

Symplectic capacities on $\mathbb{R}^{2n}$ are numerical invariants which associate with every open set $\mathcal{U} \subseteq \mathbb{R}^{2n}$ a number $c(\mathcal{U}) \in [0, \infty]$. This number, roughly speaking, measures the symplectic size of the set $\mathcal{U}$ (see e.g. [3], for a survey on symplectic capacities). We refer the reader to the Appendix of this paper for more information regarding symplectic capacities, and their role as an incentive for the current paper. Recently, the authors observed (see Theorem 1.8 in [8]) that for symmetric convex domains in $\mathbb{R}^{2n}$, a certain symplectic capacity $\overline{c}$, which is the largest possible normalized symplectic capacity and is known as the "cylindrical capacity", is asymptotically equivalent to its linearized version given by

$$\overline{c}_{\mathrm{Sp}(2n)}(\mathcal{U}) = \inf_{S \in \mathrm{Isp}(2n)} \mathrm{Area}\big(\pi(S(\mathcal{U}))\big). \tag{1}$$

E.D. Gluskin (✉) • Y. Ostrover
School of Mathematical Sciences, Tel Aviv University, Tel Aviv 69978, Israel
e-mail: gluskin@post.tau.ac.il; ostrover@post.tau.ac.il

© Springer International Publishing AG 2017
B. Klartag, E. Milman (eds.), *Geometric Aspects of Functional Analysis*,
Lecture Notes in Mathematics 2169, DOI 10.1007/978-3-319-45282-1_9

Here, π is the orthogonal projection to the complex line $E = \{z \in \mathbb{C}^n \mid z_j = 0 \text{ for } j \neq 1\}$, and the infimum is taken over all S in the affine symplectic group $\mathrm{ISp}(2n) = \mathrm{Sp}(2n) \ltimes T(2n)$, which is the semi-direct product of the linear symplectic group and the group of translations in $\mathbb{R}^{2n}$. We remark that in what follows we consider only centrally symmetric convex bodies in $\mathbb{R}^{2n}$, and hence one can take S in (1) to be a genuine symplectic matrix (i.e., $S \in \mathrm{Sp}(2n)$).

An interesting natural variation of the quantity $\bar{c}_{\mathrm{Sp}(2n)}$, which serves as an upper bound to it and is of independent interest, is obtained by restricting the infimum on the right-hand side of (1) to the unitary group $\mathrm{U}(n)$ (see the Appendix for more details). More precisely, let $L \subset \mathbb{R}^{2n}$ be a complex line, i.e., $L = \mathrm{span}\{v, Jv\}$ for some non-zero vector $v \in \mathbb{R}^{2n}$, and denote by π_L the orthogonal projection to the subspace L. For a symmetric convex body $K \subset \mathbb{R}^{2n}$, the quantity of interest is

$$\bar{c}_{\mathrm{U}(n)}(K) := \inf_{U \in \mathrm{U}(n)} \mathrm{Area}\big(\pi(U(K))\big) = \inf\Big\{\mathrm{Area}\big(\pi_L(K)\big) \mid L \subset \mathbb{R}^{2n} \text{ is a complex line}\Big\}. \tag{2}$$

In this note we focus on understanding $\bar{c}_{\mathrm{U}(n)}(OQ)$, where $O \in O(2n)$ is a random orthogonal transformation, and $Q = [-1, 1]^{2n} \subseteq \mathbb{R}^{2n}$ is the standard cube. We remark that in [8] it was shown that, in contrast with projections to arbitrary two-dimensional subspaces of $\mathbb{R}^{2n}$, there exist an orthogonal transformation $O \in O(2n)$ such that for every complex line $L \subset \mathbb{R}^{2n}$ one has that $\mathrm{Area}(\pi_L(OQ)) \geq \sqrt{n}/2$. Here we study the expectation of $\bar{c}_{\mathrm{U}(n)}(OQ)$ with respect to the Haar measure on the orthogonal group $O(2n)$. The main result of this note is the following:

Theorem 1.1 *There exist universal constants $C, c_1, c_2 > 0$ such that*

$$\mu\left\{O \in O(2n) \mid \exists \text{ a complex line } L \subset \mathbb{R}^{2n} \text{ with } \mathrm{diam}(\pi_L(OQ)) \leq c_1\sqrt{n}\right\} \leq C\exp(-c_2 n),$$

where μ is the unique normalized Haar measure on $O(2n)$.

Note that for any rotation $U \in O(2n)$, the image UQ contains the Euclidean unit ball and hence for every complex line L one has $\mathrm{Area}(\pi_L UQ) \geq \mathrm{diam}(\pi_L UQ)$. An immediate corollary from this observation, Theorem 1.1, and the easily verified fact that for every $O \in O(2n)$, the complex line $L' := \mathrm{Span}\{v, Jv\}$, where v is one of the directions where the minimal-width of OQ is obtained, satisfies $\mathrm{Area}(\pi_{L'}(OQ)) \leq 4\sqrt{2n}$, is that

Corollary 1.2 *With the above notations one has*

$$\mathbb{E}_\mu\left(\bar{c}_{\mathrm{U}(n)}(OQ)\right) \asymp \sqrt{n}, \tag{3}$$

where $\mathbb{E}_\mu$ stands for the expectation with respect to the Haar measure μ on $O(2n)$, and the symbol $\asymp$ means equality up to universal multiplicative constants.

Remark 1.3 We will see below that for every $O \in O(2n)$, the quantity $\bar{c}_{U(n)}(OQ)$ is bounded from below by the diameter of the section of the $4n$-dimensional octahedron B_1^{4n} by the subspace

$$L_O = \{(x,y) \in \mathbb{R}^{2n} \oplus \mathbb{R}^{2n} \mid y = O^*JOx\}. \tag{4}$$

This reduces the above problem of estimating $\mathbb{E}_\mu\left(\bar{c}_{U(n)}(OQ)\right)$ to estimating the diameter of a random section of the octahedron B_1^{4n} with respect to a probability measure ν on the real Grassmannian $G(4n, 2n)$ induced by the map $O \mapsto L_O$ from the Haar measure μ on $O(2n)$. By duality, the diameter of a section of the octahedron by a linear subspace is equal to the deviation of the Euclidean ball from the orthogonal subspace with respect the l_∞-norm. The right order of the minimal deviation from half-dimensional subspaces was found in the remarkable work of Kašin [11]. For this purpose, he introduced some special measure on the Grassmannian and proved that the approximation of the ball by random subspaces is almost optimal. In his exposition lecture [17], Mitjagin treated Kashin's work as a result about octahedron sections, which gave a more geometric intuition into it, and rather simplified the proof. At about the same time, the diameter of random (this time with respect to the classical Haar measure on the Grassmannian) sections of the octahedron, and more general convex bodies, was studied by Milman [14]; Figiel, Lindenstrauss and Milman [4]; Szarek [22], and many others with connection with Dvoretzky's theorem (see also [1, 5–7, 15, 19], as well as Chap. 5 of [20] and Chaps. 5 and 7 of [2] for more details). It turns out that random sections of the octahedron B_1^{4n}, with respect to the measure ν on the real Grassmannian $G(4n, 2n)$ mentioned above, also have almost optimal diameter. To prove this we use techniques which are now standard in the field. For completeness, all details will be given in Sects. 2 and 3 below.

Notations The letters $C, c, c_1, c_2, \ldots$ denote positive universal constants that take different values from one line to another. Whenever we write $\alpha \asymp \beta$, we mean that there exist universal constants $c_1, c_2 > 0$ such that $c_1\alpha \leq \beta \leq c_2\alpha$. For a finite set V, denote by $\#V$ the number of elements in V. For $a \in \mathbb{R}$ let $[a]$ be its integer part. The standard Euclidean inner product and norm on $\mathbb{R}^n$ will be denoted by $\langle \cdot, \cdot \rangle$, and $|\cdot|$, respectively. The diameter of a subset $V \subset \mathbb{R}^n$ is denoted by $\mathrm{diam}(V) = \sup\{|x-y| : x, y \in V\}$. For $1 \leq p \leq \infty$, we denote by l_p^n the space $\mathbb{R}^n$ equipped with the norm $\|\cdot\|_p$ given by $\|x\|_p = (\sum_{j=1}^n \|x_i\|^p)^{1/p}$ (where $\|x\|_\infty = \max\{|x_i| \mid i = 1, \ldots, n\}$), and the unit ball of the space l_p^n is denoted by $B_p^n = \{x \in \mathbb{R}^n \mid \|x\|_p \leq 1\}$. We denote by S^n the unit sphere in $\mathbb{R}^{n+1}$, i.e., $S^n = \{x \in \mathbb{R}^{n+1} \mid |x|^2 = 1\}$, and by σ_n the standard measure on S^n. Finally, for a measure space (X, μ) and a measurable function $\varphi : X \to \mathbb{R}$ we denote by $\mathbb{E}_\mu\varphi$ the expectation of φ with respect to the measure μ.

2 Preliminaries

Here we recall some basic notations and results required for the proof of Theorem 1.1.

Let V be a subset of a metric space (X, ρ), and let $\varepsilon > 0$. A set $\mathcal{F} \subset V$ is called an ε-net for V if for any $x \in V$ there exist $y \in \mathcal{F}$ such that $\rho(x, y) \leq \varepsilon$. It is a well known and easily verified fact that for any given set G with $V \subseteq G$, if $\mathcal{T}$ is a finite ε-net for G, then there exists a 2ε-net $\mathcal{F}$ of V with $\#\mathcal{F} \leq \#\mathcal{T}$.

Remark 2.1 From now on, unless stated otherwise, all nets are assumed to be taken with respect to the standard Euclidean metric on the relevant space.

Next, fix $n \in \mathbb{N}$ and $0 < \theta < 1$. We denote by G_θ^n the set $G_\theta^n := S^{n-1} \cap \theta\sqrt{n}B_1^n$. The following proposition goes back to Kašin [11]. The proof below follows Makovoz [12] (cf. [21] and the references therein).

Proposition 2.2 *For every ε such that $8\frac{\ln n}{n} < \varepsilon < \frac{1}{2}$, there exists a set $\mathcal{T} \subset G_\theta^n$ such that $\#\mathcal{T} \leq \exp(\varepsilon n)$, and which is a $8\theta\sqrt{\frac{\ln(1/\varepsilon)}{\varepsilon}}$-net for G_θ^n.*

For the proof of Proposition 2.2 we shall need the following lemma.

Lemma 2.3 *For $k, n \in \mathbb{N}$, the set $\mathcal{F}_{k,n} := \mathbb{Z}^n \cap kB_1^n$ is a $\sqrt{k}$-net for the set kB_1^n, and*

$$\#\mathcal{F}_{k,n} \leq (2e(1 + n/k)))^k. \tag{5}$$

Proof of Lemma 2.3 Let $x = (x_1, \ldots, x_n) \in kB_1^n$, and set $y_j = [|x_j|] \cdot \mathrm{sgn}(x_j)$, for $1 \leq j \leq n$. Note that $y = (y_1, \ldots, y_n) \in \mathcal{F}_{k,n}$, and $|x_j - y_j| \leq \min\{1, |x_j|\}$ for any $1 \leq j \leq n$. Thus, $|x - y|^2 = \sum_{j=1}^n |x_j - y_j|^2 \leq \sum_{j=1}^n |x_j| = k$. This shows that $\mathcal{F}_{k,n}$ is a $\sqrt{k}$-net for kB_1^n. In order to prove the bound (5) for the cardinality of $\mathcal{F}_{k,n}$, note that by definition

$$\#\mathcal{F}_{k,n} = \#\{v \in \mathbb{Z}^n \mid \sum_{i=1}^n |v_i| \leq k\} \leq 2^k \#\{v \in \mathbb{Z}_+^{n+1} \mid \sum_{i=1}^{n+1} v_i = k\}$$

$$= 2^k \binom{n+k}{k} \leq 2^k \left(\frac{e(n+k)}{k}\right)^k.$$

This completes the proof of the lemma.   □

Proof of Proposition 2.2 We assume $n > 1$ (the case $n = 1$ can be checked directly). Set $k = [\frac{\varepsilon n}{8 \ln(1/\varepsilon)}]$. Note that since $\varepsilon > 8\frac{\ln n}{n}$, one has that $k \geq 1$. From Lemma 2.3 it follows that $\theta\frac{\sqrt{n}}{k}\mathcal{F}_{k,n}$ is a $\theta\frac{\sqrt{n}}{k}$-net for $\theta\sqrt{n}B_1^n$. From the remark in the beginning of this section and Lemma 2.3 we conclude that there is a set

$\mathcal{T} \subset G_\theta^n \subset \theta \sqrt{n} B_1^n$ which is a $2\theta \sqrt{\frac{\pi}{k}}$-net for G_θ^n, and moreover,

$$\#\mathcal{T} \leq \#\mathcal{F}_{k,n} \leq \left(2e(1+n/k)\right)^k.$$

Finally, from our choice of ε it follows that $k \geq \frac{\varepsilon n}{16\ln(1/\varepsilon)}$, and hence $2\theta \sqrt{\frac{\pi}{k}} \leq 8\theta \sqrt{\frac{\ln(1/\varepsilon)}{\varepsilon}}$, and moreover that $\left(2e(1+n/k)\right)^{k/n} \leq e^\varepsilon$. This completes the proof of the proposition. $\square$

We conclude this section with the following well-known result regarding concentration of measure for Lipschitz functions on the sphere (see, e.g., [16], Sect. 2 and Appendix V).

Proposition 2.4 *Let* $f : S^{n-1} \to \mathbb{R}$ *be an L-Lipschitz function and set* $\mathbb{E}f = \int_{S^{n-1}} f d\sigma_{n-1}$, *where* σ_{n-1} *is the standard measure on* S^{n-1}. *Then,*

$$\sigma_{n-1}\left(\{x \in S^{n-1} \mid |f(x) - \mathbb{E}f| \geq t\}\right) \leq C\exp(-\kappa t^2 n/L^2),$$

where $C, \kappa > 0$ *are some universal constants.*

3   Proof of the Main Theorem

Proof of Theorem 1.1 Let $Q = [-1,1]^{2n} \subset \mathbb{R}^{2n}$. The proof is divided into two steps:

Step I (ε-Net Argument): Let $L \subset \mathbb{R}^{2n}$ be a complex line, and $e \in S^{2n-1} \cap L$. Note that the vectors e and Je form an orthogonal basis for L, and for every $x \in \mathbb{R}^{2n}$ one has

$$\pi_L(x) = \langle x, e \rangle e + \langle x, Je \rangle Je.$$

Thus, one has

$$\mathrm{diam}(\pi_L(UQ)) = 2 \max_{x \in Q} \sqrt{|\langle Ux, e \rangle|^2 + |\langle Ux, Je \rangle|^2}$$

$$\geq \max_{x \in Q} \max\{|\langle x, U^*e \rangle|, |\langle x, U^*Je \rangle|\} \qquad (6)$$

$$= \max\{\|U^*e\|_1, \|U^*Je\|_1\}.$$

It follows that for every $U \in O(2n)$, the minimum over all complex lines satisfies

$$\min_L \mathrm{diam}(\pi_L(UQ)) \geq \min_{v \in S^{2n-1}} \max\{\|v\|_1, \|U^*JUv\|_1\}. \qquad (7)$$

Next, for a given constant $\theta > 0$, denote $G_\theta := S^{2n-1} \cap \theta \sqrt{n} B_1^{2n}$, and

$$\mathcal{A}_\lambda := \{U \in O(2n) \mid \exists \text{ a complex line } L \subset \mathbb{R}^{2n} \text{ with diam}(\pi_L(UQ)) \leq \lambda \sqrt{n}\}. \tag{8}$$

Recall that in order to prove Theorem 1.1, we need to show that there is a constant λ for which the measure of $\mathcal{A}_\lambda \subset O(2n)$ is exponentially small, a task to which we now turn. From (7) it follows that for any $U \in \mathcal{A}_\lambda$ one has

$$G_\lambda \cap U^* J U G_\lambda \neq \emptyset.$$

Indeed, if $U \in \mathcal{A}_\lambda$, then by (6) one has that $\|U^* e\|_1 \leq \lambda \sqrt{n}$ and $\|(U^* J U) U^* e\|_1 \leq \lambda \sqrt{n}$, so $z := U^* e_1 \in G_\lambda$ and $U^* J z \in G_\lambda$. Hence, we conclude that

$$\mathcal{A}_\lambda \subseteq \{U \in O(2n) \mid G_\lambda \cap U^* J U G_\lambda \neq \emptyset\}.$$

Next, let $\mathcal{F}$ be a δ-net for G_λ for some $\delta > 0$. For any $U \in \mathcal{A}_\lambda$ there exists $x \in G_\lambda \cap U^* J U G_\lambda$, and $y \in \mathcal{F}$ for which $|y - x| \leq \delta$. Thus, one has

$$\|U^* J U y\|_1 \leq \|U^* J U x\|_1 + \|U^* J U (y - x)\|_1$$

$$\leq \lambda \sqrt{n} + \sqrt{2n} |U^* J U (y - x)| \leq \sqrt{n}(\lambda + \sqrt{2}\delta).$$

It follows that

$$\mathcal{A}_\lambda \subseteq \bigcup_{y \in \mathcal{F}} \left\{U \in O(2n) \mid U^* J U y \in G_{\lambda + \sqrt{2}\delta}\right\}. \tag{9}$$

From (9) and Proposition 2.2 from Sect. 2 it follows that for every $\lambda > 0$

$$\mu(\mathcal{A}_\lambda) \leq \sum_{y \in \mathcal{F}} \mu\{U \in O(2n) \mid U^* J U y \in G_{\lambda + \sqrt{2}\delta}\}$$

$$\leq \exp(2\varepsilon n) \sup_{y \in S^{2n-1}} \mu\{U \in O(2n) \mid U^* J U y \in G_{\lambda + \sqrt{2}\delta}\}, \tag{10}$$

where $8\frac{\ln(2n)}{2n} < \varepsilon < \frac{1}{2}$, and $\delta = 8\lambda \sqrt{\frac{\ln(1/\varepsilon)}{\varepsilon}}$.

Step II (Concentration of Measure): For $y \in S^{2n-1}$ let ν_y be the push-forward measure on S^{2n-1} induced by the Haar measure μ on $O(2n)$ through the map $f : O(2n) \to S^{2n-1}$ defined by $U \mapsto U^* J U y$. Using the measure ν_y, we can rewrite inequality (10) as

$$\mu(\mathcal{A}_\lambda) \leq \exp(2\varepsilon n) \sup_{y \in S^{2n-1}} \nu_y(G_{\lambda+\sqrt{2}\delta})$$

$$= \exp(2\varepsilon n) \sup_{y \in S^{2n-1}} \nu_y\{x \in S^{2n-1} \mid \|x\|_1 \leq \sqrt{n}(\lambda + \sqrt{2}\delta)\}. \tag{11}$$

Note that if $V \in O(2n)$ preserves y, i.e., $Vy = y$, then

$$V(f(U)) = V(U^*JUy) = (UV^*)^*J(UV^*)(Vy) = f(UV^*).$$

Thus, the measure ν_y is invariant under any rotation in $O(2n)$ that preserves y. Note also that for any $y \in S^{2n-1}$ one has

$$\langle U^*JUy, y \rangle = \langle JUy, Uy \rangle = 0.$$

This means that ν_y is supported on $S^{2n-1} \cap \{y\}^\perp$, and hence we conclude that ν_y is the standard normalized measure on $S^{2n-1} \cap \{y\}^\perp$.

Next, let $S_y = S^{2n-1} \cap \{y\}^\perp$. For $x \in S_y$ set $\varphi(x) = \|x\|_1$. Note that φ is a Lipschitz function on S_y with Lipschitz constant $\|\varphi\|_{\text{Lip}} \leq \sqrt{2n}$. Using a concentration of measure argument (see Proposition 2.4 above), we conclude that for any $\alpha > 0$

$$\nu_y\{x \in S_y \mid \varphi(x) < \mathbb{E}_{\nu_y}\varphi - \alpha\sqrt{n}\} \leq C\exp(-\kappa^2\alpha^2 n^2/\|\varphi\|_{\text{Lip}}^2) \leq C\exp(-\kappa^2\alpha^2 n), \tag{12}$$

for some universal constants C and κ.

Our next step is to estimate the expectation $\mathbb{E}_{\nu_y}\varphi$ that appear in (12). For this purpose let us take some orthogonal basis $\{z_1, \ldots, z_{2n-1}\}$ of the subspace $L = \{y\}^\perp \subset \mathbb{R}^{2n}$. For $1 \leq j \leq 2n$, denote by w_j the vector $w_j = (z_1(j), \ldots, z_{2n-1}(j))$, where $z_k(j)$ stands for the jth coordinate of the vector z_k. Then, the measure ν_y, which is the standard normalized Lebesgue measure on $S^{2n-1} \cap \{y\}^\perp$, can be described as the image of the normalized Lebesgue measure σ_{2n-2} of S^{2n-2} under the map

$$S^{2n-2} \ni a = (a_1, \ldots, a_{2n-1}) \mapsto \sum_{k=1}^{2n-1} a_k z_k = (\langle a, w_1 \rangle, \langle a, w_2 \rangle, \ldots \langle a, w_{2n} \rangle) \in S_y.$$

Consequently,

$$\mathbb{E}_{\nu_y}\varphi = \mathbb{E}_{\sigma_{2n-2}}\left(a \mapsto \sum_{j=1}^{2n} |\langle a, w_j \rangle|\right) \geq \frac{1}{\sqrt{2n-1}}\sqrt{\frac{2}{\pi}} \sum_{j=1}^{2n} |w_j|.$$

Since $\{z_1,\ldots,z_{2n-1},y\}$ is a basis of $\mathbb{R}^{2n}$, one has that $|w_j|^2 + y_j^2 = 1$ and hence

$$\mathbb{E}_{\nu_y}\varphi = \frac{1}{\sqrt{2n-1}}\sqrt{\frac{2}{\pi}}\sum_{j=1}^{2n}\sqrt{1-y_j^2} \geq \frac{1}{\sqrt{2n-1}}\sqrt{\frac{2}{\pi}}(2n-1) \geq \frac{1}{2}\sqrt{n}.$$

Thus, from inequality (12) with $\alpha = \frac{1}{4}$ we conclude that

$$\nu_y\{x \in S_y \mid \varphi(x) < \frac{1}{4}\sqrt{n}\} \leq \nu_y\{x \in S_y \mid \varphi(x) < \mathbb{E}_{\nu_y}\varphi - \frac{1}{4}\sqrt{n}\} \leq C\exp(-\frac{\kappa^2 n}{16}). \tag{13}$$

In other words, for any $\theta \leq \frac{1}{4}$ and any $y \in S^{2n-1}$ one has that

$$\nu_y(G_\theta) \leq C\exp(-\frac{\kappa^2 n}{16}),$$

for some constant κ. Thus, for every λ such that $\lambda + \sqrt{2}\delta \leq 1/4$, we conclude by (11) that

$$\mu(\mathcal{A}_\lambda) \leq C\exp(2n\varepsilon)\cdot\exp\left(-\frac{\kappa^2 n}{16}\right).$$

To complete the proof of the Theorem it is enough to take $\varepsilon = \kappa^2/64$, and λ which satisfies the inequality $\lambda\left(1 + 16\sqrt{\frac{\ln(1/\varepsilon)}{\varepsilon}}\right) \leq 1/4$. $\qquad\square$

Acknowledgements The authors would like to thank the anonymous referee for helpful comments and remarks, and in particular for his/her suggestion to elaborate more on the symplectic topology background which partially served as a motivation for the current note. The second-named author was partially supported by the European Research Council (ERC) under the European Union's Horizon 2020 research and innovation programme, starting grant No. 637386, and by the ISF grant No. 1274/14.

Appendix

Here we provide some background from symplectic topology which partially served as a motivation for the current paper. For more detailed information on symplectic topology we refer the reader e.g., to the books [10, 13] and the references therein.

A symplectic vector space is a pair (V,ω), consisting of a finite-dimensional vector space and a non-degenerate skew-symmetric bilinear form ω, called the symplectic structure. The group of linear transformations which preserve ω is denoted by $\mathrm{Sp}(V,\omega)$. The archetypal example of a symplectic vector space is the Euclidean space $\mathbb{R}^{2n}$ equipped with the skew-symmetric bilinear form ω which is the imaginary part of the standard Hermitian inner product in $\mathbb{R}^{2n} \simeq \mathbb{C}^n$. More precisely, if $\{x_1, \ldots, x_n, y_1, \ldots, y_n\}$ stands for the standard basis of $\mathbb{R}^{2n}$, then $\omega(x_i, x_j) = \omega(y_i, y_j) = 0$, and $\omega(x_i, y_j) = \delta_{ij}$. In this case the group of linear symplectomorphisms is usually denoted by $\mathrm{Sp}(2n)$. More generally, the group of diffeomorphisms φ of $\mathbb{R}^{2n}$ which preserve the symplectic structure, i.e., when the differential $d\varphi$ at each point is a linear symplectic map, is called the group of symplectomorphisms of $\mathbb{R}^{2n}$, and is denoted by $\mathrm{Symp}(\mathbb{R}^{2n}, \omega)$. In the spirit of Klein's Erlangen program, symplectic geometry can be defined as the study of transformations which preserves the symplectic structure. We remark that already in the linear case, the geometry of a skew-symmetric bilinear form is very different from that of a symmetric form, e.g., there is no natural notion of distance or angle between two vectors. We further remark that symplectic vector spaces, and more generally symplectic manifolds, provide a natural setting for Hamiltonian dynamics, as the evolution of a Hamiltonian system is known to preserve the symplectic form (see, e.g., [10]). Historically, this is one of the main motivations to study symplectic geometry.

In sharp contrast with Riemannian geometry where, e.g., curvature is an obstruction for two manifolds to be locally isometric, in the realm of symplectic geometry it is known that there are no local invariants (Darboux's theorem). Moreover, unlike the Riemannian setting, a symplectic structure has a very rich group of automorphisms. More precisely, the group of symplectomorphisms is an infinite-dimensional Lie group. The first results distinguishing (non-linear) symplectomorphisms from volume preserving transformations were discovered only in the 1980s. The most striking difference between the category of volume preserving transformations and the category of symplectomorphisms was demonstrated by Gromov [9] in his famous non-squeezing theorem. This theorem asserts that if $r < 1$, there is no symplectomorphism ψ of $\mathbb{R}^{2n}$ which maps the open unit ball $B^{2n}(1)$ into the open cylinder $Z^{2n}(r) = B^2(r) \times \mathbb{C}^{n-1}$. This result paved the way to the introduction of global symplectic invariants, called symplectic capacities, which are significantly differ from any volume related invariants, and roughly speaking measure the symplectic size of a set (see e.g., [3], for the precise definition and further discussion). Two examples, defined for open subsets of $\mathbb{R}^{2n}$, are the Gromov radius $\underline{c}(\mathcal{U}) = \sup\{\pi r^2 : B^{2n}(r) \overset{s}{\hookrightarrow} \mathcal{U}\}$, and the cylindrical capacity $\overline{c}(\mathcal{U}) = \inf\{\pi r^2 : \mathcal{U} \overset{s}{\hookrightarrow} Z^{2n}(r)\}$. Here $\overset{s}{\hookrightarrow}$ stands for symplectic embedding.

Shortly after Gromov's work [9] many other symplectic capacities were constructed, reflecting different geometrical and dynamical properties. Nowadays, these invariants play an important role in symplectic geometry, and their properties, interrelations, and applications to symplectic topology and Hamiltonian dynamics are intensively studied (see e.g., [3]). However, in spite of the rapidly accumulating knowledge regarding symplectic capacities, they are usually notoriously difficult to compute, and there are very few general methods to effectively estimate them, even within the class of convex domains in $\mathbb{R}^{2n}$ (we refer the reader to [18] for a survey of some known results and open questions regarding symplectic measurements of convex sets in $\mathbb{R}^{2n}$). In particular, a long standing central question is whether all symplectic capacities coincide on the class of convex bodies in $\mathbb{R}^{2n}$ (see, e.g., Sect. 5 in [18]). Recently, the authors proved that for centrally symmetric convex bodies, several symplectic capacities, including the Ekeland-Hofer-Zehnder capacity c_{EHZ}, spectral capacities, the cylindrical capacity $\bar{c}$, and its linearized version $c_{\mathrm{Sp}(2n)}$ given in (1), are all equivalent up to an absolute constant. More precisely, the following was proved in [8].

Theorem 3.1 *For every centrally symmetric convex body $K \subset \mathbb{R}^{2n}$*

$$\frac{1}{\|J\|_{K^\circ \to K}} \leq c_{\mathrm{EHZ}}(K) \leq \bar{c}(K) \leq \bar{c}_{\mathrm{Sp}(2n)}(K) \leq \frac{4}{\|J\|_{K^\circ \to K}},$$

where $\|J\|_{K^\circ \to K}$ is the operator norm of the complex structure J, when the latter is considered as a linear map between the normed spaces $J : (\mathbb{R}^{2n}, \|\cdot\|_{K^\circ}) \to (\mathbb{R}^{2n}, \|\cdot\|_K)$.

Theorem 3.1 implies, in particular, that despite the non-linear nature of the Ekeland-Hofer-Zehnder capacity c_{EHZ}, and the cylindrical capacity $\bar{c}$ (both, by definition, are invariant under non-linear symplectomorphisms), for centrally symmetric convex bodies they are asymptotically equivalent to a linear invariant: the linearized cylindrical capacity $\bar{c}_{\mathrm{Sp}(2n)}$. Motivated by the comparison between the capacities $\bar{c}$ and $\bar{c}_{\mathrm{Sp}(2n)}$ in Theorem 3.1, it is natural to introduce and study the following geometric quantity:

$$\bar{c}_G(K) = \inf_{g \in G} \mathrm{Area}\big(\pi(g(K))\big), \tag{14}$$

where K lies in the class of convex domains of $\mathbb{R}^{2n} \simeq \mathbb{C}^n$ (or possibly, some other class of bodies), π is the orthogonal projection to the complex line $E = \{z \in \mathbb{C}^n \mid z_j = 0 \text{ for } j \neq 1\}$, and G is some group of transformations of $\mathbb{R}^{2n}$. One possible choice is to take the group G in (14) to be the unitary group $U(n)$, which is the maximal compact subgroup of $\mathrm{Sp}(2n)$. In this case it is not hard to check (by looking at linear symplectic images of the cylinder $Z^{2n}(1)$) that the cylindrical capacity $\bar{c}$ is not asymptotically equivalent to $\bar{c}_{U(n)}$. Still, one can ask if these two quantities are asymptotically equivalent on average. More precisely,

Question 3.2 Is it true that for every convex body $K \subset \mathbb{R}^{2n}$ one has

$$\mathbb{E}_\mu\left(\bar{c}(OK)\right) \asymp \mathbb{E}_\mu\left(\bar{c}_{U(n)}(OK)\right)?,$$

where μ is the Haar measure on the orthogonal group $O(2n)$.

The answer to Question 3.2 is negative. A counterexample is given by the standard cube $Q = [-1, 1]^{2n}$ in $\mathbb{R}^{2n}$. We remark that the quantity $\mathbb{E}_\mu\left(\bar{c}_{U(n)}(OQ)\right)$ is the main objects of interest of the current paper. To be more precise, we turn now to the following proposition, which is a direct corollary of Theorem 3.1, and might be of independent interest. For completeness, we shall give a proof below.

Proposition 3.3 *For the standard cube $Q = [-1, 1]^{2n} \subset \mathbb{R}^{2n}$ one has*

$$\mathbb{E}_\mu\left(c_{\text{EHZ}}(OQ)\right) \asymp \mathbb{E}_\mu\left(\bar{c}(OQ)\right) \asymp \mathbb{E}_\mu\left(\bar{c}_{\text{Sp}(2n)}(OQ)\right) \asymp \sqrt{\frac{n}{\ln n}},$$

where μ is the Haar measure on the orthogonal group $O(2n)$.

Note that the combination of the main result of the current paper (in particular, Corollary 1.2) with Proposition 3.3 above gives a negative answer to Question 3.2, and thus further emphasizes the difference between the symplectic and complex structures on $\mathbb{R}^{2n} \simeq \mathbb{C}^n$.

Proof of Proposition 3.3 Note that by definition one has that

$$\|J\|_{(OQ)^\circ \to (OQ)} = \max_{x \in (OQ)^\circ} \|Jx\|_{OQ} = \max_{x \in B_1^{2n}} \|O^* JOx\|_\infty = \max_{i=1,\dots,2n} \|O^* JOe_i\|_\infty,$$

where $\{e_i\}_{i=1}^{2n}$ stands for the standard basis of $\mathbb{R}^{2n}$. It follows from Step II of the proof of Theorem 1.1 above that for a random rotation $O \in O(2n)$, the vector $O^* JOe_i$ is uniformly distributed on $S^{2n-2} \simeq S^{2n-1} \cap \{e_i\}^\perp$ with respect to the standard normalized measure σ_{2n-2} on S^{2n-2}. The distribution of the l_∞^k-norm on the sphere S^{k-1} is well-studied, and in particular one has (see e.g., Sects. 5.7 and 7 in [16]) that for every e_i

$$\mathbb{E}_\mu\left(\|(O^* JOe_i)\|_\infty\right) \asymp \sqrt{\frac{\ln n}{n}}, \tag{15}$$

and

$$\mathbb{P}_\mu\left\{(\|(O^* JOe_i)\|_\infty - \mathbb{E}_\mu\left(\|(O^* JOe_i)\|_\infty\right) > t\right\} \leq c_1 \exp(-c_2 t^2 n), \tag{16}$$

for some universal constants $c_1, c_2 > 0$. From (15) and (16) it immediately follows that

$$\mathbb{E}_\mu\left(\|J\|_{(OQ)^\circ \to (OQ)}\right) \asymp \sqrt{\frac{\ln n}{n}}. \tag{17}$$

Moreover, one has that for some universal constants $c_3, c_4 > 0$,

$$\mathbb{P}_\mu\left\{(\|J\|_{(OQ)^\circ \to (OQ)} \leq c_3 \sqrt{\frac{\ln n}{n}}\right\} \leq \frac{c_4}{n}. \tag{18}$$

Indeed, from the above it follows that

$$\mathbb{P}_\mu\{\|J\|_{(OQ)^\circ \to (OQ)} \leq t\} \leq \mathbb{P}_\mu\{(\|(O^* JOe_1)\|_\infty \leq t\} = \mathbb{P}_{\sigma_{2n-2}}\{\|v\|_\infty \leq t\}.$$

Using the standard Gaussian probability measure γ_{2n-1} on $\mathbb{R}^{2n-1}$, one can further estimate

$$\mathbb{P}_{\sigma_{2n-2}}\{\|v\|_\infty \leq t\} = \gamma_{2n-1}\{\|g\|_\infty \leq t\|g\|_2\}$$

$$\leq \gamma_{2n-1}\{\|g\|_\infty \leq 2\sqrt{2n-1}t\} + \gamma_{2n-1}\{\|g\|_2 \geq 2\sqrt{2n-1}\},$$

where g is a Gaussian vector in $\mathbb{R}^{2n-1}$ with independent standard Gaussian coordinates. One can directly check that (18) now follows from the above inequalities, and the following standard estimates for the Gaussian probability measure γ_k on $\mathbb{R}^k$, and $0 < \varepsilon < 1$:

$$\gamma_k\{\|g\|_\infty \leq \alpha\} \leq [1-\sqrt{\tfrac{2}{\pi}}\tfrac{\exp(-\alpha^2/2)}{\alpha}]^k, \text{ and } \gamma_k\left\{x \in \mathbb{R}^k \mid \|g\|_2^2 \geq \tfrac{k}{(1-\varepsilon)}\right\} \leq \exp(-\varepsilon^2 k/4).$$

Taking into account the fact that $\frac{1}{\sqrt{2n}} \leq \|J\|_{(OQ)^\circ \to (OQ)} \leq 1$, we conclude from (17) and (18) above that

$$\mathbb{E}_\mu\left((\|J\|_{(OQ)^\circ \to (OQ)})^{-1}\right) \asymp \sqrt{\tfrac{n}{\ln n}}.$$

Together with Theorem 3.1, this completes the proof of Proposition 3.3.   □

References

1. S. Artstein-Avidan, V.D. Milman, Logarithmic reduction of the level of randomness in some probabilistic geometric constructions. J. Funct. Anal. **235**(1), 297–329 (2006)
2. S. Artstein-Avidan, A. Giannopoulos, V.D. Milman, *Asymptotic Geometric Analysis, Part I.* Mathematical Surveys and Monographs, vol. 202 (American Mathematical Society, Providence, RI, 2015)
3. T. Cieliebak, H. Hofer, J. Latschev, F. Schlenk, Quantitative symplectic geometry, in *Dynamics, Ergodic Theory, and Geometry*, vol. 54. Mathematical Sciences Research Institute Publications (Cambridge University Press, Cambridge, 2007), pp. 1–44
4. T. Figiel, J. Lindenstrauss, V.D. Milman, The dimension of almost spherical sections of convex bodies. Bull. Am. Math. Soc. **82**(4), 575–578 (1976); expanded in "The dimension of almost spherical sections of convex bodies". Acta Math. **139**, 53–94 (1977)
5. A.Y. Garnaev, E.D. Gluskin, The widths of a Euclidean ball. Dokl. Akad. Nauk SSSR **277**(5), 1048–1052 (1984)
6. G. Giannopoulos, V.D. Milman, A. Tsolomitis, Asymptotic formula for the diameter of sections of symmetric convex bodies. JFA **233**(1), 86–108 (2005)
7. E.D. Gluskin, Norms of random matrices and diameters of finite-dimensional sets. Mat. Sb. (N.S.) **120**(162) (2), 180–189, 286 (1983)
8. E.D. Gluskin, Y. Ostrover, Asymptotic equivalence of symplectic capacities. Comm. Math. Helv. **91**(1), 131–144 (2016)
9. M. Gromov, Pseudoholomorphic curves in symplectic manifolds. Invent. Math. **82**, 307–347 (1985)
10. H. Hofer, E. Zehnder, *Symplectic Invariants and Hamiltonian Dynamics* (Birkhäuser Advanced Texts, Birkhäuser Verlag, 1994)

11. B.S. Kašin, The widths of certain finite-dimensional sets and classes of smooth functions (Russian). Izv. Akad. Nauk SSSR Ser. Mat. **41**(2), 334–351, 478 (1977)
12. Y. Makovoz, A simple proof of an inequality in the theory of n-widths, in *Constructive Theory of Functions (Varna, 1987)* (Publ. House Bulgar. Acad. Sci., Sofia, 1988), pp. 305–308
13. D. McDuff, D. Salamon, *Introduction to Symplectic Topology*. Oxford Mathematical Monographs (Oxford University Press, New York, 1995)
14. V.D. Milman, A new proof of A. Dvoretzky's theorem on cross-sections of convex bodies. Funk. Anal. i Prilozhen. **5**(4), 28–37 (1971)
15. V.D. Milman, Spectrum of a position of a convex body and linear duality relations, in *Israel Mathematical Conference Proceedings (IMCP)* vol. 3. Festschrift in Honor of Professor I. Piatetski-Shapiro (Part II) (Weizmann Science Press of Israel, 1990), pp. 151–162
16. V.D. Milman, G. Schechtman, *Asymptotic Theory of Finite-Dimensional Normed Spaces*. Lecture Notes in Mathematics, vol. 1200 (Springer, Berlin, 1986)
17. B.S. Mitjagin, Random matrices and subspaces, in *Geometry of Linear Spaces and Operator Theory* (Russian) (Yaroslav. Gos. Univ., Yaroslavl, 1977), pp. 175–202
18. Y. Ostrover, When symplectic topology meets Banach space geometry, in *Proceedings of the ICM*, Seoul, vol. II (2014), pp 959–981.
19. A. Pajor, N. Tomczak-Jaegermann, Subspaces of small codimension of finite-dimensional Banach spaces. Proc. Am. Math. Soc. **97**(4), 637–642 (1986)
20. G. Pisier, *The Volume of Convex Bodies and Banach Space Geometry*. Cambridge Tracts in Mathematics, vol. 94 (Cambridge University Press, Cambridge, 1989)
21. C. Schütt, Entropy numbers of diagonal operators between symmetric Banach spaces. J. Approx. Theory **40**(2), 121–128 (1984)
22. S. Szarek, On Kashin's almost Euclidean orthogonal decomposition of l_n^1. Bull. Acad. Polon. Sci. Sér. Sci. Math. Astronom. Phys. **26**(8), 691–694 (1978)

On the Expectation of Operator Norms of Random Matrices

Olivier Guédon, Aicke Hinrichs, Alexander E. Litvak, and Joscha Prochno

Abstract We prove estimates for the expected value of operator norms of Gaussian random matrices with independent (but not necessarily identically distributed) and centered entries, acting as operators from ℓ_{p*}^n to ℓ_q^m, $1 \leq p^* \leq 2 \leq q < \infty$.

1 Introduction and Main Results

Random matrices and their spectra are under intensive study in Statistics since the work of Wishart [28] on sample covariance matrices, in Numerical Analysis since their introduction by von Neumann and Goldstine [25] in the 1940s, and in Physics as a consequence of Wigner's work [26, 27] since the 1950s. His Semicircle Law, a fundamental theorem in the spectral theory of large random matrices describing the limit of the empirical spectral measure for what is nowadays known as Wigner matrices, is among the most celebrated results of the theory.

In Banach Space Theory and Asymptotic Geometric Analysis, random matrices appeared already in the 70s (see e.g. [2, 3, 9]). In [2], the authors obtained asymptotic bounds for the expected value of the operator norm of a random matrix $B = (b_{ij})_{i,j=1}^{m,n}$ with independent mean-zero entries with $|b_{ij}| \leq 1$ from ℓ_2^n to ℓ_q^m, $2 \leq q < \infty$. To be

O. Guédon
Laboratoire d'Analyse et de Mathématiques Appliquées, Université Paris-Est, 77454
Marne-la-Vallée, Cedex 2, France
e-mail: olivier.guedon@u-pem.fr

A. Hinrichs
Institut für Analysis, Johannes Kepler Universität Linz, Altenbergerstrasse 69, 4040 Linz, Austria
e-mail: aicke.hinrichs@jku.at

A.E. Litvak (✉)
Department of Mathematical and Statistical Sciences, University of Alberta, Edmonton, AB,
Canada T6G 2G1
e-mail: alitvak@ualberta.ca

J. Prochno
Department of Mathematics, University of Hull, Cottingham Road , Hull HU6 7RX, UK
e-mail: j.prochno@hull.ac.uk

© Springer International Publishing AG 2017   151
B. Klartag, E. Milman (eds.), *Geometric Aspects of Functional Analysis*,
Lecture Notes in Mathematics 2169, DOI 10.1007/978-3-319-45282-1_10

more precise, they proved that

$$\mathbb{E}\left\|B: \ell_2^n \to \ell_q^m\right\| \leq C_q \cdot \max\left(m^{1/q}, \sqrt{n}\right),$$

where C_q depends only on q. This was then successfully used to characterize (p, q)-absolutely summing operators on Hilbert spaces. Ever since, random matrices are extensively studied and methods of Banach spaces have produced numerous deep and new results. In particular, in many applications the spectral properties of a Gaussian matrix, whose entries are independent identically distributed (i.i.d.) standard Gaussian random variables, were used. Seginer proved in [22] that for an $m \times n$ random matrix with i.i.d. symmetric random variables the expectation of its spectral norm (that is, the operator norm from ℓ_2^n to ℓ_2^m) is of the order of the expectation of the largest Euclidean norm of its rows and columns. He also obtained an optimal result in the case of random matrices with entries $\varepsilon_{ij}a_{ij}$, where ε_{ij} are independent Rademacher random variables and a_{ij} are fixed numbers. We refer the interested reader to the surveys [6, 7] and references therein.

It is natural to ask similar questions about general random matrices, in particular about Gaussian matrices whose entries are still independent centered Gaussian random variables, but with different variances. In this structured case, where we drop the assumption of identical distributions, very little is known. It is conjectured that the expected spectral norm of such a Gaussian matrix is as in Seginer's result, that is, of the order of the expectation of the largest Euclidean norm of its rows and columns. A big step toward the solution was made by Latała in [15], who proved a bound involving fourth moments, which is of the right order $\max(\sqrt{m}, \sqrt{n})$ in the i.i.d. setting, but does not capture the right behavior in the case of, for instance, diagonal matrices. On one hand, as is mentioned in [15], in view of the classical Bai-Yin theorem, the presence of fourth moments is not surprising, on the other hand they are not needed if the conjecture is true.

Later in [20], Riemer and Schütt proved the conjecture up to a $\log n$ factor. The two results are incomparable—depending on the choice of variances, one or another gives a better bound. The Riemer-Schütt estimate was used recently in [21].

We would also like to mention that the non-commutative Khintchine inequality can be used to show that the expected spectral norm is bounded from above by the largest Euclidean norm of its rows and columns times a factor $\sqrt{\log n}$ (see e.g. (4.9) in [23]).

Another big step toward the solution was made a short while ago by Bandeira and Van Handel [1]. In particular, they proved that

$$\mathbb{E}\left\|(a_{ij}g_{ij}): \ell_2^n \to \ell_2^m\right\| \leq C\left(\|\|A\|\| + \sqrt{\log\min(n, m)} \cdot \max_{ij}|a_{ij}|\right), \tag{1}$$

where $\|\|A\|\|$ denotes the largest Euclidean norm of the rows and columns of (a_{ij}), $C > 0$ is a universal constant, and g_{ij} are independent standard Gaussian random variables (see [1, Theorem 3.1]). Under mild structural assumptions, the bound (1) is already optimal. Further progress was made by Van Handel [24] who verified the conjecture up to a $\sqrt{\log \log n}$ factor. In fact, more was proved in [24]. He computed precisely the expectation of the largest Euclidean norm of the rows and columns using Gaussian concentration. And, while the moment method is at the heart of the proofs in [22] and [1], he proposed a very nice approach based on the comparison of Gaussian processes to improve the result of Latała. His approach can be also used for our setting. We comment on this in Sect. 4.

The purpose of this work is to provide bounds for operator norms of such structured Gaussian random matrices considered as operators from $\ell_{p^*}^n$ to ℓ_q^m.

In what follows, by $g_i, g_{ij}, i \geq 1, j \geq 1$ we always denote independent standard Gaussian random variables. Let $n, m \in \mathbb{N}$ and $A = (a_{ij})_{i,j=1}^{m,n} \in \mathbb{R}^{m \times n}$. We write $G = G_A = (a_{ij}g_{ij})_{i,j=1}^{m,n}$. For $r \geq 1$, we denote by $\gamma_r \approx \sqrt{r}$ the L_r-norm of a standard Gaussian random variable. The notation $f \approx h$ means that there are two absolute positive constants c and C (that is, independent of any parameters) such that $cf \leq h \leq Cf$ and $f \approx_{p,q} h$ means that there are two positive constants $c(p, q)$ and $C(p, q)$, which depend only on the parameters p and q, such that $c(p, q)f \leq h \leq C(p, q)f$.

Our main result is the following theorem.

Theorem 1.1 *For every* $1 < p^* \leq 2 \leq q < \infty$ *one has*

$$\mathbb{E} \left\| G : \ell_{p^*}^n \to \ell_q^m \right\| \leq \left(\mathbb{E} \left\| G : \ell_{p^*}^n \to \ell_q^m \right\|^q \right)^{1/q}$$

$$\leq C p^{5/q} (\log m)^{1/q} \left[\gamma_p \max_{i \leq m} \|(a_{ij})_{j=1}^n\|_p + \gamma_q \, \mathbb{E} \max_{\substack{i \leq m \\ j \leq n}} |a_{ij}g_{ij}| \right]$$

$$+ 2^{1/q} \gamma_q \max_{j \leq n} \|(a_{ij})_{i=1}^m\|_q,$$

where C is a positive absolute constant.

We conjecture the following bound.

Conjecture 1.2 *For every* $1 \leq p^* \leq 2 \leq q \leq \infty$ *one has*

$$\mathbb{E} \left\| G : \ell_{p^*}^n \to \ell_q^m \right\| \approx \max_{i \leq m} \|(a_{ij})_{j=1}^n\|_p + \max_{j \leq n} \|(a_{ij})_{i=1}^m\|_q + \mathbb{E} \max_{\substack{i \leq m \\ j \leq n}} |a_{ij}g_{ij}|.$$

Here, as usual, p is defined via the relation $1/p + 1/p^* = 1$. This conjecture extends the corresponding conjecture for the case $p = q = 2$ and $m = n$. In this case, Bandeira and Van Handel proved in [1] an estimate with $\sqrt{\log \min(m, n)} \max |a_{ij}|$ instead of $\mathbb{E} \max |a_{ij}g_{ij}|$ (see Eq. (1)), while in [24] the corresponding bound is proved with $\sqrt{\log \log n}$ in front of the right hand side.

Remark 1.3 The lower bound in the conjecture is almost immediate and follows from standard estimates. Thus the upper bound is the only difficulty.

Remark 1.4 In the case $p^* = 1$ and $q \geq 2$, a direct computation following along the lines of Lemma 3.2 below, shows that

$$\mathbb{E}\left\|G : \ell_1^n \to \ell_q^m\right\| \lesssim \gamma_q \max_{j \leq n} \left\|(a_{ij})_{i=1}^m\right\|_q + \mathbb{E} \max_{\substack{i \leq m \\ j \leq n}} |a_{ij} g_{ij}|.$$

Remark 1.5 Note that if $1 \leq p^* \leq 2 \leq q \leq \infty$, in the case of matrices of tensor structure, that is, $(a_{ij})_{i,j=1}^n = x \otimes y = (x_j \cdot y_i)_{i,j=1}^n$, with $x, y \in \mathbb{R}^n$, Chevet's theorem [3, 4] and a direct computation show that

$$\mathbb{E}\left\|G : \ell_{p^*}^n \to \ell_q^n\right\| \approx_{p,q} \|y\|_q \|x\|_\infty + \|y\|_\infty \|x\|_p.$$

If the matrix is diagonal, that is, $(a_{ij})_{i,j=1}^n = \mathrm{diag}(a_{11}, \ldots, a_{nn})$, then we immediately obtain

$$\mathbb{E}\left\|G : \ell_{p^*}^n \to \ell_q^n\right\| = \mathbb{E}\left\|(a_{ii} g_{ii})_{i=1}^n\right\|_\infty \approx \max_{i \leq n} \sqrt{\ln(i+3)} \cdot a_{ii}^* \approx \left\|(a_{ii})_{i=1}^n\right\|_{M_g},$$

where $(a_{ii}^*)_{i \leq n}$ is the decreasing rearrangement of $(|a_{ii}|)_{i \leq n}$ and M_g is the Orlicz function given by

$$M_g(s) = \sqrt{\frac{2}{\pi}} \int_0^s e^{-\frac{1}{2t^2}}\, dt$$

(see Lemma 2.2 below and [11, Lemma 5.2] for the Orlicz norm expression).

Slightly different estimates, but of the same flavour, can also be obtained in the case $1 \leq q \leq 2 \leq p^* \leq \infty$.

2 Notation and Preliminaries

By $c, C, C_1, \ldots$ we always denote positive absolute constants, whose values may change from line to line, and we write $c_p, C_p, \ldots$ if the constants depend on some parameter p.

Given $p \in [1, \infty]$, p^* denotes its conjugate and is given by the relation $1/p + 1/p^* = 1$. For $x = (x_i)_{i \leq n} \in \mathbb{R}^n$, $\|x\|_p$ denotes its ℓ_p-norm, that is $\|x\|_\infty = \max_{i \leq n} |x_i|$ and, for $p < \infty$,

$$\|x\|_p = \left(\sum_{i=1}^n |x_i|^p\right)^{1/p}.$$

The corresponding space $(\mathbb{R}^n, \|\cdot\|_p)$ is denoted by ℓ_p^n, its unit ball by B_p^n.

If E is a normed space, then E^* denotes its dual space and B_E its closed unit ball. The modulus of convexity of E is defined for any $\varepsilon \in (0, 2)$ by

$$\delta_E(\varepsilon) := \inf \left\{ 1 - \left\| \frac{x+y}{2} \right\|_E : \|x\|_E = 1, \ \|y\|_E = 1, \ \|x - y\|_E > \varepsilon \right\}.$$

We say that E has modulus of convexity of power type 2 if there exists a positive constant c such that for all $\varepsilon \in (0, 2)$, $\delta_E(\varepsilon) \geq c\varepsilon^2$. It is well known that this property (see e.g. [8] or [18, Proposition 2.4]) is equivalent to the fact that

$$\left\| \frac{x+y}{2} \right\|_E^2 + \lambda^{-2} \left\| \frac{x-y}{2} \right\|_E^2 \leq \frac{\|x\|_E^2 + \|y\|_E^2}{2}$$

holds for all $x, y \in E$, where $\lambda > 0$ is a constant depending only on c. In that case, we say that E has modulus of convexity of power type 2 with constant λ. We clearly have $\delta_E(\varepsilon) \geq \varepsilon^2/(2\lambda^2)$.

Recall that a Banach space E is of Rademacher type r for some $1 \leq r \leq 2$ if there is $C > 0$ such that for all $n \in \mathbb{N}$ and for all $x_1, \ldots, x_n \in E$,

$$\left(\mathbb{E}_\varepsilon \left\| \sum_{i=1}^n \varepsilon_i x_i \right\|^2 \right)^{1/2} \leq C \left(\sum_{i=1}^n \|x_i\|^r \right)^{1/r},$$

where $(\varepsilon_i)_{i=1}^\infty$ is a sequence of independent random variables defined on some probability space $(\Omega, \mathbb{P})$ such that $\mathbb{P}(\varepsilon_i = 1) = \mathbb{P}(\varepsilon_i = -1) = \frac{1}{2}$ for every $i \in \mathbb{N}$. The smallest C is called type-r constant of E, denoted by $T_r(E)$. This concept was introduced into Banach space theory by Hoffmann-Jørgensen [14] in the early 1970s and the basic theory was developed by Maurey and Pisier [17].

We will need the following theorem.

Theorem 2.1 *Let E be a Banach space with modulus of convexity of power type 2 with constant λ. Let $X_1, \ldots, X_m \in E^*$ be independent random vectors, $q \geq 2$ and define*

$$B := C\lambda^4 T_2(E^*) \sqrt{\frac{\log m}{m}} \left(\mathbb{E} \max_{i \leq m} \|X_i\|_{E^*}^q \right)^{1/2},$$

and

$$\sigma := \sup_{y \in B_E} \left(\frac{1}{m} \sum_{i=1}^m \mathbb{E} |\langle X_i, y \rangle|^q \right)^{1/q}.$$

Then

$$\mathbb{E} \sup_{y \in B_E} \left| \frac{1}{m} \sum_{i=1}^m |\langle X_i, y \rangle|^q - \mathbb{E} |\langle X_i, y \rangle|^q \right| \leq B^2 + B \cdot \sigma^{q/2}.$$

Its proof is done following the argument "proof of condition (H)" of [13] in combination with the improvement on covering numbers established in [12, Lemma 2]. Indeed, in [12], the argument is only made in the simpler case $q = 2$, but it can be extended verbatim to the case $q \geq 2$.

We also recall known facts about Gaussian random variables. The next lemma is well-known (see e.g. Lemmas 2.3, 2.4 in [24]).

Lemma 2.2 *Let $a = (a_i)_{i \leq n} \in \mathbb{R}^n$ and $(a_i^*)_{i \leq n}$ be the decreasing rearrangement of $(|a_i|)_{i \leq n}$. Then*

$$\mathbb{E} \max_{i \leq n} |a_i g_i| \approx \max_{i \leq n} \sqrt{\ln(i + 3)} \cdot a_i^*.$$

Note that in general the maximum of i.i.d. random variables weighted by coordinates of a vector a is equivalent to a certain Orlicz norm $\|a\|_M$, where the function M depends only on the distribution of random variables (see [10, Corollary 2] and Lemma 5.2 in [11]).

The following theorem is the classical Gaussian concentration inequality (see e.g. [5] or inequality (2.35) and Proposition 2.18 in [16]).

Theorem 2.3 *Let $n \in \mathbb{N}$ and $(Y, \|\cdot\|_Y)$ be a Banach space. Let $y_1, \dots, y_n \in Y$ and $X = \sum_{i=1}^n g_i y_i$. Then, for every $t > 0$,*

$$\mathbb{P}\left(\left| \|X\|_Y - \mathbb{E}\|X\|_Y \right| \geq t \right) \leq 2 \exp\left(-\frac{t^2}{2\sigma_Y(X)^2} \right), \tag{2}$$

where $\sigma_Y(X) = \sup_{\|\xi\|_{Y^*}=1} \left(\sum_{i=1}^n |\xi(y_i)|^2 \right)^{1/2}$.

Remark 2.4 Let $p \geq 2$. Let $a = (a_j)_{j \leq n} \in \mathbb{R}^n$ and $X = (a_j g_j)_{j \leq n}$. Then we clearly have

$$\sigma_{\ell_p^n}(X) = \max_{j \leq n} |a_j|.$$

Thus, Theorem 2.3 implies for $X = (a_j g_j)_{j \leq n}$

$$\mathbb{P}\left(\left| \|X\|_p - \mathbb{E}\|X\|_p \right| > t \right) \leq 2 \exp\left(-\frac{t^2}{2 \max_{j \leq n} |a_j|^2} \right). \tag{3}$$

Note also that

$$\mathbb{E}\|X\|_p \leq \left(\sum_{j=1}^n |a_j|^p \, \mathbb{E}|g_j|^p \right)^{1/p} = \gamma_p \|a\|_p. \tag{4}$$

3   Proof of the Main Result

We will apply Theorem 2.1 with $E = \ell_{p*}^n$, $1 < p^* \le 2$ and $X_1, \ldots, X_m$ being the rows of the matrix $G = (a_{ij}g_{ij})_{i,j=1}^{m,n}$. We start with two lemmas in which we estimate the quantity σ and the expectation, appearing in that theorem.

Lemma 3.1 *Let* $m, n \in \mathbb{N}$, $1 < p^* \le 2 \le q$, *and for* $i \le m$ *let* $X_i = (a_{ij}g_{ij})_{j=1}^n$. *Then*

$$\sigma = \sup_{y \in B_{p*}^n} \left(\frac{1}{m} \sum_{i=1}^m \mathbb{E} |\langle X_i, y \rangle|^q \right)^{1/q} = \frac{\gamma_q}{m^{1/q}} \max_{j \le n} \|(a_{ij})_{i=1}^m\|_q.$$

Proof For every $i \le m$, $\langle X_i, y \rangle = \sum_{j=1}^n a_{ij}y_jg_{ij}$, is a Gaussian random variable with variance $\|(a_{ij}y_j)_{j=1}^n\|_2$. Hence,

$$\sigma^q = \sup_{y \in B_{p*}^n} \frac{1}{m} \sum_{i=1}^m \mathbb{E} |\langle X_i, y \rangle|^q = \frac{\gamma_q^q}{m} \sup_{y \in B_{p*}^n} \sum_{i=1}^m \left(\sum_{j=1}^n |a_{ij}y_j|^2 \right)^{q/2}.$$

Since $p^* \le 2 \le q$, the function

$$\phi(z) = \sum_{i=1}^m \left(\sum_{j=1}^n |a_{ij}|^2 |z_j|^{2/p^*} \right)^{q/2}$$

is a convex function on the simplex $S = \{z \in \mathbb{R}^n \mid \sum_{j=1}^n \le 1, \forall j : z_j \ge 0\}$. Therefore, it attains its maximum on extreme points, that is, on vectors of the canonical unit basis of $\mathbb{R}^n$, $e_1, \ldots, e_n$. Thus,

$$\sup_{y \in B_{p*}^n} \sum_{i=1}^m \left(\sum_{j=1}^n |a_{ij}y_j|^2 \right)^{q/2} = \sup_{z \in S} \phi(z) = \sup_{k \le n} \phi(e_k) = \max_{j \le n} \|(a_{ij})_{i=1}^m\|_q^q,$$

which completes the proof.   □

Now we estimate the expectation in Theorem 2.1. The proof is based on the Gaussian concentration, Theorem 2.3, and is similar to Theorem 2.1 and Remark 2.2 in [24].

Lemma 3.2 *Let* $m, n \in \mathbb{N}$, $1 < p^* \le 2 \le q$, *and for* $i \le m$ *let* $X_i = (a_{ij}g_{ij})_{j=1}^n$. *Then*

$$\left(\mathbb{E} \max_{i \le m} \|X_i\|_p^q \right)^{1/q} \le \max_{i \le m} \mathbb{E} \|X_i\|_p + C \gamma_q \, \mathbb{E} \max_{\substack{i \le m \\ j \le n}} |a_{ij}g_{ij}|$$

$$\le \gamma_p \max_{i \le m} \|(a_{ij})_{j=1}^n\|_p + C \gamma_q \, \mathbb{E} \max_{\substack{i \le m \\ j \le n}} |a_{ij}g_{ij}|,$$

where C is a positive absolute constant.

Proof We have

$$\left(\mathbb{E} \max_{i \leq m} \|X_i\|_p^q\right)^{1/q} \leq \left\| \max_{i \leq m} \left| \|X_i\|_p - \mathbb{E}\|X_i\|_p \right| + \max_{i \leq m} \mathbb{E}\|X_i\|_p \right\|_{L_q}$$

$$\leq \left(\mathbb{E} \max_{i \leq m} \left| \|X_i\|_p - \mathbb{E}\|X_i\|_p \right|^q\right)^{1/q} + \max_{i \leq m} \mathbb{E}\|X_i\|_p.$$

For all $i \leq m$ and $t > 0$ by (3) we have

$$\mathbb{P}\left(\left| \|X_i\|_p - \mathbb{E}\|X_i\|_p \right| > t \right) \leq 2 \exp\left(- \frac{t^2}{2 \max_{j \leq n} |a_{ij}|^2} \right). \tag{5}$$

By permuting the rows of $(a_{ij})_{i,j=1}^{m,n}$, we can assume that

$$\max_{j \leq n} |a_{1j}| \geq \cdots \geq \max_{j \leq n} |a_{mj}|.$$

For each $i \leq m$, choose $j(i) \leq n$ such that $|a_{ij(i)}| = \max_{j \leq n} |a_{ij}|$. Clearly,

$$\max_{\substack{i \leq m \\ j \leq n}} |a_{ij} g_{ij}| \geq \max_{i \leq m} |a_{ij(i)}| \cdot |g_{ij(i)}|$$

and hence, by independence of g_{ij}'s and Lemma 2.2,

$$b := \mathbb{E} \max_{\substack{i \leq m \\ j \leq n}} |a_{ij} g_{ij}| \geq \mathbb{E} \max_{i \leq m} |a_{ij(i)}| \cdot |g_i| \geq c \max_{i \leq m} \sqrt{\log(i+3)} \cdot |a_{ij(i)}|,$$

where the latter inequality follows since $|a_{1j(1)}| \geq \cdots \geq |a_{nj(n)}|$. Thus, for $i \leq m$,

$$\max_{j \leq n} |a_{ij}|^2 = a_{ij(i)}^2 \leq \frac{b^2}{c \log(i+3)}.$$

By (5) we observe for every $t > 0$,

$$\mathbb{P}\left(\max_{i \leq m} \left| \|X_i\|_p - \mathbb{E}\|X_i\|_p \right| > t \right) \leq 2 \sum_{i=1}^m \exp\left(- \frac{ct^2 \log(i+3)}{2b^2} \right)$$

$$= 2 \sum_{i=1}^m \left(\frac{1}{i+3} \right)^{ct^2/2b^2} \leq 2 \int_3^\infty x^{-ct^2/2b^2} \, dx$$

$$\leq 6 \cdot 3^{-ct^2/2b^2},$$

whenever $ct^2/b^2 \geq 4$. Integrating the tail inequality proves that

$$\left(\mathbb{E} \max_{i \leq m} \left| \|X_i\|_p - \mathbb{E}\|X_i\|_p \right|^q \right)^{1/q} \leq C_1 \sqrt{q} \, b \leq C_2 \gamma_q \, \mathbb{E} \max_{\substack{i \leq m \\ j \leq n}} |a_{ij} g_{ij}|.$$

By the triangle inequality, we obtain the first desired inequality, the second one follows by (4). $\qquad\square$

We are now ready to present the proof of the main theorem.

Proof of Theorem 1.1 First observe that

$$\mathbb{E}\left\|G:\ell_{p*}^n \to \ell_q^m\right\| \leq \left(\mathbb{E}\left\|G:\ell_{p*}^n \to \ell_q^m\right\|^q\right)^{1/q} = \left(\mathbb{E}\sup_{y\in B_{p*}^n}\sum_{i=1}^m|\langle X_i,y\rangle|^q\right)^{1/q}.$$

We have

$$\mathbb{E}\sup_{y\in B_{p*}^n}\sum_{i=1}^m|\langle X_i,y\rangle|^q \leq \mathbb{E}\sup_{y\in B_{p*}^n}\left[\sum_{i=1}^m|\langle X_i,y\rangle|^q - \mathbb{E}|\langle X_i,y\rangle|^q\right] + \sup_{y\in B_{p*}^n}\sum_{i=1}^m\mathbb{E}|\langle X_i,y\rangle|^q$$

$$= m\cdot\mathbb{E}\sup_{y\in B_{p*}^n}\left[\frac{1}{m}\sum_{i=1}^m|\langle X_i,y\rangle|^q - \mathbb{E}|\langle X_i,y\rangle|^q\right] + m\cdot\sigma^q.$$

Hence, Theorem 2.1 applied with $E = \ell_{p*}^n$ implies

$$\mathbb{E}\left\|G:\ell_{p*}^n \to \ell_q^m\right\|^q \leq m\cdot\left[B^2 + B\sigma^{q/2}\right] + m\cdot\sigma^q \leq 2m\left(B^2 + \sigma^q\right),$$

where B and σ are defined in that theorem. Therefore,

$$\left(\mathbb{E}\left\|G:\ell_{p*}^n \to \ell_q^m\right\|^q\right)^{1/q} \leq 2^{1/q}m^{1/q}\left(B^{2/q} + \sigma\right).$$

Now, recall that $T_2(\ell_p^n) \approx \sqrt{p}$ and that B_{p*}^n has modulus of convexity of power type 2 with $\lambda^{-2} \approx 1/p$ (see, e.g., [19, Theorem 5.3]). Therefore,

$$B^{2/q} = C^{2/q}\lambda^{8/q}T_2^{2/q}(\ell_p^n)\left(\frac{\log m}{m}\right)^{1/q}\left(\mathbb{E}\max_{i\leq m}\|X_i\|_p^q\right)^{1/q}$$

$$= C^{2/q}p^{5/q}(\log m)^{1/q}m^{-1/q}\left(\mathbb{E}\max_{i\leq m}\|X_i\|_p^q\right)^{1/q}.$$

Applying Lemma 3.1, we obtain

$$\left(\mathbb{E}\left\|G:\ell_{p*}^n \to \ell_q^m\right\|^q\right)^{1/q}$$

$$\leq (2C^2)^{1/q}\cdot p^{5/q}\cdot(\log m)^{1/q}\left(\mathbb{E}\max_{i\leq m}\|X_i\|_p^q\right)^{1/q}$$

$$+ 2^{1/q}\gamma_q\cdot\max_{j\leq n}\|(a_{ij})_{i=1}^m\|_q.$$

The desired bound follows now from Lemma 3.2. $\qquad\square$

Remark 3.3 This proof can be extended to the case of random matrices whose rows are centered independent vectors with multivariate Gaussian distributions. We leave the details to the interested reader.

4 Concluding Remarks

In this section, we briefly outline what can be obtained using the approach of [24]. We use a standard trick to pass to a symmetric matrix. The matrix G_A being given, define S as

$$S = \frac{1}{2}\begin{pmatrix} 0 & G_A^T \\ G_A & 0 \end{pmatrix}.$$

Then, S is a random symmetric matrix and

$$\sup_w \langle Sw, w \rangle = \sup_{u \in B_{p*}^n} \sup_{v \in B_{q*}^m} \langle G_A u, v \rangle = \| G_A : \ell_{p*}^n \to \ell_q^m \|,$$

where the supremum in w is taken over all vectors of the form $(u, v)^T$ with $u \in B_{p*}^n$ and $v \in B_{q*}^m$. Repeating verbatim the proof of Theorem 4.1 in [24] one gets

$$\mathbb{E} \| G_A : \ell_{p*}^n \to \ell_q^m \| \lesssim_{p,q} \mathbb{E} \max_{i \le m} \left(\sum_{j=1}^n |g_j|^p |a_{ij}|^p \right)^{1/p}$$

$$+ \mathbb{E} \max_{j \le n} \left(\sum_{i=1}^m |g_i|^q |a_{ij}|^g \right)^{1/q} + \mathbb{E} \max_i Y_i,$$

where $Y \sim N(0, A^-)$ and A^- is a positive definite matrix whose diagonal elements are bounded by

$$\max \left(\max_{i \le m} \sqrt{\sum_j a_{ij}^4}, \max_{j \le n} \sqrt{\sum_i a_{ij}^4} \right).$$

However, the bounds obtained here and in Theorem 1.1 are incomparable. Depending on the situation one may be better than the other.

Acknowledgements Part of this work was done while Alexander E. Litvak visited Joscha Prochno at the Johannes Kepler University in Linz (supported by FWFM 1628000). Alexander E. Litvak thanks the support of the Bézout Research Foundation (Labex Bézout) for the invitation to the University Marne la Vallée (France).

We would also like to thank our colleague R. Adamczak for helpful comments. We are grateful to the anonymous referee for many useful comments and remarks helping us to improve the presentation as well as for showing the argument outlined in the last section.

J. Prochno was supported in parts by the Austrian Science Fund, FWFM 1628000.

References

1. A. Bandeira, R. van Handel, Sharp nonasymptotic bounds on the norm of random matrices with independent entries. Ann. Probab. **44**, 2479–2506 (2016)
2. G. Bennett, V. Goodman, C.M. Newman, Norms of random matrices. Pac. J. Math. **59**(2), 359–365 (1975)
3. Y. Benyamini, Y. Gordon, Random factorization of operators between Banach spaces. J. Anal. Math. **39**, 45–74 (1981)
4. S. Chevet, Séries de variables aléatoires gaussiennes à valeurs dans $E \hat{\otimes}_\varepsilon F$. Application aux produits d'espaces de Wiener abstraits, in *Séminaire sur la Géométrie des Espaces de Banach (1977–1978)*, pages Exp. No. 19, 15 (École Polytech., Palaiseau, 1978)
5. B.S. Cirel'son, I.A. Ibragimov, V.N. Sudakov, Norms of Gaussian sample functions, in *Proceedings of the Third Japan-USSR Symposium on Probability Theory (Tashkent, 1975)*. Lecture Notes in Mathematics, vol. 550 (Springer, Berlin, 1976), pp. 20–41
6. K.R. Davidson, S.J. Szarek, Addenda and corrigenda to: "Local operator theory, random matrices and Banach spaces", in *Handbook of the Geometry of Banach Spaces*, vol. 2 (North-Holland, Amsterdam, 2003)
7. K.R. Davidson, S.J. Szarek, Local operator theory, random matrices and Banach spaces, in *Handbook of the Geometry of Banach Spaces*, vol. 1 (North-Holland, Amsterdam, 2003)
8. T. Figiel, On the moduli of convexity and smoothness. Stud. Math. **56**(2), 121–155 (1976)
9. Y. Gordon, Some inequalities for gaussian processes and applications. Isr. J. Math. **50**, 265–289 (1985)
10. Y. Gordon, A.E. Litvak, C. Schütt, E. Werner, Orlicz norms of sequences of random variables. Ann. Probab. **30**(4), 1833–1853 (2002)
11. Y. Gordon, A.E. Litvak, C. Schütt, E. Werner, Uniform estimates for order statistics and Orlicz functions. Positivity **16**(1), 1–28 (2012)
12. O. Guédon, S. Mendelson, A. Pajor, N. Tomczak-Jaegermann, Majorizing measures and proportional subsets of bounded orthonormal systems. Rev. Mat. Iberoam. **24**(3), 1075–1095 (2008)
13. O. Guédon, M. Rudelson, Moments of random vectors via majorizing measures. Adv. Math. **208**(2), 798–823 (2007)
14. J. Hoffmann-Jørgensen, Sums of independent Banach space valued random variables. Stud. Math. **52**, 159–186 (1974)
15. R. Latała, Some estimates of norms of random matrices. Proc. Am. Math. Soc. **133**(5), 1273–1282 (electronic) (2005)
16. M. Ledoux, *The Concentration of Measure Phenomenon*. Mathematical Surveys and Monographs, vol. 89 (American Mathematical Society, Providence, RI, 2001)
17. B. Maurey, G. Pisier, Séries de variables aléatoires vectorielles indépendantes et propriétés géométriques des espaces de Banach. Stud. Math. **58**(1), 45–90 (1976)
18. G. Pisier, Martingales with values in uniformly convex spaces. Isr. J. Math. **20**(3–4), 326–350 (1975)
19. G. Pisier, Q. Xu, Non-commutative L^p-spaces, in *Handbook of the Geometry of Banach Spaces*, vol. 2 (North-Holland, Amsterdam, 2003), pp. 1459–1517
20. S. Riemer, C. Schütt, On the expectation of the norm of random matrices with non-identically distributed entries. Electron. J. Probab. **18**(29), 1–13 (2013)

21. M. Rudelson, O. Zeitouni, Singular values of gaussian matrices and permanent estimators. Random Struct. Algorithm. **48**, 183–212 (2016)
22. Y. Seginer, The expected norm of random matrices. Combin. Probab. Comput. **9**, 149–166 (2000)
23. J.A. Tropp, User-friendly tail bounds for sums of random matrices. Found. Comput. Math. **12**(4), 389–434 (2012)
24. R. Van Handel, On the spectral norm of gaussian random matrices. Trans. Am. Math. Soc. (to appear)
25. J. von Neumann, H.H. Goldstine, Numerical inverting of matrices of high order. Bull. Am. Math. Soc. **53**(11), 1021–1099 (1947)
26. E.P. Wigner, Characteristic vectors of bordered matrices with infinite dimensions. Ann. Math. **62**(3), 548–564 (1955)
27. E.P. Wigner, On the distribution of the roots of certain symmetric matrices. Ann. Math. **67**(2), 325–327 (1958)
28. J. Wishart, The generalised product moment distribution in samples from a normal multivariate population. Biometrika **20A**(1/2), 32–52 (1928)

The Restricted Isometry Property of Subsampled Fourier Matrices

Ishay Haviv and Oded Regev

Abstract A matrix $A \in \mathbb{C}^{q \times N}$ satisfies the *restricted isometry property* of order k with constant ϵ if it preserves the ℓ_2 norm of all k-sparse vectors up to a factor of $1 \pm \epsilon$. We prove that a matrix A obtained by randomly sampling $q = O(k \cdot \log^2 k \cdot \log N)$ rows from an $N \times N$ Fourier matrix satisfies the restricted isometry property of order k with a fixed ϵ with high probability. This improves on Rudelson and Vershynin (Comm Pure Appl Math, 2008), its subsequent improvements, and Bourgain (GAFA Seminar Notes, 2014).

1 Introduction

A matrix $A \in \mathbb{C}^{q \times N}$ satisfies the *restricted isometry property* of order k with constant $\epsilon > 0$ if for every k-sparse vector $x \in \mathbb{C}^N$ (i.e., a vector with at most k nonzero entries), it holds that

$$(1 - \epsilon) \cdot \|x\|_2^2 \leq \|Ax\|_2^2 \leq (1 + \epsilon) \cdot \|x\|_2^2 . \tag{1}$$

Intuitively, this means that every k columns of A are nearly orthogonal. This notion, due to Candès and Tao [9], was intensively studied during the last decade and found various applications and connections to several areas of theoretical computer science, including sparse recovery [8, 20, 27], coding theory [14], norm embeddings [6, 22], and computational complexity [4, 25, 31].

The original motivation for the restricted isometry property comes from the area of compressed sensing. There, one wishes to compress a high-dimensional sparse vector $x \in \mathbb{C}^N$ to a vector Ax, where $A \in \mathbb{C}^{q \times N}$ is a measurement matrix that enables

A preliminary version appeared in Proceedings of the 27th Annual ACM-SIAM Symposium on Discrete Algorithms (SODA), 2016, pages 288–297.

I. Haviv (✉)
School of Computer Science, The Academic College of Tel Aviv-Yaffo, Tel Aviv 61083, Israel

O. Regev
Courant Institute of Mathematical Sciences, New York University, New York, NY, USA

© Springer International Publishing AG 2017
B. Klartag, E. Milman (eds.), *Geometric Aspects of Functional Analysis*,
Lecture Notes in Mathematics 2169, DOI 10.1007/978-3-319-45282-1_11

reconstruction of x from Ax. Typical goals in this context include minimizing the number of measurements q and the running time of the reconstruction algorithm. It is known that the restricted isometry property of A, for $\epsilon < \sqrt{2} - 1$, is a sufficient condition for reconstruction. In fact, it was shown in [8, 9, 11, 12] that under this condition, reconstruction is equivalent to finding the vector of least ℓ_1 norm among all vectors that agree with the given measurements, a task that can be formulated as a linear program [13, 16], and thus can be solved efficiently.

The above application leads to the challenge of finding matrices $A \in \mathbb{C}^{q \times N}$ that satisfy the restricted isometry property and have a small number of rows q as a function of N and k. (For simplicity, we ignore for now the dependence on ϵ.) A general lower bound of $q = \Omega(k \cdot \log(N/k))$ is known to follow from [18] (see also [17]). Fortunately, there are matrices that match this lower bound, e.g., random matrices whose entries are chosen independently according to the normal distribution [10]. However, in many applications the measurement matrix cannot be chosen arbitrarily but is instead given by a random sample of rows from a unitary matrix, typically the discrete Fourier transform. This includes, for instance, various tests and experiments in medicine and biology (e.g., MRI [28] and ultrasound imaging [21]) and applications in astronomy (e.g., radio telescopes [32]). An advantage of subsampled Fourier matrices is that they support fast matrix-vector multiplication, and as such, are useful for efficient compression as well as for efficient reconstruction based on iterative methods (see, e.g., [26]).

In recent years, with motivation from both theory and practice, an intensive line of research has aimed to study the restricted isometry property of random submatrices of unitary matrices. Letting $A \in \mathbb{C}^{q \times N}$ be a (normalized) matrix whose rows are chosen uniformly and independently from the rows of a unitary matrix $M \in \mathbb{C}^{N \times N}$, the goal is to prove an upper bound on q for which A is guaranteed to satisfy the restricted isometry property with high probability. Note that the fact that the entries of every row of A are not independent makes this question much more difficult than in the case of random matrices with independent entries.

The first upper bound on the number of rows of a subsampled Fourier matrix that satisfies the restricted isometry property was $O(k \cdot \log^6 N)$, which was proved by Candès and Tao [10]. This was then improved by Rudelson and Vershynin [30] to $O(k \cdot \log^2 k \cdot \log(k \log N) \cdot \log N)$ (see also [15, 29] for a simplified analysis with better success probability). A modification of their analysis led to an improved bound of $O(k \cdot \log^3 k \cdot \log N)$ by Cheraghchi, Guruswami, and Velingker [14], who related the problem to a question on the list-decoding rate of random linear codes over finite fields. Interestingly, replacing the $\log(k \log N)$ term in the bound of [30] by $\log k$ was crucial for their application.[1] Recently, Bourgain [7] proved a bound of $O(k \cdot \log k \cdot \log^2 N)$, which is incomparable to those of [14, 30] (and has a worse dependence on ϵ; see below). We finally mention that the best known lower bound on the number of rows is $\Omega(k \cdot \log N)$ [5].

---

[1]Note that the list-decoding result of [14] was later improved by Wootters [33] using different techniques.

1.1 Our Contribution

In this work, we improve the previous bounds and prove the following.

Theorem 1.1 (Simplified) *Let $M \in \mathbb{C}^{N \times N}$ be a unitary matrix with entries of absolute value $O(1/\sqrt{N})$, and let $\epsilon > 0$ be a fixed constant. For some $q = O(k \cdot \log^2 k \cdot \log N)$, let $A \in \mathbb{C}^{q \times N}$ be a matrix whose q rows are chosen uniformly and independently from the rows of M, multiplied by $\sqrt{N/q}$. Then, with high probability, the matrix A satisfies the restricted isometry property of order k with constant ϵ.*

The main idea in our proof is described in Sect. 1.3. We arrived at the proof from our recent work on list-decoding [19], where a baby version of the idea was used to bound the sample complexity of learning the class of Fourier-sparse Boolean functions.[2] Like all previous work on this question, our proof can be seen as a careful union bound applied to a sequence of progressively finer nets, a technique sometimes known as chaining. However, unlike the work of Rudelson and Vershynin [30] and its improvements [14, 15], we avoid the use of Gaussian processes, the "symmetrization process," and Dudley's inequality. Instead, we follow and refine Bourgain's proof [7], and apply the chaining argument directly to the problem at hand using only elementary arguments. It would be interesting to see if our proof can be cast in the Gaussian framework of Rudelson and Vershynin.

We remark that the bounds obtained in the previous works [14, 30] have a multiplicative $O(\epsilon^{-2})$ term, whereas a much worse term of $O(\epsilon^{-6})$ was obtained in [7]. In our proof of Theorem 1.1 we nearly obtain the best known dependence on ϵ. For simplicity of presentation we first prove in Sect. 3 our bound with a weaker multiplicative term of $O(\epsilon^{-4})$, and then, in Sect. 4, we modify the analysis and decrease the dependence on ϵ to $O(\epsilon^{-2})$ up to logarithmic terms.

1.2 Related Literature

As mentioned before, one important advantage of using subsampled Fourier matrices in compressed sensing is that they support fast, in fact nearly linear time, matrix-vector multiplication. In certain scenarios, however, one is not restricted to using subsampled Fourier matrices as the measurement matrix. The question then is whether one can decrease the number of rows using another measurement matrix, while still keeping the near-linear multiplication time. For $k < N^{1/2-\gamma}$ where $\gamma > 0$ is an arbitrary constant, the answer is yes: a construction with the *optimal* number

---

[2]The result in [19] is weaker in two main respects. First, it is restricted to the case that Ax is in $\{0, 1\}^q$. This significantly simplifies the analysis and leads to a better bound on the number of rows of A. Second, the order of quantifiers is switched, namely it shows that for any sparse x, a random subsampled A works with high probability, whereas for the restricted isometry property we need to show that a random A works for all sparse x.

$O(k \cdot \log N)$ of rows follows from works by Ailon and Chazelle [1] and Ailon and Liberty [2] (see [6]). For general k, Nelson, Price, and Wootters [27] suggested taking subsampled Fourier matrices and "tweaking" them by bunching together rows with random signs. Using the Gaussian-process-based analysis of [14, 30] and introducing further techniques from [23], they showed that with this construction one can reduce the number of rows by a logarithmic factor to $O(k \cdot \log^2(k \log N) \cdot \log N)$ while still keeping the nearly linear multiplication time. Our result shows that the same number of rows (in fact, a slightly smaller number) can be achieved already with the original subsampled Fourier matrices without having to use the "tweak." A natural open question is whether the "tweak" from [27] and their techniques can be combined with ours to further reduce the number of rows. An improvement in the regime of parameters of $k = \omega(\sqrt{N})$ would lead to more efficient low-dimensional embeddings based on Johnson–Lindenstrauss matrices (see, e.g., [1–3, 22, 27]).

1.3 Proof Overview

Recall from Theorem 1.1 and from (1) that our goal is to prove that a matrix A given by a random sample Q of q rows of M satisfies with high probability that for all k-sparse x, $\|Ax\|_2^2 \approx \|x\|_2^2$. Since M is unitary, the latter is equivalent to saying that $\|Ax\|_2^2 \approx \|Mx\|_2^2$. Yet another way of expressing this condition is as

$$\mathop{\mathbb{E}}_{j \in Q} \left[(|Mx|^2)_j \right] \approx \mathop{\mathbb{E}}_{j \in [N]} \left[(|Mx|^2)_j \right] ,$$

i.e., that a sample $Q \subseteq [N]$ of q coordinates of the vector $|Mx|^2$ gives a good approximation to the average of all its coordinates. Here, $|Mx|^2$ refers to the vector obtained by taking the squared absolute value of Mx coordinate-wise. For reasons that will become clear soon, it will be convenient to assume without loss of generality that $\|x\|_1 = 1$. With this scaling, the sparsity assumption implies that $\|Mx\|_2^2$ is not too small (namely at least $1/k$), and this will determine the amount of additive error we can afford in the approximation above. This is the only way we use the sparsity assumption.

At a high level, the proof proceeds by defining a finite set of vectors $\mathcal{H}$ that forms a *net*, i.e., a set satisfying that any vector $|Mx|^2$ is close to one of the vectors in $\mathcal{H}$. We then argue using the Chernoff-Hoeffding bound that for any fixed vector $h \in \mathcal{H}$, a sample of q coordinates gives a good approximation to the average of h. Finally, we complete the proof by a union bound over all $h \in \mathcal{H}$.

In order to define the set $\mathcal{H}$ we notice that since $\|x\|_1 = 1$, Mx can be seen as a weighted average of the columns of M (possibly with signs). In other words, we can think of Mx as the *expectation* of a vector-valued random variable given by a certain probability distribution over the columns of M. Using the Chernoff-Hoeffding bound again, this implies that we can approximate Mx well by taking the average over a small number of samples from this distribution. We then let $\mathcal{H}$ be the set of all possible such averages, and a bound on the cardinality of $\mathcal{H}$ follows easily (basically N raised to the number of samples). This technique is sometimes referred to as Maurey's empirical method.

The argument above is actually oversimplified, and carrying it out leads to rather bad bounds on q. As a result, our proof in Sect. 3 is slightly more delicate. Namely, instead of just one set $\mathcal{H}$, we have a sequence of sets, $\mathcal{H}_1, \mathcal{H}_2, \ldots$, each being responsible for approximating a different scale of $|Mx|^2$. The first set $\mathcal{H}_1$ approximates $|Mx|^2$ on coordinates on which its value is highest; since the value is high, we need less samples in order to approximate it well, as a result of which the set $\mathcal{H}_1$ is small. The next set $\mathcal{H}_2$ approximates $|Mx|^2$ on coordinates on which its value is somewhat smaller, and is therefore a bigger set, and so on and so forth. The end result is that any vector $|Mx|^2$ can be approximately decomposed into a sum $\sum_i h^{(i)}$, with $h^{(i)} \in \mathcal{H}_i$. To complete the proof, we argue that a random choice of q coordinates approximates all the vectors in all the $\mathcal{H}_i$ well. The reason working with several $\mathcal{H}_i$ leads to the better bound stated in Theorem 1.1 is this: even though as i increases the number of vectors in $\mathcal{H}_i$ grows, the quality of approximation that we need the q coordinates to provide decreases, since the value of $|Mx|^2$ there is small and so errors are less significant. It turns out that these two requirements on q balance each other perfectly, leading to the desired bound on q.

2 Preliminaries

Notation The notation $x \approx_{\epsilon, \alpha} y$ means that $x \in [(1 - \epsilon)y - \alpha, (1 + \epsilon)y + \alpha]$. For a matrix M, we denote by $M^{(\ell)}$ the ℓth column of M and define $\|M\|_\infty = \max_{i,j} |M_{i,j}|$.

The Restricted Isometry Property The restricted isometry property is defined as follows.

Definition 2.1 We say that a matrix $A \in \mathbb{C}^{q \times N}$ satisfies the *restricted isometry property* of order k with constant ϵ if for every k-sparse vector $x \in \mathbb{C}^N$ it holds that

$$(1 - \epsilon) \cdot \|x\|_2^2 \leq \|Ax\|_2^2 \leq (1 + \epsilon) \cdot \|x\|_2^2.$$

Chernoff-Hoeffding Bounds We now state the Chernoff-Hoeffding bound (see, e.g., [24]) and derive several simple corollaries that will be used extensively later.

Theorem 2.2 *Let $X_1, \ldots, X_N$ be N identically distributed independent random variables in $[0, a]$ satisfying $\mathbb{E}[X_i] = \mu$ for all i, and denote $\overline{X} = \frac{1}{N} \cdot \sum_{i=1}^N X_i$. Then there exists a universal constant C such that for every $0 < \epsilon \leq 1/2$, the probability that $\overline{X} \approx_{\epsilon, 0} \mu$ is at least $1 - 2e^{-C \cdot N \mu \epsilon^2 / a}$.*

Corollary 2.3 *Let $X_1, \ldots, X_N$ be N identically distributed independent random variables in $[0, a]$ satisfying $\mathbb{E}[X_i] = \mu$ for all i, and denote $\overline{X} = \frac{1}{N} \cdot \sum_{i=1}^N X_i$.*

Then there exists a universal constant C such that for every $0 < \epsilon \leq 1/2$ and $\alpha > 0$, the probability that $\overline{X} \approx_{\epsilon,\alpha} \mu$ is at least $1 - 2e^{-C \cdot N\alpha\epsilon/a}$.

Proof If $\mu \geq \frac{\alpha}{\epsilon}$ then by Theorem 2.2 the probability that $\overline{X} \approx_{\epsilon,0} \mu$ is at least $1 - 2e^{-C \cdot N\mu\epsilon^2/a}$, which is at least $1 - 2e^{-C \cdot N\alpha\epsilon/a}$. Otherwise, Theorem 2.2 for $\tilde{\epsilon} = \frac{\alpha}{\mu} > \epsilon$ implies that the probability that $\overline{X} \approx_{\tilde{\epsilon},0} \mu$, hence $\overline{X} \approx_{0,\alpha} \mu$, is at least $1 - 2e^{-C \cdot N\mu\tilde{\epsilon}^2/a}$, and the latter is at least $1 - 2e^{-C \cdot N\alpha\epsilon/a}$. ∎

Corollary 2.4 *Let $X_1, \ldots, X_N$ be N identically distributed independent random variables in $[-a, +a]$ satisfying $\mathbb{E}[X_i] = \mu$ and $\mathbb{E}[|X_i|] = \tilde{\mu}$ for all i, and denote $\overline{X} = \frac{1}{N} \cdot \sum_{i=1}^{N} X_i$. Then there exists a universal constant C such that for every $0 < \epsilon' \leq 1/2$ and $\alpha > 0$, the probability that $\overline{X} \approx_{0, \epsilon' \cdot \tilde{\mu} + \alpha} \mu$ is at least $1 - 4e^{-C \cdot N\alpha\epsilon'/a}$.*

Proof The corollary follows by applying Corollary 2.3 to $\max(X_i, 0)$ and to $-\min(X_i, 0)$. ∎

We end with the additive form of the bound, followed by an easy extension to the complex case.

Corollary 2.5 *Let $X_1, \ldots, X_N$ be N identically distributed independent random variables in $[-a, +a]$ satisfying $\mathbb{E}[X_i] = \mu$ for all i, and denote $\overline{X} = \frac{1}{N} \cdot \sum_{i=1}^{N} X_i$. Then there exists a universal constant C such that for every $b > 0$, the probability that $\overline{X} \approx_{0,b} \mu$ is at least $1 - 4e^{-C \cdot Nb^2/a^2}$.*

Proof We can assume that $b \leq 2a$. The corollary follows by applying Corollary 2.4 to, say, $\alpha = 3b/4$ and $\epsilon' = b/(4a)$. ∎

Corollary 2.6 *Let $X_1, \ldots, X_N$ be N identically distributed independent complex-valued random variables satisfying $|X_i| \leq a$ and $\mathbb{E}[X_i] = \mu$ for all i, and denote $\overline{X} = \frac{1}{N} \cdot \sum_{i=1}^{N} X_i$. Then there exists a universal constant C such that for every $b > 0$, the probability that $|\overline{X}| \approx_{0,b} |\mu|$ is at least $1 - 8e^{-C \cdot Nb^2/a^2}$.*

Proof By Corollary 2.5 applied to the real and imaginary parts of the random variables $X_1, \ldots, X_N$ it follows that for a universal constant C, the probability that $\mathsf{Re}(\overline{X}) \approx_{0, b/\sqrt{2}} \mathsf{Re}(\mu)$ and $\mathsf{Im}(\overline{X}) \approx_{0, b/\sqrt{2}} \mathsf{Im}(\mu)$ is at least $1 - 8e^{-C \cdot Nb^2/a^2}$. By triangle inequality, it follows that with such probability we have $|\overline{X}| \approx_{0,b} |\mu|$, as required. ∎

3 The Simpler Analysis

In this section we prove our result with a multiplicative term of $O(\epsilon^{-4})$ in the bound. This will be obtained in Theorem 3.7 as an easy corollary of the following theorem.

Theorem 3.1 *For a sufficiently large N, a matrix $M \in \mathbb{C}^{N \times N}$, and sufficiently small $\epsilon, \eta > 0$, the following holds. For some $q = O(\epsilon^{-3}\eta^{-1} \log N \cdot \log^2(1/\eta))$, let Q*

be a multiset of q uniform and independent random elements of $[N]$. *Then, with probability* $1 - 2^{-\Omega(\epsilon^{-2} \cdot \log N \cdot \log(1/\eta))}$, *it holds that for every* $x \in \mathbb{C}^N$,

$$\mathop{\mathbb{E}}_{j \in Q} \left[|(Mx)_j|^2 \right] \approx_{\epsilon, \eta \cdot \|x\|_1^2 \cdot \|M\|_\infty^2} \mathop{\mathbb{E}}_{j \in [N]} \left[|(Mx)_j|^2 \right].$$

Throughout the proof we assume without loss of generality that the matrix $M \in \mathbb{C}^{N \times N}$ satisfies $\|M\|_\infty = 1$. For $\epsilon, \eta > 0$, we denote $t = \log_2(1/\eta)$, $r = \log_2(1/\epsilon^2)$, and $\gamma = \eta/(2t)$.

We now define the approximating vector sets $\mathcal{H}_i$, $i = 1, \ldots, t$, each responsible for coordinates of $|Mx|^2$ of a different scale (the larger the i the smaller the scale). We start by defining the "raw approximations" $\mathcal{G}_i$, which are essentially vectors obtained by averaging a certain number of columns of M. We then define the vectors in $\mathcal{H}_i$ by restricting the vectors in $\mathcal{G}_i$ (actually $\mathcal{G}_{i+r}$) to the set of coordinates B_i where there is a clear "signal" and not just noise. This is necessary in order to make sure that the small coordinates of $|Mx|^2$ are not flooded by noise from the coarse approximations. Details follow.

The Vector Sets $\mathcal{G}_i$ For every $1 \leq i \leq t + r$, let $\mathcal{G}_i$ denote the set of all vectors $g^{(i)} \in \mathbb{C}^N$ that can be represented as

$$g^{(i)} = \frac{\sqrt{2}}{|F|} \cdot \sum_{(\ell, s) \in F} (-1)^{s/2} \cdot M^{(\ell)} \tag{2}$$

for a multiset F of $O(2^i \cdot \log(1/\gamma))$ pairs in $[N] \times \{0, 1, 2, 3\}$. A trivial counting argument gives the following.

Claim 3.2 *For every* $1 \leq i \leq t + r$, $|\mathcal{G}_i| \leq N^{O(2^i \cdot \log(1/\gamma))}$.

The Vector Sets $\mathcal{H}_i$ For a t-tuple of vectors $(g^{(1+r)}, \ldots, g^{(t+r)}) \in \mathcal{G}_{1+r} \times \cdots \times \mathcal{G}_{t+r}$ and for $1 \leq i \leq t$, let B_i be the set of all $j \in [N]$ for which i is the smallest index satisfying $|g_j^{(i+r)}| \geq 2 \cdot 2^{-i/2}$. For such i, define the vector $h^{(i)}$ by

$$h_j^{(i)} = \min(|g_j^{(i+r)}|^2 \cdot \mathbf{1}_{j \in B_i}, 9 \cdot 2^{-i}). \tag{3}$$

Let $\mathcal{H}_i$ be the set of all vectors $h^{(i)}$ that can be obtained in this way.

Claim 3.3 *For every* $1 \leq i \leq t$, $|\mathcal{H}_i| \leq N^{O(\epsilon^{-2} \cdot 2^i \cdot \log(1/\gamma))}$.

Proof Observe that every $h^{(i)} \in \mathcal{H}_i$ is fully defined by some $(g^{(1+r)}, \ldots, g^{(i+r)}) \in \mathcal{G}_{1+r} \times \cdots \times \mathcal{G}_{i+r}$. Hence

$$|\mathcal{H}_i| \leq |\mathcal{G}_{1+r}| \cdots |\mathcal{G}_{i+r}| \leq N^{O(\log(1/\gamma)) \cdot (2^{1+r} + 2^{2+r} + \cdots + 2^{i+r})} \leq N^{O(\log(1/\gamma)) \cdot 2^{i+r+1}}.$$

Using the definition of r, the claim follows. ∎

Lemma 3.4 *For every $\tilde{\eta} > 0$ and some $q = O(\epsilon^{-3}\tilde{\eta}^{-1}\log N \cdot \log(1/\gamma))$, let Q be a multiset of q uniform and independent random elements of $[N]$. Then, with probability $1 - 2^{-\Omega(\epsilon^{-2}\cdot\log N \cdot\log(1/\gamma))}$, it holds that for all $1 \le i \le t$ and $h^{(i)} \in \mathcal{H}_i$,*

$$\mathop{\mathbb{E}}_{j \in Q}\left[h_j^{(i)}\right] \approx_{\epsilon,\tilde{\eta}} \mathop{\mathbb{E}}_{j \in [N]}\left[h_j^{(i)}\right].$$

Proof Fix an $1 \le i \le t$ and a vector $h^{(i)} \in \mathcal{H}_i$, and denote $\mu = \mathbb{E}_{j \in [N]}[h_j^{(i)}]$. By Corollary 2.3, applied with $\alpha = \tilde{\eta}$ and $a = 9 \cdot 2^{-i}$ (recall that $h_j^{(i)} \le a$ for every j), with probability $1 - 2^{-\Omega(2^i \cdot q \epsilon \tilde{\eta})}$, it holds that $\mathbb{E}_{j \in Q}[h_j^{(i)}] \approx_{\epsilon,\tilde{\eta}} \mu$. Using Claim 3.3, the union bound over all the vectors in $\mathcal{H}_i$ implies that the probability that some $h^{(i)} \in \mathcal{H}_i$ does not satisfy $\mathbb{E}_{j \in Q}[h_j^{(i)}] \approx_{\epsilon,\tilde{\eta}} \mu$ is at most

$$N^{O(\epsilon^{-2}\cdot 2^i \cdot \log(1/\gamma))} \cdot 2^{-\Omega(2^i \cdot q\epsilon\tilde{\eta})} \le 2^{-\Omega(\epsilon^{-2}\cdot 2^i \cdot \log N \cdot \log(1/\gamma))}.$$

We complete the proof by a union bound over i.   ∎

Approximating the Vectors *Mx*

Lemma 3.5 *For every vector $x \in \mathbb{C}^N$ with $\|x\|_1 = 1$, every multiset $Q \subseteq [N]$, and every $1 \le i \le t + r$, there exists a vector $g \in \mathcal{G}_i$ that satisfies $|(Mx)_j| \approx_{0,2^{-i/2}} |g_j|$ for all but at most γ fraction of $j \in [N]$ and for all but at most γ fraction of $j \in Q$.*

Proof Observe that for every $\ell \in [N]$ there exist $p_{\ell,0}, p_{\ell,1}, p_{\ell,2}, p_{\ell,3} \ge 0$ that satisfy

$$\sum_{s=0}^{3} p_{\ell,s} = |x_\ell| \quad \text{and} \quad \sqrt{2} \cdot \sum_{s=0}^{3} p_{\ell,s} \cdot (-1)^{s/2} = x_\ell.$$

Notice that the assumption $\|x\|_1 = 1$ implies that the numbers $p_{\ell,s}$ form a probability distribution. Thus, the vector Mx can be represented as

$$Mx = \sum_{\ell=1}^{N} x_\ell \cdot M^{(\ell)} = \sqrt{2} \cdot \sum_{\ell=1}^{N} \sum_{s=0}^{3} p_{\ell,s} \cdot (-1)^{s/2} \cdot M^{(\ell)} = \mathop{\mathbb{E}}_{(\ell,s)\sim D}[\sqrt{2} \cdot (-1)^{s/2} \cdot M^{(\ell)}],$$

where D is the distribution that assigns probability $p_{\ell,s}$ to the pair (ℓ, s).

Let F be a multiset of $O(2^i \cdot \log(1/\gamma))$ independent random samples from D, and let $g \in \mathcal{G}_i$ be the vector corresponding to F as in (2). By Corollary 2.6, applied with $a = \sqrt{2}$ (recall that $\|M\|_\infty = 1$) and $b = 2^{-i/2}$, for every $j \in [N]$ the probability that

$$|(Mx)_j| \approx_{0,2^{-i/2}} |g_j| \tag{4}$$

is at least $1 - \gamma/4$. It follows that the expected number of $j \in [N]$ that do not satisfy (4) is at most $\gamma N/4$, so by Markov's inequality the probability that the number of $j \in [N]$ that do not satisfy (4) is at most γN is at least $3/4$. Similarly, the expected number of $j \in Q$ that do not satisfy (4) is at most $\gamma|Q|/4$, so by Markov's inequality, with probability at least $3/4$ it holds that the number of $j \in Q$ that do not satisfy (4) is at most $\gamma|Q|$. It follows that there exists a vector $g \in \mathcal{G}_i$ for which (4) holds for all but at most γ fraction of $j \in [N]$ and for all but at most γ fraction of $j \in Q$, as required. ∎

Lemma 3.6 *For every multiset $Q \subseteq [N]$ and every vector $x \in \mathbb{C}^N$ with $\|x\|_1 = 1$ there exists a t-tuple of vectors $(h^{(1)}, \ldots, h^{(t)}) \in \mathcal{H}_1 \times \cdots \times \mathcal{H}_t$ for which*

1. $\mathbb{E}_{j \in Q}\left[|(Mx)_j|^2\right] \approx_{O(\epsilon), O(\eta)} \mathbb{E}_{j \in Q}\left[\sum_{i=1}^{t} h_j^{(i)}\right]$ *and*

2. $\mathbb{E}_{j \in [N]}\left[|(Mx)_j|^2\right] \approx_{O(\epsilon), O(\eta)} \mathbb{E}_{j \in [N]}\left[\sum_{i=1}^{t} h_j^{(i)}\right]$.

Proof By Lemma 3.5, for every $1 \le i \le t$ there exists a vector $g^{(i+r)} \in \mathcal{G}_{i+r}$ that satisfies

$$|(Mx)_j| \approx_{0, 2^{-(i+r)/2}} |g_j^{(i+r)}| \tag{5}$$

for all but at most γ fraction of $j \in [N]$ and for all but at most γ fraction of $j \in Q$. We say that $j \in [N]$ is *good* if (5) holds for every $1 \le i \le t$, and otherwise that it is *bad*. Notice that all but at most $t\gamma$ fraction of $j \in [N]$ are good and that all but at most $t\gamma$ fraction of $j \in Q$ are good. Let $(h^{(1)}, \ldots, h^{(t)})$ and $(B_1, \ldots, B_t)$ be the vectors and sets associated with $(g^{(1+r)}, \ldots, g^{(t+r)})$ as defined in (3). We claim that $h^{(1)}, \ldots, h^{(t)}$ satisfy the requirements of the lemma.

We first show that for every good j it holds that $|(Mx)_j|^2 \approx_{3\epsilon, 9\eta} \sum_{i=1}^{t} h_j^{(i)}$. To obtain it, we observe that if $j \in B_i$ for some i, then

$$2 \cdot 2^{-i/2} \le |g_j^{(i+r)}| \le 3 \cdot 2^{-i/2}. \tag{6}$$

The lower bound follows simply from the definition of B_i. For the upper bound, which trivially holds for $i = 1$, assume that $i \ge 2$, and notice that the definition of B_i implies that $|g_j^{(i+r-1)}| < 2 \cdot 2^{-(i-1)/2}$. Using (5), and assuming that ϵ is sufficiently small, we obtain that

$$|g_j^{(i+r)}| \le |(Mx)_j| + 2^{-(i+r)/2} \le |g_j^{(i+r-1)}| + 2^{-(i+r-1)/2} + 2^{-(i+r)/2}$$
$$\le 2^{-i/2}(2^{3/2} + 2^{1/2} \cdot \epsilon + \epsilon) \le 3 \cdot 2^{-i/2}.$$

Hence, by the upper bound in (6), for a good $j \in B_i$ we have $h_j^{(i)} = |g_j^{(i+r)}|^2$ and $h_j^{(i')} = 0$ for $i' \ne i$. Observe that by the lower bound in (6),

$$|(Mx)_j| \in [|g_j^{(i+r)}| - 2^{-(i+r)/2}, |g_j^{(i+r)}| + 2^{-(i+r)/2}] \subseteq [(1-\epsilon) \cdot |g_j^{(i+r)}|, (1+\epsilon) \cdot |g_j^{(i+r)}|],$$

and that this implies that $|(Mx)_j|^2 \approx_{3\epsilon,0} \sum_{i=1}^{t} h_j^{(i)}$. On the other hand, in case that j is good but does not belong to any B_i, recalling that $t = \log_2(1/\eta)$, it follows that

$$|(Mx)_j| \leq |g_j^{(t+r)}| + 2^{-(t+r)/2} \leq 2 \cdot 2^{-t/2} + 2^{-(t+r)/2} \leq 3 \cdot 2^{-t/2} \leq 3\sqrt{\eta},$$

and thus $|(Mx)_j|^2 \approx_{0,9\eta} 0 = \sum_{i=1}^{t} h_j^{(i)}$.

Finally, for every bad j we have

$$\left| |(Mx)_j|^2 - \sum_{i=1}^{t} h_j^{(i)} \right| \leq \max\left(|(Mx)_j|^2, \sum_{i=1}^{t} h_j^{(i)} \right) \leq 2.$$

Since at most $t\gamma$ fraction of the elements in $[N]$ and in Q are bad, their effect on the difference between the expectations in the lemma can be bounded by $2t\gamma$. By our choice of γ, this is η, completing the proof of the lemma. ∎

Finally, we are ready to prove Theorem 3.1.

Proof of Theorem 3.1 By Lemma 3.4, applied with $\bar{\eta} = \eta/(2t)$, a random multiset Q of size

$$q = O\left(\epsilon^{-3}\eta^{-1} \cdot t \cdot \log N \cdot \log(1/\gamma)\right) = O\left(\epsilon^{-3}\eta^{-1} \log N \cdot \log^2(1/\eta)\right)$$

satisfies with probability $1 - 2^{-\Omega(\epsilon^{-2} \cdot \log N \cdot \log(1/\eta))}$ that for all $1 \leq i \leq t$ and $h^{(i)} \in \mathcal{H}_i$,

$$\mathop{\mathbb{E}}_{j \in Q}\left[h_j^{(i)} \right] \approx_{\epsilon,\eta/t} \mathop{\mathbb{E}}_{j \in [N]}\left[h_j^{(i)} \right],$$

in which case we also have

$$\mathop{\mathbb{E}}_{j \in Q}\left[\sum_{i=1}^{t} h_j^{(i)} \right] \approx_{\epsilon,\eta} \mathop{\mathbb{E}}_{j \in [N]}\left[\sum_{i=1}^{t} h_j^{(i)} \right].$$

We show that a Q with the above property satisfies the requirement of the theorem. Let $x \in \mathbb{C}^N$ be a vector, and assume without loss of generality that $\|x\|_1 = 1$. By Lemma 3.6, there exists a t-tuple of vectors $(h^{(1)}, \ldots, h^{(t)}) \in \mathcal{H}_1 \times \cdots \times \mathcal{H}_t$ satisfying Items 1 and 2 there. As a result,

$$\mathop{\mathbb{E}}_{j \in Q}\left[|(Mx)_j|^2 \right] \approx_{O(\epsilon),O(\eta)} \mathop{\mathbb{E}}_{j \in [N]}\left[|(Mx)_j|^2 \right],$$

and we are done. ∎

3.1 The Restricted Isometry Property

Equipped with Theorem 3.1, it is easy to derive our result on the restricted isometry property (see Definition 2.1) of random sub-matrices of unitary matrices.

Theorem 3.7 *For sufficiently large N and k, a unitary matrix $M \in \mathbb{C}^{N \times N}$ satisfying $\|M\|_\infty \leq O(1/\sqrt{N})$, and a sufficiently small $\epsilon > 0$, the following holds. For some $q = O(\epsilon^{-4} \cdot k \cdot \log^2(k/\epsilon) \cdot \log N)$, let $A \in \mathbb{C}^{q \times N}$ be a matrix whose q rows are chosen uniformly and independently from the rows of M, multiplied by $\sqrt{N/q}$. Then, with probability $1 - 2^{-\Omega(\epsilon^{-2} \cdot \log N \cdot \log(k/\epsilon))}$, the matrix A satisfies the restricted isometry property of order k with constant ϵ.*

Proof Let Q be a multiset of q uniform and independent random elements of $[N]$, defining a matrix A as above. Notice that by the Cauchy-Schwarz inequality, any k-sparse vector $x \in \mathbb{C}^N$ with $\|x\|_2 = 1$ satisfies $\|x\|_1 \leq \sqrt{k}$. Applying Theorem 3.1 with $\epsilon/2$ and some $\eta = \Omega(\epsilon/k)$, we get that with probability $1 - 2^{-\Omega(\epsilon^{-2} \cdot \log N \cdot \log(k/\epsilon))}$, it holds that for every $x \in \mathbb{C}^N$ with $\|x\|_2 = 1$,

$$\|Ax\|_2^2 = N \cdot \mathop{\mathbb{E}}_{j \in Q} \left[|(Mx)_j|^2\right] \approx_{\epsilon/2, \epsilon/2} N \cdot \mathop{\mathbb{E}}_{j \in [N]} \left[|(Mx)_j|^2\right] = \|Mx\|_2^2 = 1 \,.$$

It follows that every vector $x \in \mathbb{C}^N$ satisfies $\|Ax\|_2^2 \approx_{\epsilon, 0} \|x\|_2^2$, hence A satisfies the restricted isometry property of order k with constant ϵ. ∎

4 The Improved Analysis

In this section we prove the following theorem, which improves the bound of Theorem 3.1 in terms of the dependence on ϵ.

Theorem 4.1 *For a sufficiently large N, a matrix $M \in \mathbb{C}^{N \times N}$, and sufficiently small $\epsilon, \eta > 0$, the following holds. For some $q = O(\log^2(1/\epsilon) \cdot \epsilon^{-1} \eta^{-1} \log N \cdot \log^2(1/\eta))$, let Q be a multiset of q uniform and independent random elements of $[N]$. Then, with probability $1 - 2^{-\Omega(\log N \cdot \log(1/\eta))}$, it holds that for every $x \in \mathbb{C}^N$,*

$$\mathop{\mathbb{E}}_{j \in Q} \left[|(Mx)_j|^2\right] \approx_{\epsilon, \eta \cdot \|x\|_1^2 \cdot \|M\|_\infty^2} \mathop{\mathbb{E}}_{j \in [N]} \left[|(Mx)_j|^2\right] \,. \tag{7}$$

We can assume that $\epsilon \geq \eta$, as otherwise, one can apply the theorem with parameters $\eta/2, \eta/2$ and derive (7) for ϵ, η as well (because the right-hand size is bounded from above by $\|x\|_1^2 \cdot \|M\|_\infty^2$). As before, we assume without loss of generality that $\|M\|_\infty = 1$. For $\epsilon \geq \eta > 0$, we define $t = \log_2(1/\eta)$ and $r = \log_2(1/\epsilon^2)$. For the analysis given in this section, we define $\gamma = \eta/(60(t+r))$. Throughout the proof, we use the vector sets $\mathcal{G}_i$ from Sect. 3 and Lemma 3.5 for this value of γ.

The Vector Sets $\mathcal{D}_{i,m}$ For a $(t + r)$-tuple of vectors $(g^{(1)}, \ldots, g^{(t+r)}) \in \mathcal{G}_1 \times \cdots \times \mathcal{G}_{t+r}$ and for $1 \leq i \leq t$, let C_i be the set of all $j \in [N]$ for which i is the smallest index satisfying $|g_j^{(i)}| \geq 2 \cdot 2^{-i/2}$. For $m = i, \ldots, i + r$ define the vector $h^{(i,m)}$ by

$$h_j^{(i,m)} = |g_j^{(m)}|^2 \cdot \mathbf{1}_{j \in C_i}, \tag{8}$$

and for other values of m define $h^{(i,m)} = 0$. Now, for every m, let $\Delta^{(i,m)}$ be the vector defined by

$$\Delta_j^{(i,m)} = \begin{cases} h_j^{(i,m)} - h_j^{(i,m-1)}, & \text{if } |h_j^{(i,m)} - h_j^{(i,m-1)}| \leq 30 \cdot 2^{-(i+m)/2}; \\ 0, & \text{otherwise.} \end{cases} \tag{9}$$

Note that the support of $\Delta^{(i,m)}$ is contained in C_i. Let $\mathcal{D}_{i,m}$ be the set of all vectors $\Delta^{(i,m)}$ that can be obtained in this way.

Claim 4.2 *For every* $1 \leq i \leq t$ *and* $i \leq m \leq i + r$, $|\mathcal{D}_{i,m}| \leq N^{O(2^m \cdot \log(1/\gamma))}$.

Proof Observe that every vector in $\mathcal{D}_{i,m}$ is fully defined by some $(g^{(1)}, \ldots, g^{(m)}) \in \mathcal{G}_1 \times \cdots \times \mathcal{G}_m$. Hence

$$|\mathcal{D}_{i,m}| \leq |\mathcal{G}_1| \cdots |\mathcal{G}_m| \leq N^{O(\log(1/\gamma)) \cdot (2^1 + 2^2 + \cdots + 2^m)} \leq N^{O(\log(1/\gamma)) \cdot 2^{m+1}},$$

and the claim follows. ∎

Lemma 4.3 *For every* $\tilde{\varepsilon}, \tilde{\eta} > 0$ *and some* $q = O(\tilde{\varepsilon}^{-1} \tilde{\eta}^{-1} \log N \cdot \log(1/\gamma))$, *let* Q *be a multiset of* q *uniform and independent random elements of* $[N]$. *Then, with probability* $1 - 2^{-\Omega(\log N \cdot \log(1/\gamma))}$, *it holds that for every* $1 \leq i \leq t$, m, *and a vector* $\Delta^{(i,m)} \in \mathcal{D}_{i,m}$ *associated with a set* C_i,

$$\mathop{\mathbb{E}}_{j \in Q}\left[\Delta_j^{(i,m)}\right] \approx_{0,b} \mathop{\mathbb{E}}_{j \in [N]}\left[\Delta_j^{(i,m)}\right] \quad \text{for} \quad b = O\left(\tilde{\varepsilon} \cdot 2^{-i} \cdot \frac{|C_i|}{N} + \tilde{\eta}\right). \tag{10}$$

Proof Fix i, m, and a vector $\Delta^{(i,m)} \in \mathcal{D}_{i,m}$ associated with a set C_i as in (9). Notice that

$$\mathop{\mathbb{E}}_{j \in [N]}[|\Delta_j^{(i,m)}|] \leq 30 \cdot 2^{-(i+m)/2} \cdot \frac{|C_i|}{N}.$$

By Corollary 2.4, applied with

$$\epsilon' = \tilde{\varepsilon} \cdot 2^{(m-i)/2}, \quad \alpha = \tilde{\eta}, \quad \text{and} \quad a = 30 \cdot 2^{-(i+m)/2},$$

we have that (10) holds with probability $1 - 2^{-\Omega(2^m \cdot q \tilde{\varepsilon} \tilde{\eta})}$. Using Claim 4.2, the union bound over all the vectors in $\mathcal{D}_{i,m}$ implies that the probability that some $\Delta^{(i,m)} \in \mathcal{D}_{i,m}$

The Restricted Isometry Property of Subsampled Fourier Matrices

does not satisfy (10) is at most

$$N^{O(2^m \cdot \log(1/\gamma))} \cdot 2^{-\Omega(2^m \cdot q\bar{\epsilon}\bar{\eta})} \leq 2^{-\Omega(2^m \cdot \log N \cdot \log(1/\gamma))} \,.$$

The result follows by a union bound over i and m. ∎

Approximating the Vectors Mx

Lemma 4.4 *For every multiset $Q \subseteq [N]$ and every vector $x \in \mathbb{C}^N$ with $\|x\|_1 = 1$ there exist vector collections $(\Delta^{(i,m)} \in \mathcal{D}_{i,m})_{m=i,\dots,i+r}$ associated with sets C_i $(1 \leq i \leq t)$, for which*

1. *$\mathbb{E}_{j \in [N]} \left[|(Mx)_j|^2 \right] \geq \sum_{i=1}^{t} 2^{-i} \cdot \frac{|C_i|}{N} - \eta$,*
2. *$\mathbb{E}_{j \in Q} \left[|(Mx)_j|^2 \right] \approx_{O(\epsilon),O(\eta)} \mathbb{E}_{j \in Q} \left[\sum_{i=1}^{t} \sum_{m=i}^{i+r} \Delta_j^{(i,m)} \right]$, and*
3. *$\mathbb{E}_{j \in [N]} \left[|(Mx)_j|^2 \right] \approx_{O(\epsilon),O(\eta)} \mathbb{E}_{j \in [N]} \left[\sum_{i=1}^{t} \sum_{m=i}^{i+r} \Delta_j^{(i,m)} \right]$.*

Proof By Lemma 3.5, for every $1 \leq i \leq t + r$ there exists a vector $g^{(i)} \in \mathcal{G}_i$ that satisfies

$$|(Mx)_j| \approx_{0,2^{-i/2}} |g_j^{(i)}| \tag{11}$$

for all but at most γ fraction of $j \in [N]$ and for all but at most γ fraction of $j \in Q$. We say that $j \in [N]$ is *good* if (11) holds for every i, and otherwise that it is *bad*. Notice that all but at most $(t + r)\gamma$ fraction of $j \in [N]$ are good and that all but at most $(t + r)\gamma$ fraction of $j \in Q$ are good. Consider the sets C_i and vectors $h^{(i,m)}$, $\Delta^{(i,m)}$ associated with $(g^{(1)}, \dots, g^{(t+r)})$ as defined in (8). We claim that $\Delta^{(i,m)}$ satisfy the requirements of the lemma.

Fix some $1 \leq i \leq t$. For every good $j \in C_i$, the definition of C_i implies that $|g_j^{(i)}| \geq 2 \cdot 2^{-i/2}$, so using (11) it follows that

$$|(Mx)_j| \geq |g_j^{(i)}| - 2^{-i/2} \geq 2^{-i/2}. \tag{12}$$

We also claim that $|(Mx)_j| \leq 3 \cdot 2^{-(i-1)/2}$. This trivially holds for $i = 1$, so assume that $i \geq 2$, and notice that the definition of C_i implies that $|g_j^{(i-1)}| < 2 \cdot 2^{-(i-1)/2}$, so using (11), it follows that

$$|(Mx)_j| \leq |g_j^{(i-1)}| + 2^{-(i-1)/2} \leq 3 \cdot 2^{-(i-1)/2}. \tag{13}$$

Since at most $(t + r)\gamma$ fraction of $j \in [N]$ are bad, (12) yields that

$$\mathbb{E}_{j \in [N]} \left[|(Mx)_j|^2 \right] \geq \sum_{i=1}^{t} 2^{-i} \cdot \frac{|C_i|}{N} - (t+r)\gamma/2 \geq \sum_{i=1}^{t} 2^{-i} \cdot \frac{|C_i|}{N} - \eta,$$

as required for Item 1.

Next, we claim that every good j satisfies

$$|(Mx)_j|^2 \approx_{O(\epsilon),O(\eta)} \sum_{i=1}^{t} h_j^{(i,i+r)} .\tag{14}$$

For a good $j \in C_i$ and $m \geq i$,

$$\left| |(Mx)_j|^2 - h_j^{(i,m)} \right| \leq 2 \cdot |(Mx)_j| \cdot 2^{-m/2} + 2^{-m} \leq 10 \cdot 2^{-(i+m)/2},\tag{15}$$

where the first inequality follows from (11) and the second from (13). In particular, for $m = i + r$ (recall that $r = \log_2(1/\epsilon^2)$), we have

$$\left| |(Mx)_j|^2 - h_j^{(i,i+r)} \right| \leq 10 \cdot \epsilon \cdot 2^{-i} \leq 10 \cdot \epsilon \cdot |(Mx)_j|^2 ,$$

and thus $|(Mx)_j|^2 \approx_{O(\epsilon),0} h_j^{(i,i+r)}$. Since every good j belongs to at most one of the sets C_i, for every good $j \in \bigcup C_i$ we have $|(Mx)_j|^2 \approx_{O(\epsilon),0} \sum_{i=1}^{t} h_j^{(i,i+r)}$. On the other hand, if j is good but does not belong to any C_i, by our choice of t, it satisfies

$$|(Mx)_j| \leq |g_j^{(t)}| + 2^{-t/2} \leq 3 \cdot 2^{-t/2} = 3\sqrt{\eta} ,$$

and thus $|(Mx)_j|^2 \approx_{0,9\eta} 0 = \sum_{i=1}^{t} h_j^{(i,i+r)}$. This establishes that (14) holds for every good j.

Next, we claim that for every good j,

$$|(Mx)_j|^2 \approx_{O(\epsilon),O(\eta)} \sum_{i=1}^{t} \sum_{m=i}^{i+r} \Delta_j^{(i,m)} .\tag{16}$$

This follows since for every $1 \leq i \leq t$, the vector $h^{(i,i+r)}$ can be written as the telescopic sum

$$h^{(i,i+r)} = \sum_{m=i}^{i+r} (h^{(i,m)} - h^{(i,m-1)}) ,$$

where we used that $h^{(i,i-1)} = 0$. We claim that for every good j, these differences satisfy

$$|h_j^{(i,m)} - h_j^{(i,m-1)}| \leq 30 \cdot 2^{-(i+m)/2},$$

thus establishing that (16) holds for every good j. Indeed, for $m \geq i+1$, (15) implies that

$$|h_j^{(i,m)} - h_j^{(i,m-1)}| \leq 10 \cdot (2^{-(i+m)/2} + 2^{-(i+m-1)/2}) \leq 30 \cdot 2^{-(i+m)/2},\tag{17}$$

and for $m = i$ it follows from (11) combined with (13).

Finally, for every bad j we have

$$\left| |(Mx)_j|^2 - \sum_{i=1}^{t}\sum_{m=i}^{i+r}\Delta_j^{(i,m)} \right| \leq 1 + 30 \cdot \max_{1\leq i\leq t}\left(\sum_{m=i}^{i+r}2^{-(i+m)/2} \right) \leq 60 .$$

Since at most $(t+r)\gamma$ fraction of the elements in $[N]$ and in Q are bad, their effect on the difference between the expectations in Items 2 and 3 can be bounded by $60(t+r)\gamma$. By our choice of γ this is η, as required.   ∎

Finally, we are ready to prove Theorem 4.1.

Proof of Theorem 4.1 Recall that it can be assumed that $\epsilon \geq \eta$. By Lemma 4.3, applied with $\tilde{\varepsilon} = \epsilon/r$ and $\tilde{\eta} = \eta/(rt)$, a random multiset Q of size

$$q = O\left(\epsilon^{-1}\eta^{-1}\cdot r^2 \cdot t \cdot \log N \cdot \log(1/\gamma) \right)$$

$$= O\left(\log^2(1/\epsilon)\cdot \epsilon^{-1}\eta^{-1}\log N \cdot \log^2(1/\eta) \right)$$

satisfies with probability $1 - 2^{-\Omega(\log N \cdot \log(1/\eta))}$, that for every $1 \leq i \leq t$, m, and $\Delta^{(i,m)} \in \mathcal{D}_{i,m}$ associated with a set C_i,

$$\mathbb{E}_{j\in Q}\left[\Delta_j^{(i,m)} \right] \approx_{0,b_i} \mathbb{E}_{j\in[N]}\left[\Delta_j^{(i,m)} \right] \quad \text{for} \quad b_i = O\left(\frac{\varepsilon}{r}\cdot 2^{-i}\cdot \frac{|C_i|}{N} + \frac{\eta}{rt} \right),$$

in which case we also have

$$\mathbb{E}_{j\in Q}\left[\sum_{i=1}^{t}\sum_{m=i}^{i+r}\Delta_j^{(i,m)} \right] \approx_{0,b} \mathbb{E}_{j\in[N]}\left[\sum_{i=1}^{t}\sum_{m=i}^{i+r}\Delta_j^{(i,m)} \right] \quad \text{for} \quad b = O\left(\epsilon\cdot\sum_{i=1}^{t}2^{-i}\cdot\frac{|C_i|}{N} + \eta \right).$$

$$(18)$$

We show that a Q with the above property satisfies the requirement of the theorem. Let $x \in \mathbb{C}^N$ be a vector, and assume without loss of generality that $\|x\|_1 = 1$. By Lemma 4.4, there exist vector collections $(\Delta^{(i,m)} \in \mathcal{D}_{i,m})_{m=i,...,i+r}$ associated with sets C_i $(1 \leq i \leq t)$, satisfying Items 1, 2, and 3 there. Combined with (18), this gives

$$\mathbb{E}_{j\in Q}\left[|(Mx)_j|^2 \right] \approx_{O(\epsilon),O(\eta)} \mathbb{E}_{j\in[N]}\left[|(Mx)_j|^2 \right] ,$$

and we are done.   ∎

4.1 The Restricted Isometry Property

It is easy to derive now the following theorem. The proof is essentially identical to that of Theorem 3.7, using Theorem 4.1 instead of Theorem 3.1.

Theorem 4.5 *For sufficiently large N and k, a unitary matrix $M \in \mathbb{C}^{N \times N}$ satisfying $\|M\|_\infty \leq O(1/\sqrt{N})$, and a sufficiently small $\epsilon > 0$, the following holds. For some $q = O(\log^2(1/\epsilon)\epsilon^{-2} \cdot k \cdot \log^2(k/\epsilon) \cdot \log N)$, let $A \in \mathbb{C}^{q \times N}$ be a matrix whose q rows are chosen uniformly and independently from the rows of M, multiplied by $\sqrt{N/q}$. Then, with probability $1 - 2^{-\Omega(\log N \cdot \log(k/\epsilon))}$, the matrix A satisfies the restricted isometry property of order k with constant ϵ.*

Acknowledgements We thank Afonso S. Bandeira, Mahdi Cheraghchi, Michael Kapralov, Jelani Nelson, and Eric Price for useful discussions, and anonymous reviewers for useful comments.

Oded Regev was supported by the Simons Collaboration on Algorithms and Geometry and by the National Science Foundation (NSF) under Grant No. CCF-1320188. Any opinions, findings, and conclusions or recommendations expressed in this material are those of the authors and do not necessarily reflect the views of the NSF.

References

1. N. Ailon, B. Chazelle, The fast Johnson–Lindenstrauss transform and approximate nearest neighbors. SIAM J. Comput. **39**(1), 302–322 (2009). Preliminary version in STOC'06
2. N. Ailon, E. Liberty, Fast dimension reduction using Rademacher series on dual BCH codes. Discrete Comput. Geom. **42**(4), 615–630 (2009). Preliminary version in SODA'08
3. N. Ailon, E. Liberty, An almost optimal unrestricted fast Johnson–Lindenstrauss transform. ACM Trans. Algorithms **9**(3), 21 (2013). Preliminary version in SODA'11
4. A.S. Bandeira, E. Dobriban, D.G. Mixon, W.F. Sawin, Certifying the restricted isometry property is hard. IEEE Trans. Inform. Theory **59**(6), 3448–3450 (2013)
5. A.S. Bandeira, M.E. Lewis, D.G. Mixon, Discrete uncertainty principles and sparse signal processing. CoRR abs/1504.01014 (2015)
6. R. Baraniuk, M. Davenport, R. DeVore, M. Wakin, A simple proof of the restricted isometry property for random matrices. Constr. Approx. **28**(3), 253–263 (2008)
7. J. Bourgain, An improved estimate in the restricted isometry problem, in *Geometric Aspects of Functional Analysis*. Lecture Notes in Mathematics, vol. 2116, pp. 65–70 (Springer, Berlin, 2014)
8. E.J. Candès, The restricted isometry property and its implications for compressed sensing. C. R. Math. **346**(9–10), 589–592 (2008)
9. E.J. Candès, T. Tao, Decoding by linear programming. IEEE Trans. Inform. Theory **51**(12), 4203–4215 (2005)
10. E.J. Candès, T. Tao, Near-optimal signal recovery from random projections: universal encoding strategies? IEEE Trans. on Inform. Theory **52**(12), 5406–5425 (2006)
11. E.J. Candès, M. Rudelson, T. Tao, R. Vershynin, Error correction via linear programming, in *46th Annual IEEE Symposium on Foundations of Computer Science, FOCS*, pp. 295–308 (2005)
12. E.J. Candès, J.K. Romberg, T. Tao, Stable signal recovery from incomplete and inaccurate measurements. Commun. Pure Appl. Math. **59**(8), 1207–1223 (2006)
13. S.S. Chen, D.L. Donoho, M.A. Saunders, Atomic decomposition by basis pursuit. SIAM J. Comput. **20**(1), 33–61 (1998)
14. M. Cheraghchi, V. Guruswami, A. Velingker, Restricted isometry of Fourier matrices and list decodability of random linear codes. SIAM J. Comput. **42**(5), 1888–1914 (2013). Preliminary version in SODA'13
15. S. Dirksen, Tail bounds via generic chaining. Electron. J. Prob. **20**(53), 1–29 (2015)

16. D.L. Donoho, M. Elad, V.N. Temlyakov, Stable recovery of sparse overcomplete representations in the presence of noise. IEEE Trans. Inform. Theory **52**(1), 6–18 (2006)
17. S. Foucart, A. Pajor, H. Rauhut, T. Ullrich, The Gelfand widths of ℓ_p-balls for $0 < p \leq 1$. J. Complex. **26**(6), 629–640 (2010)
18. A.Y. Garnaev, E.D. Gluskin, On the widths of Euclidean balls. Sov. Math. Dokl. **30**, 200–203 (1984)
19. I. Haviv, O. Regev, The list-decoding size of Fourier-sparse boolean functions, in *Proceedings of the 30th Conference on Computational Complexity, CCC*, pp. 58–71 (2015)
20. P. Indyk, I. Razenshteyn, On model-based RIP-1 matrices, in *Automata, Languages, and Programming - 40th International Colloquium, ICALP*, pp. 564–575 (2013)
21. A.C. Kak, M. Slaney, *Principles of Computerized Tomographic Imaging* (Society of Industrial and Applied Mathematics, Philadelphia, 2001)
22. F. Krahmer, R. Ward, New and improved Johnson-Lindenstrauss embeddings via the restricted isometry property. SIAM J. Math. Anal. **43**(3), 1269–1281 (2011)
23. F. Krahmer, S. Mendelson, H. Rauhut, Suprema of chaos processes and the restricted isometry property. CoRR abs/1207.0235 (2012)
24. C. McDiarmid, Concentration, in *Probabilistic Methods for Algorithmic Discrete Mathematics*. Algorithms Combination, vol. 16 (Springer, Berlin, 1998), pp. 195–248
25. A. Natarajan, Y. Wu, Computational complexity of certifying restricted isometry property, in *Approximation, Randomization, and Combinatorial Optimization. Algorithms and Techniques, APPROX*, pp. 371–380 (2014)
26. D. Needell, J.A. Tropp, CoSaMP: iterative signal recovery from incomplete and inaccurate samples. Commun. ACM **53**(12), 93–100 (2010)
27. J. Nelson, E. Price, M. Wootters, New constructions of RIP matrices with fast multiplication and fewer rows, in *Proceedings of the 25th Annual ACM-SIAM Symposium on Discrete Algorithms, SODA*, pp. 1515–1528 (2014)
28. D.G. Nishimura, *Principles of Magnetic Resonance Imaging* (Stanford University, Stanford, CA, 2010)
29. H. Rauhut, Compressive sensing and structured random matrices, in *Theoretical Foundations and Numerical Methods for Sparse Recovery*, vol. 9, ed. by M. Fornasier (De Gruyter, Berlin, 2010), pp. 1–92
30. M. Rudelson, R. Vershynin, On sparse reconstruction from Fourier and Gaussian measurements. Commun. Pure Appl. Math. **61**(8), 1025–1045 (2008). Preliminary version in CISS'06
31. A.M. Tillmann, M.E. Pfetsch, The computational complexity of the restricted isometry property, the nullspace property, and related concepts in compressed sensing. IEEE Trans. Inform. Theory **60**(2), 1248–1259 (2014)
32. S. Wenger, S. Darabi, P. Sen, K. Glassmeier, M.A. Magnor, Compressed sensing for aperture synthesis imaging, in *Proceedings of the International Conference on Image Processing, ICIP*, pp. 1381–1384 (2010)
33. M. Wootters, On the list decodability of random linear codes with large error rates, in *Proceedings of the 45th Annual ACM Symposium on Theory of Computing, STOC*, pp. 853–860 (2013)

Upper Bound for the Dvoretzky Dimension in Milman-Schechtman Theorem

Han Huang and Feng Wei

Abstract For a symmetric convex body $K \subset \mathbb{R}^n$, the Dvoretzky dimension $k(K)$ is the largest dimension for which a random central section of K is almost spherical. A Dvoretzky-type theorem proved by V.D. Milman in 1971 provides a lower bound for $k(K)$ in terms of the average $M(K)$ and the maximum $b(K)$ of the norm generated by K over the Euclidean unit sphere. Later, V.D. Milman and G. Schechtman obtained a matching upper bound for $k(K)$ in the case when $\frac{M(K)}{b(K)} > c(\frac{\log(n)}{n})^{\frac{1}{2}}$. In this paper, we will give an elementary proof of the upper bound in Milman-Schechtman theorem which does not require any restriction on $M(K)$ and $b(K)$.

1 Introduction

Given a symmetric convex body K in $\mathbb{R}^n$, we have a corresponding norm $\|x\|_K = \inf\{r > 0, x \in rK\}$. Let $|\cdot|$ denote the Euclidean norm, ν_n denote the normalized Haar measure on the Euclidean sphere, S^{n-1}, and $\nu_{n,k}$ denote the normalized Haar measure on the Grassmannian manifold $Gr_{n,k}$. Let $M = M(K) := \int_{S^{n-1}} \|x\|_K d\nu_n$ and $b = b(K) := \sup\{\|x\|_K, x \in S^{n-1}\}$ be the mean and the maximum of the norm over the unit sphere.

In 1971, V.D. Milman proved the following Dvoretzky-type theorem [3]:

Theorem 1 *Let K be a symmetric convex body in $\mathbb{R}^n$. Assume that $\|x\|_K \leq b|x|$ for all $x \in \mathbb{R}^n$. For any $\epsilon \in (0, 1)$, there is $k \geq C_\epsilon (M/b)^2 n$ such that*

$$\nu_{n,k}\{F \in G_{n,k} : (1 - \varepsilon)M < \|\cdot\|_{K \cap F} < (1 + \varepsilon)M\} > 1 - \exp(-\tilde{c}k)$$

where $\tilde{c} > 0$ is a universal constant, $C_\epsilon > 0$ is a constant depending only on ϵ.

The quantity C_ϵ was of the order $\epsilon^2 \log^{-1}(\frac{1}{\epsilon})$ in the original proof of V.D. Milman. It was improved to the order of ϵ^2 by Gordon [2] and later, with a simpler argument, by Schechtman [6].

H. Huang (✉) • F. Wei
Department of Mathematics, University of Michigan, Ann Arbor, MI, USA
e-mail: sthhan@umich.edu; weifeng@umich.edu

© Springer International Publishing AG 2017
B. Klartag, E. Milman (eds.), *Geometric Aspects of Functional Analysis*,
Lecture Notes in Mathematics 2169, DOI 10.1007/978-3-319-45282-1_12

In 1997, Milman and Schechtman [5] found that the bound on k appearing in Theorem 1 is essentially optimal. More precisely, they proved the following theorem.

Theorem A (Milman–Schechtman, See e.g., Sect. 5.3 in [1]) *Let K be a symmetric convex body in $\mathbb{R}^n$. For $\epsilon \in (0, 1)$, define $k(K)$ to be the largest dimension k such that*

$$\nu_{n,k}\left(\{F \in G_{n,k} : \forall x \in S^{n-1} \cap F, (1-\varepsilon)M < \|x\|_K < (1+\varepsilon)M\}\right) > p_{n,k} = \frac{n}{n+k}.$$

Then,

$$\tilde{C}_\epsilon n(M/b)^2 \geq k(K) \geq \bar{C}_\epsilon n(M/b)^2$$

when $\frac{M}{b} > c(\frac{\log(n)}{n})^{\frac{1}{2}}$ for some universal constant $c > 0$, where $\| \cdot \|_F$ denotes the norm corresponding to the convex body $K \cap F$ in F, and $\tilde{C}_\epsilon, \bar{C}_\epsilon > 0$ are constants depending only on ϵ.

Because the Dvoretzky-Milman theorem cannot guarantee the lower bound with small $\frac{M}{b}$ for $p_{n,k} = \frac{n}{n+k}$, the original proof required an assumption that $\frac{M}{b} > c(\frac{\log(n)}{n})^{\frac{1}{2}}$ for some c. In [1, p. 197], S. Artstein-Avidan, A.A. Giannopoulos, and V.D. Milman addressed it as an open question whether one can prove the same result when $p_{n,k}$ is a constant, such as $\frac{1}{2}$. When $p_{n,k} = \frac{1}{2}$, the lower estimate on $k(K)$ is a direct result of Dvoretzky-Milman theorem [3], but the upper bound was unknown. In this paper, we are going to give upper bound estimate with $p_{n,k} = \frac{1}{2}$, our main result is the following theorem:

Theorem B *Let K be a symmetric convex body in $\mathbb{R}^n$. Fix a constant $\epsilon \in (0, 1)$, let $k(K)$ be the largest dimension k such that*

$$\nu_{n,k}\{F \in G_{n,k} : (1-\varepsilon)M < \| \cdot \|_{K \cap F} < (1+\varepsilon)M\} > \frac{1}{2}.$$

Then,

$$Cn(M/b)^2 \geq k(X) \geq \bar{C}_\epsilon n(M/b)^2$$

where $C > 0$ is a universal constant and $\bar{C}_\epsilon > 0$ is a constant depending only on ϵ.

In the next section, we will provide a proof of Theorem B with no restriction on $\frac{M}{b}$. In fact, from the proof, one can see that $\frac{1}{2}$ can be replaced by any $c \in (0, 1)$ or $1 - \exp(-\tilde{c}k)$, which is the probability appearing in Milman-Dvoretzky theorem.

2   Proof of Theorem B

Let P_k be the orthogonal projection from S^{n-1} to some fixed k-dimensional subspace, and $|\cdot|$ be the Euclidean norm. The upper estimate is related to the distribution of $|P_k(x)|$, where x is uniformly distributed on S^{n-1}.

Recall the concentration inequality for Lipschitz functions on the sphere (see, e.g., [4]):

Theorem 2 (Measure Concentration on S^{n-1}) *Let $f : S^{n-1} \to \mathbb{R}$ be a Lipschitz continuous function with Lipschitz constant b. Then, for every $t > 0$,*

$$\nu_n(\{x \in S^{n-1} : |f(x) - \mathbb{E}(f)| \geq bt\}) \leq 4\exp(-c_0 t^2 n)$$

where $c_0 > 0$ is a universal constant.

Theorem 2 implies the following elementary lemma.

Lemma 3 *Fix any $c_1 > 0$, let P_k be an orthogonal projection from $\mathbb{R}^n$ to some subspace $\mathbb{R}^k$. If $t > \frac{c_1}{\sqrt{n}}$ and $\nu_n(\{x \in S^{n-1} : |P_k(x)| < t\}) > \frac{1}{2}$, then $k < c_2 t^2 n$, where $c_2 > 0$ is a constant depending only on c_1.*

Proof $|P_k(x)|$ is a 1-Lipschitz function on S^{n-1} with $\mathbb{E}|P_k(x)|$ about $\sqrt{\frac{k}{n}}$. If we want the measure of $\{x : |P_k(x)| < t\}$ to be greater than $1/2$, then measure concentration will force $\mathbb{E}|P_k|$ to be bounded by the size of t, which means $k < c_2 t^2 n$ for some universal constant c_2. Since $t^2 n > c_1^2$, we may and shall assume k is bigger than some absolute constant in our proof, then adjust c_2.

To make it precise, we will first give a lower bound on $\mathbb{E}|P_k|$. By Theorem 2,

$$\nu_n(||P_k(x)| - \mathbb{E}|P_k(x)||^2 > t) \leq 4\exp(-c_0 t n).$$

Thus,

$$\mathbb{E}|P_k|^2 - (\mathbb{E}|P_k|)^2 = \mathbb{E}(|P_k|(x) - \mathbb{E}|P_k|)^2$$

$$< \int_0^\infty \nu_n(||P_k(x)| - \mathbb{E}|P_k(x)||^2 > t)dt$$

$$\leq \int_0^\infty 4\exp(-c_0 t n)dt = \frac{4}{c_0 n}.$$

With $\mathbb{E}|P_k|^2 = \mathbb{E}\sum_{i=1}^k |x_i|^2 = \frac{k}{n}$, we get $\mathbb{E}(|P_k|) > \sqrt{\frac{k}{n} - \frac{4}{c_0 n}}$. If we assume that $k > \frac{24}{c_0}$, then we have

$$\mathbb{E}(|P_k|) > \sqrt{\frac{1}{2}\frac{k}{n}}.$$

Assuming $k > 8t^2 n$, we have

$$\mathbb{E}(|P_k|) - t > \sqrt{\frac{1}{2}\frac{k}{n}} - t \geq \frac{1}{2}\sqrt{\frac{1}{2}\frac{k}{n}} > 0.$$

Applying Theorem 2 again, we obtain

$$\nu_n(|P_k| < t) < \nu_n\left(||P_k| - \mathbb{E}|P_k|| > \mathbb{E}(|P_k|) - t\right) \leq 4\exp(-c_0(\mathbb{E}(|P_k|) - t)^2 n)$$

$$\leq 4\exp\left(-c_0\left(\frac{1}{2}\sqrt{\frac{1}{2}\frac{k}{n}}\right)^2 n\right) \leq 4\exp\left(-\frac{c_0}{8}k\right) \leq 4\exp(-3) < \frac{1}{2},$$

which proves our result by contradiction.   □

Theorem 4 *Let K be a symmetric convex body with inradius $\frac{1}{b}$. For $\epsilon \in (0,1)$, let k be the largest integer such that*

$$\nu_{n,k}\{F \in G_{n,k} : (1-\epsilon)M < \|\cdot\|_{K \cap F} < (1+\epsilon)M\} > \frac{1}{2}.$$

Then $k < Cn(\frac{M}{b})^2$ where $C > 0$ is an absolute constant.

Proof We may assume $\|e_1\|_K = b$, then $K \subset S = \{x \in \mathbb{R}^n : |x_1| < \frac{1}{b}\}$, thus $\|x\|_K \geq \|x\|_S = b|\langle x, e_1 \rangle|$. This implies

$$\{V \in G_{n,k} : \forall x \in V \cap S^{n-1}, (1-\epsilon)M < \|x\|_K < (1+\epsilon)M\}$$

$$\subset \{V \in G_{n,k} : \forall x \in V \cap S^{n-1}, \|x\|_S < (1+\epsilon)M\}$$

$$= \{V \in G_{n,k} : \sup_{x \in V \cap S^{n-1}} \langle x, e_1 \rangle < (1+\epsilon)\frac{M}{b}\}$$

$$= \{V \in G_{n,k} : |P_V(e_1)| < (1+\epsilon)\frac{M}{b}\} \qquad (1)$$

where P_V is the orthogonal projection from $\mathbb{R}^n$ to V. If V is uniformly distributed on $G_{n,k}$ and x is uniformly distributed on S^{n-1}, then $|P_{V_0}(x)|$ and $|P_V(e_1)|$ are equi-distributed for any fixed k-dimensional subspace V_0. Therefore,

$$\nu_{n,k}(\{V \in G_{n,k} : |P_V(e_1)| < (1+\epsilon)\frac{M}{b}\}) = \nu_n(\{x \in S^{n-1} : |P_{V_0}(x)| < (1+\epsilon)\frac{M}{b}\}).$$

As shown in the Remark 5.2.2(iii) of [1, p. 164], the ratio $\frac{M}{b}$ has a lower bound $\frac{c'}{\sqrt{n}}$. Setting $c_1 = c'$ and $t = (1+\epsilon)\frac{M}{b}$, it is easy to see that if

$$\nu_{n,k}\{F \in G_{n,k} : (1-\epsilon)M < \|\cdot\|_{K \cap F} < (1+\epsilon)M\} > \frac{1}{2},$$

then $k \leq c_1(1+\epsilon)^2 \left(\frac{M}{b}\right)^2 n < Cn(\frac{M}{b})^2$ by Lemma 3 and (1).   □

Now we can prove Theorem B as a corollary of Theorems 4 and 1:

Proof of Theorem B Theorem 1 shows that if $C_\epsilon (M/b)^2 n > \frac{\log(2)}{\tilde{c}}$, then there is $k \geq C_\epsilon (M/b)^2 n$ such that

$$\nu_{n,k}\{F \in G_{n,k} : (1-\varepsilon)M < \|\cdot\|_F < (1+\varepsilon)M\} > 1 - \exp(-\tilde{c}k) > \frac{1}{2}.$$

Otherwise, $(M/b)^2 n < \frac{\log(2)}{\tilde{c}C_\epsilon}$. Therefore, $k(K) \geq \min\{\frac{\tilde{c}C_\epsilon}{\log(2)}, C_\epsilon\}(M/b)^2 n$. Combining it with Theorem 4, we get

$$C(\frac{M}{b})^2 n \geq k(K) \geq \min\{\frac{\tilde{c}C_\epsilon}{\log(2)}, C_\epsilon\}(M/b)^2 n.$$

$\square$

Remark

(1) It is worth noticing that the number $\frac{1}{2}$ plays no special role in our proof. Thus, if we define the Dvoretzky dimension to be the largest dimension such that

$$\nu_{n,k}\{F \in G_{n,k} : (1-\varepsilon)M < \|\cdot\|_{K \cap F} < (1+\varepsilon)M\} > c$$

for some $c \in (0,1)$, then exactly the same proof will work. We will still have $k(K) \sim (\frac{M}{b})^2 n$. Similarly, if we fix ϵ and replace $\frac{1}{2}$ by $1 - \exp(-\tilde{c}k)$, then the lower bound of $k(K)$ is the one from Theorem 1. For k bigger than some absolute constant, we have $1 - \exp(-\tilde{c}k) > \frac{1}{2}$. Thus, the upper bound is still of order $\left(\frac{M}{b}\right)^2 n$. Therefore, we can replace $\frac{1}{2}$ by $1 - \exp(-\tilde{c}k)$ in Theorem A. With this probability choice, it also shows Theorem 1 provides an optimal k depending on M, b.

(2) Usually, we are only interested in $\epsilon \in (0,1)$. In the lower bound, $\bar{C}_\epsilon = o_\epsilon(1)$. It is a natural question to ask if we could improve the upper bound from a universal constant C to $o_\epsilon(1)$. Unfortunately, it is not possible due to the following observation. Let $K = \mathrm{conv}(B_2^n, Re_1)^\circ$. By passing from the intersection on K to the projection of K°, one can show that $k(K)$ does not exceed the maximum dimension k such that $\nu_n(P_k(Rx) < 1 + \epsilon) > \frac{1}{2}$. Choosing $R = \sqrt{\frac{\pi}{l}}$, we get $n(\frac{M}{b})^2 \sim l$ and $k(X) \sim l$ by Theorem 2 and a similar argument to that of Lemma 3. This example shows that no matter what $\frac{M}{b}$ is, one can not improve the upper bound in Theorem A from an absolute constant C to $o_\epsilon(1)$.

Acknowledgements We want to thank our advisor Professor Mark Rudelson for his advise and encouragement on solving this problem. And thank both Professor Mark Rudelson and Professor Vitali Milman for encouraging us to organize our result as this paper.

Partially supported by M. Rudelson's NSF Grant DMS-1464514, and USAF Grant FA9550-14-1-0009.

References

1. S. Artstein-Avidan, A.A. Giannopoulos, V.D. Milman, *Asymptotic Geometric Analysis. Part I*. Mathematical Surveys and Monographs, vol. 202 (American Mathematical Society, Providence, RI, 2015)
2. Y. Gordon, Some inequalities for Gaussian processes and applications. Isr. J. Math. **50**, 265–289 (1985)
3. V.D. Milman, New proof of the theorem of A. Dvoretzky on sections of convex bodies. Funct. Anal. Appl. **5**(4), 288–295 (1971)
4. V.D. Milman, G. Schechtman, *Asymptotic Theory of Finite-Dimensional Normed Spaces*. With an appendix by M. Gromov. Lecture Notes in Mathematics, vol. 1200 (Springer, Berlin, 1986)
5. V.D. Milman, G. Schechtman, Global versus local asymptotic theories of finite dimensional normed spaces. Duke Math. J. **90**, 73–93 (1997)
6. G. Schechtman, A remark concerning the dependence on ϵ in Dvoretzky's theorem, in *Geometric Aspects of Functional Analysis (1987–88)*. Lecture Notes in Mathematics, vol. 1376 (Springer, Berlin, 1989), pp. 274–277

Super-Gaussian Directions of Random Vectors

Bo'az Klartag

Abstract We establish the following universality property in high dimensions: Let X be a random vector with density in $\mathbb{R}^n$. The density function can be arbitrary. We show that there exists a fixed unit vector $\theta \in \mathbb{R}^n$ such that the random variable $Y = \langle X, \theta \rangle$ satisfies

$$\min\{\mathbb{P}(Y \geq tM), \mathbb{P}(Y \leq -tM)\} \geq ce^{-Ct^2} \qquad \text{for all } 0 \leq t \leq \tilde{c}\sqrt{n},$$

where $M > 0$ is any median of $|Y|$, i.e., $\min\{\mathbb{P}(|Y| \geq M), \mathbb{P}(|Y| \leq M)\} \geq 1/2$. Here, $c, \tilde{c}, C > 0$ are universal constants. The dependence on the dimension n is optimal, up to universal constants, improving upon our previous work.

1 Introduction

Consider a random vector X that is distributed uniformly in some Euclidean ball centered at the origin in $\mathbb{R}^n$. For any fixed vector $0 \neq \theta \in \mathbb{R}^n$, the density of the random variable $\langle X, \theta \rangle = \sum_i \theta_i X_i$ may be found explicitly, and in fact it is proportional to the function

$$t \mapsto \left(1 - \frac{t^2}{A^2 n}\right)_+^{(n-1)/2} \qquad (t \in \mathbb{R}) \tag{1}$$

where $x_+ = \max\{x, 0\}$ and $A > 0$ is a parameter depending on the length of θ and the radius of the Euclidean ball. It follows that when the dimension n is large, the density in (1) is close to a Gaussian density, and the random variable $Y = \langle X, \theta \rangle$ has a tail of considerable size:

$$\mathbb{P}(Y \geq tM) \geq c \exp(-Ct^2) \qquad \text{for all } 0 \leq t \leq \tilde{c}\sqrt{n}. \tag{2}$$

B. Klartag (✉)
Department of Mathematics, Weizmann Institute of Science, Rehovot 76100, Israel

School of Mathematical Sciences, Tel Aviv University, Tel Aviv 69978, Israel
e-mail: boaz.klartag@weizmann.ac.il

© Springer International Publishing AG 2017
B. Klartag, E. Milman (eds.), *Geometric Aspects of Functional Analysis*,
Lecture Notes in Mathematics 2169, DOI 10.1007/978-3-319-45282-1_13

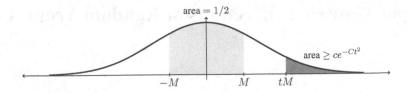

Fig. 1 An example of a density of a Super-Gaussian random variable

Here, $M = \text{Median}(|Y|)$ is any median of $|Y|$, i.e., $\min\{\mathbb{P}(|Y| \geq M), \mathbb{P}(|Y| \leq M)\} \geq 1/2$, and $c, \tilde{c}, C > 0$ are universal constants. Both the median and the expectation of $|Y|$ differ from A by a factor which is at most a universal constant. We prefer to work with a median since in the cases we will consider shortly, the expectation of $|Y|$ is not guaranteed to be finite. The inequality in (2) expresses the property that the tail distribution of Y/M is at least as heavy as the standard Gaussian tail distribution, for $\sqrt{n}$ standard deviations. The dependence on the dimension n is optimal, since for $t > \tilde{C}\sqrt{n}$, the probability on the left-hand side of (2) vanishes (Fig. 1).

Our goal in this paper is to show that a similar phenomenon occurs for essentially any random vector in $\mathbb{R}^n$, and not only for the uniform distribution on the high-dimensional Euclidean ball. Recall that when n is large and the random vector $X = (X_1, \ldots, X_n)$ has independent coordinates, the classical central limit theorem implies that under mild assumptions, there exists $0 \neq \theta \in \mathbb{R}^n$ for which $\langle X, \theta \rangle$ is approximately Gaussian. It is curious to note that a Gaussian lower bound on the tail persists, even when the independence assumption is completely dropped.

Let Y be a real-valued random variable and let $L > 0$. We say that Y is *Super-Gaussian of length L* with parameters $\alpha, \beta > 0$ if $\mathbb{P}(Y = 0) = 0$ and for any $0 \leq t \leq L$,

$$\min\{\mathbb{P}(Y \geq tM), \mathbb{P}(Y \leq -tM)\} \geq \alpha e^{-t^2/\beta},$$

where $M = \text{Median}(|Y|)$ is any median of $|Y|$. The requirement that $\mathbb{P}(Y = 0) = 0$ is necessary only to avoid trivialities. A Gaussian random variable is certainly super-Gaussian of infinite length, as well as a symmetric exponential random variable. Write $|x| = \sqrt{\langle x, x \rangle}$ for the standard Euclidean norm of $x \in \mathbb{R}^n$, and denote $S^{n-1} = \{x \in \mathbb{R}^n ; |x| = 1\}$.

Theorem 1.1 *Let X be a random vector with density in $\mathbb{R}^n$. Then there exists a fixed vector $\theta \in S^{n-1}$ such that $\langle X, \theta \rangle$ is Super-Gaussian of length $c_1 \sqrt{n}$ with parameters $c_2, c_3 > 0$, where $c_1, c_2, c_3 > 0$ are universal constants.*

Theorem 1.1 improves upon Corollary 1.4 from [5], in which the dependence on the dimension n was logarithmic. In the case where X is distributed uniformly in a 1-unconditional convex body in $\mathbb{R}^n$, Theorem 1.1 goes back to Pivovarov [9] up to logarithmic factors. In the case where X is distributed uniformly in a convex body satisfying the hyperplane conjecture with a uniform constant, Theorem 1.1 is due to

Paouris [8]. Theorem 1.1 provides a universal lower bound on the tail distribution, which is tight up to constants in the case where X is uniformly distributed in a Euclidean ball centered at the origin. In particular, the dependence on the dimension in Theorem 1.1 is optimal, up to the value of the universal constants.

The assumption that the random vector X has a density in $\mathbb{R}^n$ may be somewhat relaxed. The following definition appears in [2, 5] with minor modifications:

Definition 1.2 Let X be a random vector in a finite-dimensional vector space $\mathcal{B}$ and let $d > 0$. We say that "the effective rank of X is at least d", or in short that X is of class eff.rank$_{\geq d}$ if for any linear subspace $E \subseteq \mathcal{B}$,

$$\mathbb{P}(X \in E) \leq \dim(E)/d, \tag{3}$$

with equality if and only if there is a subspace $F \subseteq \mathcal{B}$ with $E \oplus F = \mathcal{B}$ and $\mathbb{P}(X \in E \cup F) = 1$.

Intuitively, when X is of class eff.rank$_{\geq d}$ we think of the support of X as effectively spanning a subspace whose dimension is at least d. Note, however, that d is not necessarily an integer. By substituting $E = \mathcal{B}$ in (3), we see that there are no random vectors in $\mathbb{R}^n$ of class eff.rank$_{\geq d}$ with $d > n$. We say that the effective rank of X is d when X is of class eff.rank$_{\geq d}$, but for any $\varepsilon > 0$ the random vector X is not of class eff.rank$_{\geq d+\varepsilon}$. The effective rank of X is d^- if X is of class eff.rank$_{\geq d-\varepsilon}$ for all $0 < \varepsilon < d$ but X is not of class eff.rank$_{\geq d}$. In the terminology of [5], the random vector X has an effective rank greater than d if and only if it is ε-decent for some $\varepsilon < 1/d$.

There are many random vectors in $\mathbb{R}^n$ whose effective rank is precisely n. For example, any random vector with density in $\mathbb{R}^n$, or any random vector X that is distributed uniformly on a finite set that spans $\mathbb{R}^n$ and does not contain the origin. It was shown by Böröczky et al. [1] and by Henk and Linke [4] that the cone volume measure of any convex body in $\mathbb{R}^n$ with barycenter at the origin is of class eff.rank$_{\geq n}$ as well. Note that a random variable Y is Super-Gaussian of length L with parameters $\alpha, \beta > 0$ if and only if for any number $0 \neq r \in \mathbb{R}$, also rY is Super-Gaussian of length L with the same parameters $\alpha, \beta > 0$. Theorem 1.1 is thus a particular case of the following:

Theorem 1.3 *Let $d \geq 1$ and let $\mathcal{B}$ be a finite-dimensional linear space. Let X be a random vector in $\mathcal{B}$ whose effective rank is at least d. Then there exists a non-zero, fixed, linear functional $\ell : \mathcal{B} \to \mathbb{R}$ such that the random variable $\ell(X)$ is Super-Gaussian of length $c_1 \sqrt{d}$ with parameters $c_2, c_3 > 0$, where $c_1, c_2, c_3 > 0$ are universal constants.*

Theorem 1.3 admits the following corollary, pertaining to infinite-dimensional spaces:

Corollary 1.4 *Let $\mathcal{B}$ be a topological vector space with a countable family of continuous linear functionals that separates points in $\mathcal{B}$. Let X be a random vector, distributed according to a Borel probability measure in $\mathcal{B}$. Assume that $d \geq 1$ is such that $\mathbb{P}(X \in E) \leq \dim(E)/d$ for any finite-dimensional subspace $E \subseteq \mathcal{B}$.*

Then there exists a non-zero, fixed, continuous linear functional $\ell : \mathcal{B} \to \mathbb{R}$ such that the random variable $\ell(X)$ is Super-Gaussian of length $c_1\sqrt{d}$ with parameters $c_2, c_3 > 0$, where $c_1, c_2, c_3 > 0$ are universal constants.

The remainder of this paper is devoted to the proof of Theorem 1.3 and Corollary 1.4. We use the letters $c, C, \tilde{C}, c_1, C_2$ etc. to denote various positive universal constants, whose value may change from one line to the next. We use upper-case C to denote universal constants that we think of as "sufficiently large", and lower-case c to denote universal constants that are "sufficiently small". We write $\#(A)$ for the cardinality of a set A. When we write that a certain set or a certain number are fixed, we intend to emphasize that they are non-random.

We denote by σ_{n-1} the uniform probability measure on the sphere S^{n-1}, which is the unique rotationally-invariant probability measure on S^{n-1}. When we say that a random vector θ is distributed uniformly on S^{n-1}, we refer to the probability measure σ_{n-1}. Similarly, when we write that a random subspace E is distributed uniformly over the Grassmannian $G_{n,k}$ of k-dimensional subspaces of $\mathbb{R}^n$, we refer to the unique rotationally-invariant probability measure on $G_{n,k}$.

2 Proof Strategy

The main ingredient in the proof of Theorem 1.3 is the following proposition:

Proposition 2.1 *Let X be a random vector in $\mathbb{R}^n$ with $\mathbb{P}(X = 0) = 0$ such that*

$$\mathbb{E}\left\langle \frac{X}{|X|}, \theta \right\rangle^2 \le \frac{5}{n} \qquad \text{for all } \theta \in S^{n-1}. \tag{4}$$

Then there exists a fixed vector $\theta \in S^{n-1}$ such that the random variable $\langle X, \theta \rangle$ is Super-Gaussian of length $c_1\sqrt{n}$ with parameters $c_2, c_3 > 0$, where $c_1, c_2, c_3 > 0$ are universal constants.

The number 5 in Proposition 2.1 does not play any particular role, and may be replaced by any other universal constant, at the expense of modifying the values of c_1, c_2 and c_3. Let us explain the key ideas in the proof of Proposition 2.1. In our previous work [5], the unit vector $\theta \in S^{n-1}$ was chosen randomly, uniformly on S^{n-1}. In order to improve the dependence on the dimension, here we select θ a bit differently. We shall define θ_1 and θ_2 via the following procedure:

(i) Let $M > 0$ be a $1/3$-quantile of $|X|$, i.e., $\mathbb{P}(|X| \ge M) \ge 1/3$ and $\mathbb{P}(|X| \le M) \ge 2/3$. We fix a vector $\theta_1 \in S^{n-1}$ such that

$$\mathbb{P}\left(|X| \ge M \text{ and } \left| \frac{X}{|X|} - \theta_1 \right| \le \frac{1}{5} \right) \ge \frac{1}{2} \cdot \sup_{\eta \in S^{n-1}} \mathbb{P}\left(|X| \ge M \text{ and } \left| \frac{X}{|X|} - \eta \right| \le \frac{1}{5} \right).$$

(ii) Next, we fix a vector $\theta_2 \in S^{n-1}$ with $|\langle \theta_1, \theta_2 \rangle| \leq 1/10$ such that

$$\mathbb{P}\left(|X| \geq M \text{ and } \left|\frac{X}{|X|} - \theta_2\right| \leq \frac{1}{5}\right) \geq \frac{1}{2} \cdot \sup_{\substack{\eta \in S^{n-1} \\ |\langle \eta, \theta_1 \rangle| \leq 1/10}} \mathbb{P}\left(|X| \geq M \text{ and } \left|\frac{X}{|X|} - \eta\right| \leq \frac{1}{5}\right).$$

In the following pages we will describe a certain subset $\mathcal{F}_3 \subseteq S^{n-1}$ which satisfies $\sigma_{n-1}(\mathcal{F}_3) \geq 1 - C/n^c$ and $\theta_2 - \theta_1 \notin \mathcal{F}_3$. We will show that for any $\theta_3 \in \mathcal{F}_3$, the random variable $\langle X, \theta \rangle$ is Super-Gaussian of length $c\sqrt{n}$ with parameters $c_1, c_2 > 0$, where θ is defined as follows:

$$\theta = \frac{\theta_1 - \theta_2 + \theta_3}{|\theta_1 - \theta_2 + \theta_3|}. \tag{5}$$

Thus, θ_1 and θ_2 are fixed vectors, while most choices of θ_3 will work for us, where by "most" we refer to the uniform measure on S^{n-1}. The first step the proof below is to show that for any unit vector $\theta \in S^{n-1}$,

$$\text{Median}\,(|\langle X, \theta \rangle|) \leq CM/\sqrt{n}, \tag{6}$$

that is, any median of $|\langle X, \theta \rangle|$ is at most $CM/\sqrt{n}$. Then we need to show that when $\theta_3 \in \mathcal{F}_3$ and θ is defined as in (5), for all $0 \leq t \leq c\sqrt{n}$,

$$\min\left\{\mathbb{P}\left(Y \geq \frac{tM}{\sqrt{n}}\right), \mathbb{P}\left(Y \leq -\frac{tM}{\sqrt{n}}\right)\right\} \geq \tilde{c}e^{-\tilde{c}t^2}. \tag{7}$$

The proof of (7) is divided into three sections. The case where $t \in [0, \sqrt{\log n}]$ may essentially be handled by using the methods of [5], see Sect. 3. Let $t_0 > 0$ be defined via

$$e^{-t_0^2} = \mathbb{P}\left(|X| \geq M \text{ and } \left|\frac{X}{|X|} - \theta_2\right| \leq \frac{1}{5}\right). \tag{8}$$

In order to prove (7) in the range $t \in [\sqrt{\log n}, t_0]$, we will use tools from the local theory of Banach spaces, such as Sudakov's inequality as well as the concentration of measure on the sphere. Details in Sect. 4 below. The remaining interval $t \in [t_0, c\sqrt{n}]$ is analyzed in Sect. 5. In Sect. 6 we deduce Theorem 1.3 and Corollary 1.4 from Proposition 2.1 by using the angularly-isotropic position, along the lines of [5].

3   Central Limit Regime

This section is the first in a sequence of three sections that are dedicated to the proof of Proposition 2.1. Thus, we are given a random vector X in $\mathbb{R}^n$ with $\mathbb{P}(X = 0) = 0$ such that (4) holds true. We fix a number $M > 0$ with the property that

$$\mathbb{P}(|X| \geq M) \geq 1/3, \qquad \mathbb{P}(|X| \leq M) \geq 2/3. \tag{9}$$

That is, M is a $1/3$-quantile of $|X|$. Our first lemma verifies (6), as it states that for any choice of a unit vector θ, any median of the random variable $|\langle X, \theta \rangle|$ is at most $CM/\sqrt{n}$.

Lemma 3.1 *For any $\theta \in S^{n-1}$,*

$$\mathbb{P}\left(|\langle X, \theta \rangle| \geq CM/\sqrt{n}\right) < 1/2,$$

where $C > 0$ is a universal constant.

Proof It follows from (4) that for any $\theta \in S^{n-1}$,

$$\mathbb{E}\left[\langle X, \theta \rangle^2 \, \mathbb{1}_{\{|X| \leq M\}}\right] \leq \mathbb{E}\left[\langle X, \theta \rangle^2 \cdot \frac{M^2}{|X|^2}\right] = M^2 \cdot \mathbb{E}\left\langle \frac{X}{|X|}, \theta \right\rangle^2 \leq \frac{5M^2}{n}.$$

By the Markov-Chebyshev inequality,

$$\mathbb{P}\left(\langle X, \theta \rangle^2 \, \mathbb{1}_{\{|X| \leq M\}} \geq 35M^2/n\right) \leq 1/7.$$

Since $\mathbb{P}(|X| > M) \leq 1/3$, we obtain

$$\mathbb{P}\left(|\langle X, \theta \rangle| \geq \frac{6M}{\sqrt{n}}\right) \leq \mathbb{P}(|X| > M) + \mathbb{P}\left(|\langle X, \theta \rangle| \geq \frac{6M}{\sqrt{n}} \text{ and } |X| \leq M\right) \leq \frac{1}{3}$$

$$+ \frac{1}{7} < \frac{1}{2}.$$

The lemma follows with $C = 6$. $\square$

The rest of this section is devoted to the proof of (7) in the range $t \in [0, \sqrt{\log n}]$. The defining properties of $\theta_1, \theta_2 \in S^{n-1}$ from the previous section will not be used here, the entire analysis in this section applies for arbitrary unit vectors θ_1 and θ_2.

Lemma 3.2 *Let $\theta_1, \theta_2 \in S^{n-1}$ be any two fixed vectors. Then,*

$$\mathbb{P}\left(|X| \geq M, \ |\langle X, \theta_1 \rangle| \leq \frac{10|X|}{\sqrt{n}} \ \text{and} \ |\langle X, \theta_2 \rangle| \leq \frac{10|X|}{\sqrt{n}}\right) > \frac{1}{5}.$$

Proof By (4) and the Markov-Chebyshev inequality, for $j = 1, 2$,

$$\mathbb{P}\left(|\langle X, \theta_j \rangle| \geq \frac{10|X|}{\sqrt{n}}\right) \leq \frac{n}{100} \cdot \mathbb{E}\left\langle \frac{X}{|X|}, \theta_j \right\rangle^2 \leq \frac{n}{100} \cdot \frac{5}{n} = \frac{1}{20}.$$

Thanks to (9), we conclude that

$$\mathbb{P}\left(|X| \geq M, \ |\langle X, \theta_1 \rangle| \leq \frac{10|X|}{\sqrt{n}}, \ |\langle X, \theta_2 \rangle| \leq \frac{10|X|}{\sqrt{n}}\right) \geq 1 - \left(\frac{2}{3} + \frac{1}{20} + \frac{1}{20}\right) > \frac{1}{5}.$$

$\square$

Let $1 \leq k \leq n$. Following [5], we write $\mathcal{O}_k \subseteq (\mathbb{R}^n)^k$ for the collection of all k-tuples $(v_1, \ldots, v_k)$ with the following property: There exist orthonormal vectors $w_1, \ldots, w_k \in \mathbb{R}^n$ and real numbers $(a_{ij})_{i,j=0,\ldots,k}$ such that $|a_{ij}| < a_{ii}/k^2$ for $j < i$, and

$$v_i = \sum_{j=1}^{i} a_{ij} w_j \quad \text{for } i = 1, \ldots, k. \tag{10}$$

In other words, $\mathcal{O}_k$ consists of k-tuples of vectors that are almost orthogonal. By recalling the Gram-Schmidt process from linear algebra, we see that $(v_1, \ldots, v_k) \in \mathcal{O}_k$ assuming that

$$|\text{Proj}_{E_{i-1}} v_i| < |v_i|/k^2 \quad \text{for } i = 1, \ldots, k, \tag{11}$$

where E_i is the subspace spanned by the vectors $v_1, \ldots, v_i \in \mathbb{R}^n$ and Proj_{E_i} is the orthogonal projection operator onto E_i in $\mathbb{R}^n$. Here, $E_0 = \{0\}$.

Lemma 3.3 *Assume that $1 \leq k \leq n$ and fix $(v_1, \ldots, v_k) \in \mathcal{O}_k$. Then there exists $\mathcal{F} \subseteq S^{n-1}$ with $\sigma_{n-1}(\mathcal{F}) \geq 1 - C \exp(-c\sqrt{k})$ such that for any $\theta \in \mathcal{F}$ and $0 \leq t \leq \sqrt{\log k}$,*

$$\#\left\{1 \leq i \leq k; \ \langle v_i, \theta \rangle \geq c_1 \frac{|v_i|}{\sqrt{n}} \cdot t\right\} \geq c_2 e^{-C_3 t^2} \cdot k,$$

where $c_1, c_2, C_3, c, C > 0$ are universal constants.

Proof Let $w_1, \ldots, w_k$ and (a_{ij}) be as in (10). By applying an orthogonal transformation in $\mathbb{R}^n$, we may assume that $w_i = e_i$, the standard ith unit vector. Let $\Gamma = (\Gamma_1, \ldots, \Gamma_n) \in \mathbb{R}^n$ be a standard Gaussian random vector in $\mathbb{R}^n$. For $i = 1, \ldots, n$ and $t > 0$, it is well-known that

$$\mathbb{P}(\Gamma_i \geq t) = \frac{1}{\sqrt{2\pi}} \int_t^\infty e^{-s^2/2} ds \in [ce^{-t^2}, Ce^{-t^2/2}].$$

Therefore, by the Chernoff large deviations bound (e.g., [3, Chap. 2]), for any $t > 0$,

$$\mathbb{P}\left(\#\{1 \leq i \leq k; \Gamma_i \geq t\} \geq \frac{c}{2} \cdot e^{-t^2} \cdot k\right) \geq 1 - \tilde{C}\exp\left(-\tilde{c}e^{-t^2}k\right). \tag{12}$$

From the Bernstein large deviation inequality (e.g., [3, Chap. 2]),

$$\mathbb{P}\left(|\Gamma| \leq 2\sqrt{n}\right) \geq 1 - Ce^{-cn}, \qquad \mathbb{P}\left(\sum_{i=1}^{k} |\Gamma_i| \leq 2k\right) \geq 1 - \hat{C}e^{-\hat{c}k}. \tag{13}$$

Note that when $\sum_{i=1}^{k} |\Gamma_i| \leq 2k$, for any $i = 1, \ldots, k$,

$$\langle \Gamma, v_i \rangle = a_{ii} \cdot \left\langle \Gamma, e_i + \sum_{j=2}^{i} \frac{a_{ij}}{a_{ii}} e_j \right\rangle \geq a_{ii}\left(\Gamma_i - \frac{\sum_{j=1}^{k} |\Gamma_j|}{k^2}\right) \geq a_{ii}\left(\Gamma_i - \frac{2}{k}\right). \tag{14}$$

Moreover, $a_{ii} = |v_i - \sum_{j \leq 2} a_{ij}e_j| \geq |v_i| - a_{ii}/k$ for all $i = 1, \ldots, k$. Therefore $a_{ii} \geq |v_i|/2$ for all i. It thus follows from (14) that when $\sum_{i=1}^{k} |\Gamma_i| \leq 2k$, for any i,

$$\Gamma_i \geq t \quad \Longrightarrow \quad \langle \Gamma, v_i \rangle \geq a_{ii}\Gamma_i/2 \geq |v_i|t/4 \qquad \text{for all } t \geq 4/k.$$

Hence we deduce from (12) and (13) that for all $t \geq 4/k$,

$$\mathbb{P}\left(\#\left\{i; \langle \Gamma, v_i \rangle \geq \frac{t|v_i|}{4}\right\} \geq \frac{c}{2} \cdot e^{-t^2} \cdot k\right) \geq 1 - \tilde{C}\exp\left(-\tilde{c}e^{-t^2}k\right). \tag{15}$$

Write $I = \{\ell \in \mathbb{Z}; \ell \geq 2, 2^\ell \leq \sqrt{\log k}/5\}$. By substituting $t = 2^\ell$ into (15) we see that

$$\mathbb{P}\left(\forall \ell \in I, \#\{i; \langle \Gamma, v_i \rangle \geq 2^{\ell-2}|v_i|\} \geq \frac{c}{2} \cdot e^{-(2^\ell)^2} \cdot k\right) \geq 1 - \tilde{C}\sum_{\ell \in I}\exp\left(-\tilde{c}e^{-(2^\ell)^2}k\right).$$

The latter sum is at most $\overline{C}\exp(-\overline{c}\sqrt{k})$. Moreover, suppose that $x \in \mathbb{R}^n$ is a fixed vector such that $\#\{i; \langle x, v_i \rangle \geq t|v_i|/4\} \geq (c/2)e^{-t^2}k$ for all $1 \leq t \leq \sqrt{\log k}/5$ of the form $t = 2^\ell$ for an integer $\ell \geq 2$. By adjusting the constants, we see that for any real number t with $0 \leq t \leq \sqrt{\log k}$,

$$\#\{i; \langle x, v_i \rangle \geq c_1 t|v_i|\} \geq \tilde{c}e^{-\tilde{C}t^2}k.$$

Consequently,

$$\mathbb{P}\left(\forall t \in [0, \sqrt{\log k}], \#\{i; \langle \Gamma, v_i \rangle \geq c_1 t|v_i|\} \geq \tilde{c}e^{-\tilde{C}t^2} \cdot k\right) \geq 1 - \overline{C}e^{-\overline{c}\sqrt{k}}.$$

Recall that $|\Gamma| \leq 2\sqrt{n}$ with a probability of at least $1 - Ce^{-cn}$. Therefore, as $k \leq n$,

$$\mathbb{P}\left(\forall t \in [0, \sqrt{\log k}], \ \#\left\{i; \ \left\langle \frac{\Gamma}{|\Gamma|}, v_i \right\rangle \geq c_1 \frac{t|v_i|}{2\sqrt{n}}\right\} \geq \tilde{c}e^{-\tilde{C}t^2} \cdot k\right) \geq 1 - \hat{C}e^{-\hat{c}\sqrt{k}}. \tag{16}$$

Since $\Gamma/|\Gamma|$ is distributed uniformly on S^{n-1}, the lemma follows from (16). □

Let $E \subseteq \mathbb{R}^n$ be an arbitrary subspace. It follows from (4) that

$$\mathbb{E}\left|\mathrm{Proj}_E \frac{X}{|X|}\right|^2 = \mathbb{E}\sum_{i=1}^{\dim(E)}\left\langle \frac{X}{|X|}, u_i \right\rangle^2 \leq 5\frac{\dim(E)}{n}, \tag{17}$$

where $u_1, \ldots, u_m$ is an orthonormal basis of the subspace E for $m = \dim(E)$.

Lemma 3.4 *Set* $\ell = \lfloor n^{1/8} \rfloor$ *and let* $\theta_1, \theta_2 \in S^{n-1}$ *be any fixed vectors. Let* $X_1, \ldots, X_\ell$ *be independent copies of the random vector* X. *Then with a probability of at least* $1 - C/\ell$ *of selecting* $X_1, \ldots, X_\ell$, *there exists a subset* $I \subseteq \{1, \ldots, \ell\}$ *with the following three properties:*

(i) $k := \#(I) \geq \ell/10$.
(ii) *We may write* $I = \{i_1, \ldots, i_k\}$ *such that* $(X_{i_1}, \ldots, X_{i_k}) \in \mathcal{O}_k$.
(iii) *For* $j = 1, \ldots, k$,

$$|X_{i_j}| \geq M, \quad |\langle X_{i_j}, \theta_1 \rangle| \leq 10|X_{i_j}|/\sqrt{n} \quad and \quad |\langle X_{i_j}, \theta_2 \rangle| \leq 10|X_{i_j}|/\sqrt{n}.$$

Here, $C > 0$ *is a universal constant.*

Proof We may assume that $\ell \geq 10$, as otherwise the lemma trivially holds with any $C \geq 10$. Define

$$I = \left\{1 \leq i \leq \ell; \ |X_i| \geq M, \ |\langle X_i, \theta_1 \rangle| \leq 10|X_i|/\sqrt{n}, \ |\langle X_i, \theta_2 \rangle| \leq 10|X_i|/\sqrt{n}\right\}.$$

Denote $k = \#(I)$ and let $i_1 < i_2 < \ldots < i_k$ be the elements of I. We conclude from Lemma 3.2 and the Chernoff large deviation bound that

$$\mathbb{P}(\#(I) \geq \ell/10) \geq 1 - C\exp(-c\ell). \tag{18}$$

Thus (i) holds with a probability of at least $1 - C\exp(-c\ell)$. Clearly (iii) holds true with probability one, by the definition of I. All that remains is to show that (ii) holds true with a probability of at least $1 - 1/\ell$. Write F_i for the subspace spanned by $X_1, \ldots, X_i$, with $F_0 = \{0\}$. It follows from (17) that for $i = 1, \ldots, \ell$,

$$\mathbb{E}\left|\mathrm{Proj}_{F_{i-1}} \frac{X_i}{|X_i|}\right|^2 \leq \frac{5 \cdot \dim(F_{i-1})}{n} \leq \frac{5(i-1)}{n} \leq \frac{5\ell}{n} < \frac{1}{\ell^6},$$

as $10 \leq \ell \leq n^{1/8}$. It follows from the Markov-Chebyshev inequality that with a probability of at least $1 - 1/\ell$,

$$\left| \text{Proj}_{F_{i-1}} \frac{X_i}{|X_i|} \right| < \frac{1}{\ell^2} \qquad \text{for all } i = 1, \ldots, \ell.$$

Write E_j for the subspace spanned by $X_{i_1}, \ldots, X_{i_j}$. Then $E_{j-1} \subseteq F_{i_j-1}$. Therefore, with a probability of at least $1 - 1/\ell$,

$$\left| \text{Proj}_{E_{j-1}} \frac{X_{i_j}}{|X_{i_j}|} \right| \leq \left| \text{Proj}_{F_{i_j-1}} \frac{X_{i_j}}{|X_{i_j}|} \right| < \frac{1}{\ell^2} \leq \frac{1}{k^2} \qquad \text{for all } j = 1, \ldots, k.$$

In view of (11), we see that (ii) holds true with a probability of at least $1 - 1/\ell$, thus completing the proof of the lemma. $\square$

By combining Lemmas 3.3 and 3.4 we arrive at the following:

Lemma 3.5 *Let ℓ, θ_1, θ_2 be as in Lemma 3.4. Then there exists a fixed subset $\mathcal{F} \subseteq S^{n-1}$ with $\sigma_{n-1}(\mathcal{F}) \geq 1 - C/\sqrt{\ell}$ such that for any $\theta_3 \in \mathcal{F}$ the following holds: Define θ via (5). Let $X_1, \ldots, X_\ell$ be independent copies of the random vector X. Then with a probability of at least $1 - C/\sqrt{\ell}$ of selecting $X_1, \ldots, X_\ell$,*

$$\# \left\{ 1 \leq i \leq \ell ; \, \langle X_i, \theta \rangle \geq c_1 \frac{M}{\sqrt{n}} \cdot t \right\} \geq c_2 e^{-C_3 t^2} \cdot \ell, \qquad \text{for all } 0 \leq t \leq \sqrt{\log \ell}, \tag{19}$$

and

$$\# \left\{ 1 \leq i \leq \ell ; \, \langle X_i, \theta \rangle \leq -c_1 \frac{M}{\sqrt{n}} \cdot t \right\} \geq c_2 e^{-C_3 t^2} \cdot \ell, \qquad \text{for all } 0 \leq t \leq \sqrt{\log \ell}. \tag{20}$$

Here, $c_1, c_2, C_3, c, C > 0$ are universal constants.

Proof Let Θ be a random vector, distributed uniformly on S^{n-1}. According to Lemma 3.4, with a probability of at least $1 - C/\ell$ of selecting $X_1, \ldots, X_\ell$, there exists a subset

$$I = \{i_1, \ldots, i_k\} \subseteq \{1, \ldots, \ell\}$$

such that properties (i)–(iii) of Lemma 3.4 hold true. Let us apply Lemma 3.3. Then under the event where properties (i)–(iii) hold true, with a probability of at least $1 - \tilde{C} \exp(-\tilde{c}\sqrt{\ell})$ of selecting $\Theta \in S^{n-1}$,

$$\# \left\{ 1 \leq j \leq k ; \, \langle X_{i_j}, \Theta \rangle \geq c_1 \frac{|X_{i_j}|}{\sqrt{n}} \cdot t \right\} \geq c_2 e^{-C_3 t^2} \cdot k \qquad \text{for all } 0 \leq t \leq \sqrt{\log k},$$

and moreover $k \geq \ell/10$ with

$$\max\left\{\left|\left\langle \frac{X_{ij}}{|X_{ij}|}, \theta_1 \right\rangle\right|, \left|\left\langle \frac{X_{ij}}{|X_{ij}|}, \theta_2 \right\rangle\right|\right\} \leq \frac{10}{\sqrt{n}} \qquad \text{for } j = 1, \ldots, k.$$

Consequently, under the event where properties (i)–(iii) hold true, with a probability of at least $1 - \tilde{C}\exp(-\tilde{c}\sqrt{\ell})$ of selecting $\Theta \in S^{n-1}$,

$$\#\left\{1 \leq j \leq k; \left\langle \frac{X_{ij}}{|X_{ij}|}, \theta_1 - \theta_2 + \Theta \right\rangle \geq \frac{c_1}{2}\frac{t}{\sqrt{n}}\right\} \geq c_2 e^{-C_3 t^2} \cdot k \qquad \text{for } t \in [80/c_1, \sqrt{\log k}].$$

Since $k \geq \ell/10$, the condition $t \in [80/c_1, \sqrt{\log k}]$ can be upgraded to $t \in [0, \sqrt{\log \ell}]$ at the cost of modifying the universal constants. Recall that by Lemma 3.3(iii), we have that $|X_{ij}| \geq M$ for all j. By the triangle inequality, with probability one, $0 < |\theta_1 - \theta_2 + \Theta| \leq 3$. Hence,

$$|X_{ij}|/|\theta_1 - \theta_2 + \Theta| \geq M/3.$$

Therefore, under the event where properties (i)–(iii) hold true, with a probability of at least $1 - \tilde{C}\exp(-\tilde{c}\sqrt{\ell})$ of selecting $\Theta \in S^{n-1}$,

$$\forall t \in [0, \sqrt{\log \ell}], \quad \#\left\{1 \leq i \leq \ell; \left\langle X_i, \frac{\theta_1 - \theta_2 + \Theta}{|\theta_1 - \theta_2 + \Theta|} \right\rangle \geq \bar{c}_1 \frac{M}{\sqrt{n}} \cdot t\right\} \geq \bar{c}_2 e^{-\bar{C}_3 t^2} \cdot \ell. \tag{21}$$

Write $\mathcal{A}$ for the event that the statement in (21) holds true. Denoting $\vec{X} = (X_1, \ldots, X_\ell)$, we have shown that

$$\mathbb{P}((\Theta, \vec{X}) \in \mathcal{A}) \geq 1 - \tilde{C}\exp(-\tilde{c}\sqrt{\ell}) - C/\ell \geq 1 - \bar{C}/\ell.$$

Denote

$$\mathcal{F} = \left\{\theta \in S^{n-1}; \mathbb{P}_{\vec{X}}((\theta, \vec{X}) \in \mathcal{A}) \geq 1 - \bar{C}/\sqrt{\ell}\right\}.$$

Then,

$$1 - \frac{\bar{C}}{\ell} \leq \mathbb{P}((\Theta, \vec{X}) \in \mathcal{A}) \leq \mathbb{P}(\Theta \in \mathcal{F}) + \left(1 - \frac{\bar{C}}{\sqrt{\ell}}\right)\mathbb{P}(\Theta \notin \mathcal{F}). \tag{22}$$

It follows from (22) that $\sigma_{n-1}(\mathcal{F}) = \mathbb{P}(\Theta \in \mathcal{F}) \geq 1 - 1/\sqrt{\ell}$. By the definition of $\mathcal{F} \subseteq S^{n-1}$, for any $\theta_3 \in \mathcal{F}$, with a probability of at least $1 - \bar{C}\sqrt{\ell}$ of selecting $X_1, \ldots, X_\ell$,

$$\forall t \in [0, \sqrt{\log \ell}], \quad \#\left\{1 \leq i \leq \ell; \left\langle X_i, \frac{\theta_1 - \theta_2 + \theta_3}{|\theta_1 - \theta_2 + \theta_3|} \right\rangle \geq \bar{c}_1 \frac{M}{\sqrt{n}} \cdot t\right\} \geq \bar{c}_2 e^{-\bar{C}_3 t^2} \cdot \ell.$$

This completes the proof of (19). The argument for (20) requires only the most trivial modifications, and we leave it for the reader to complete. $\qquad\square$

We will use the well-known fact that for any random variable Y and measurable sets $A_1, \ldots, A_\ell$, by the Markov-Chebyshev inequality,

$$\frac{1}{s} \cdot \sum_{i=1}^{\ell} \mathbb{P}(Y \in A_i) = \frac{1}{s} \cdot \mathbb{E} \sum_{i=1}^{\ell} 1_{\{Y \in A_i\}} \geq \mathbb{P}\big(\# \{i \,;\, Y \in A_i\} \geq s\big) \qquad (s > 0).$$

Corollary 3.6 *Let $\theta_1, \theta_2 \in S^{n-1}$ be any fixed vectors. Then there exists a fixed subset $\mathcal{F} \subseteq S^{n-1}$ with $\sigma_{n-1}(\mathcal{F}) \geq 1 - C/n^c$ such that for any $\theta_3 \in \mathcal{F}$, defining θ via (5),*

$$\forall t \in [0, 5\sqrt{\log n}], \quad \min \left\{ \mathbb{P}\left(\langle X, \theta \rangle \geq c_1 \frac{M}{\sqrt{n}} \cdot t \right), \mathbb{P}\left(\langle X, \theta \rangle \leq -c_1 \frac{M}{\sqrt{n}} \cdot t \right) \right\} \geq c_2 e^{-C_3 t^2},$$

where $c, C, c_1, c_2, C_3 > 0$ are universal constants.

Proof We may assume that n exceeds a certain fixed universal constant, as otherwise the conclusion of the lemma trivially holds for $\mathcal{F} = \emptyset$. Set $\ell = \lfloor n^{1/8} \rfloor$ and let $\mathcal{F}$ be the set from Lemma 3.5. Let $\theta_3 \in \mathcal{F}$ and define θ via (5). Suppose that $X_1, \ldots, X_\ell$ are independent copies of the random vector X. Then for any $0 \leq t \leq \sqrt{\log \ell}$,

$$\mathbb{P}\left(\langle X, \theta \rangle \geq c_1 \frac{M}{\sqrt{n}} \cdot t \right) = c_2 e^{-C_3 t^2} \frac{1}{c_2 e^{-C_3 t^2} \cdot \ell} \sum_{i=1}^{\ell} \mathbb{P}\left(\langle X_i, \theta \rangle \geq c_1 \frac{M}{\sqrt{n}} \cdot t \right)$$

$$\geq c_2 e^{-C_3 t^2} \cdot \mathbb{P}\left(\# \left\{ i \,;\, \langle X_i, \theta \rangle \geq c_1 \frac{M}{\sqrt{n}} \cdot t \right\} \geq c_2 e^{-C_3 t^2} \cdot \ell \right) \geq c_2 e^{-C_3 t^2} \cdot (1 - C/\sqrt{\ell}),$$

where the last passage is the content of Lemma 3.5. We may similarly obtain a corresponding lower bound for $\mathbb{P}\big(\langle X, \theta \rangle \leq -c_1 t M/\sqrt{n} \big)$. Since $\ell = \lfloor n^{1/8} \rfloor$, the desired conclusion follows by adjusting the constants. $\qquad\square$

4   Geometry of the High-Dimensional Sphere

This is the second section dedicated to the proof of Proposition 2.1. A few geometric properties of the high-dimensional sphere will be used here. For example, the sphere S^{n-1} does not contain more than n mutually orthogonal vectors, yet it contains $e^{\varepsilon n}$ mutually almost-orthogonal vectors. Moreover, for the purpose of computing the expectation of the supremum, a family of $e^{\varepsilon n}$ standard Gaussians which are almost-orthogonal in pairs behaves approximately like a collection of independent Gaussians.

While Corollary 3.6 takes care of the interval $t \in [0, 5\sqrt{\log n}]$, in this section we deal with the range $t \in [5\sqrt{\log n}, t_0]$ where t_0 is defined in (8). We begin with some background on Sudakov's minoration theorem and the concentration of measure inequality on the sphere. Given a bounded, non-empty subset $S \subseteq \mathbb{R}^n$, its *supporting functional* is defined via

$$h_S(\theta) = \sup_{x \in S} \langle x, \theta \rangle \qquad (\theta \in \mathbb{R}^n).$$

The supporting functional h_S is a convex function on $\mathbb{R}^n$ whose Lipschitz constant is bounded by $R(S) = \sup_{x \in S} |x|$. The *mean width* of S is $2M^*(S)$ where

$$M^*(S) = \int_{S^{n-1}} h_S(\theta) d\sigma_{n-1}(\theta).$$

The concentration inequality for Lipschitz functions on the sphere (see, e.g., [7, Appendix V]) states that for any $r > 0$,

$$\sigma_{n-1}\left(\{v \in S^{n-1} ; |h_S(v) - M^*(S)| \geq r \cdot R(S)\}\right) \leq Ce^{-cr^2n}. \tag{23}$$

A lower bound for $M^*(S)$ is provided by the following Sudakov's minoration theorem (see, e.g., [6, Sect. 3.3]):

Theorem 4.1 (Sudakov) *Let $N \geq 1, \alpha > 0$ and let $x_1, \ldots, x_N \in \mathbb{R}^n$. Set $S = \{x_1, \ldots, x_N\}$ and assume that $|x_i - x_j| \geq \alpha$ for any $i \neq j$. Then,*

$$M^*(S) \geq c\alpha\sqrt{\frac{\log N}{n}},$$

where $c > 0$ is a universal constant.

We shall need the following elementary lemma:

Lemma 4.2 *Let $Z_1, \ldots, Z_N$ be random variables attaining values in $\{0, 1\}$. Let $1 \leq k \leq N, 0 \leq \varepsilon \leq 1$, and assume that for any $A \subseteq \{1, \ldots, N\}$ with $\#(A) = k$,*

$$\mathbb{P}(\exists i \in A, Z_i = 1) \geq 1 - \varepsilon. \tag{24}$$

Then,

$$\mathbb{P}\left(\sum_{i=1}^{N} Z_i \geq \frac{N}{3k}\right) \geq 1 - 2\varepsilon. \tag{25}$$

Proof If $k \geq N/3$ then (25) holds true, since it follows from (24) that with a probability of at least $1 - \varepsilon$, there is a non-zero element among $Z_1, \ldots, Z_N$. Suppose now that $k < N/3$. The number of k-elements subsets $A \subseteq \{1, \ldots, N\}$ with $\max_{i \in A} Z_i = 0$ equals

$$\binom{N - \sum_{i=1}^{N} Z_i}{k}.$$

Write $\mathcal{E}$ for the event that $\sum_{i=1}^{N} Z_i \leq N/(3k)$. Conditioning on the event $\mathcal{E}$,

$$\frac{1}{\binom{N}{k}} \sum_{\#(A)=k} \mathbb{P}\left(\forall i \in A, \, Z_i = 0 \,|\, \mathcal{E}\right) \geq \frac{\binom{N - \lfloor N/(3k) \rfloor}{k}}{\binom{N}{k}} \geq \left(1 - \frac{N/(3k)}{N-k}\right)^k$$

$$> \left(1 - \frac{1}{2k}\right)^k \geq \frac{1}{2}.$$

However, by (24),

$$\varepsilon \geq \frac{1}{\binom{N}{k}} \sum_{\#(A)=k} \mathbb{P}\left(\forall i \in A, \, Z_i = 0\right)$$

$$\geq \frac{1}{\binom{N}{k}} \sum_{\#(A)=k} \mathbb{P}(\mathcal{E}) \cdot \mathbb{P}\left(\forall i \in A, \, Z_i = 0 \,|\, \mathcal{E}\right) \geq \mathbb{P}(\mathcal{E})/2.$$

Hence $\mathbb{P}(\mathcal{E}) \leq 2\varepsilon$ and the lemma is proven. $\square$

Sudakov's theorem is used in the following lemma:

Lemma 4.3 *Let $N \geq n$ and let $x_1, \ldots, x_N \in S^{n-1}$ be such that $\langle x_i, x_j \rangle \leq 49/50$ for any $i \neq j$. Then there exists $\mathcal{F} \subseteq S^{n-1}$ with $\sigma_{n-1}(\mathcal{F}) \geq 1 - C/n^c$ such that for any $\theta \in \mathcal{F}$,*

$$\frac{\#\left\{1 \leq i \leq N ; \, \langle x_i, \theta \rangle \geq c_1 t/\sqrt{n}\right\}}{N} \geq c_2 e^{-C_3 t^2}, \qquad \text{for all } t \in [\sqrt{\log n}, \sqrt{\log N}],$$

$$(26)$$

where $c_1, c_2, C_3, c, C > 0$ are universal constants.

Proof Denote $S = \{x_1, \ldots, x_N\} \subset S^{n-1}$ and note that $|x_i - x_j| \geq \sqrt{2 - 49/25} = 1/5$ for all $i \neq j$. Fix a number $t \in [\sqrt{\log n}, \sqrt{\log N}]$. Let $A \subseteq \{x_1, \ldots, x_N\}$ be any subset with $\#(A) \geq \exp(t^2)$. By Theorem 4.1,

$$M^*(A) \geq ct/\sqrt{n}. \tag{27}$$

Next we will apply the concentration inequality (23) with $r = M^*(A)/(2R(A))$. Since $R(A) = 1$, it follows from (23) and (27) that

$$\sigma_{n-1}\left(\{\theta \in S^{n-1} ; \, h_A(\theta) \geq M^*(A)/2\}\right) \geq 1 - C \exp\left(-cn\left(\frac{M^*(A)}{R(A)}\right)^2\right) \geq 1 - \tilde{C} e^{-\tilde{c} t^2}.$$

Let Θ be a random vector, distributed uniformly over S^{n-1}. By combining the last inequality with (27), we see that for any fixed subset $\tilde{A} \subseteq \{1, \ldots, N\}$ with $\#(\tilde{A}) = \lceil \exp(t^2) \rceil$,

$$\mathbb{P}\left(\exists i \in \tilde{A} ; \langle x_i, \Theta \rangle \geq ct/\sqrt{n}\right) \geq 1 - \tilde{C}e^{-\tilde{c}t^2}.$$

Let us now apply Lemma 4.2 for $Z_i = 1_{\{\langle x_i, \Theta \rangle \geq ct/\sqrt{n}\}}$. Lemma 4.2 now implies that with a probability of at least $1 - 2\tilde{C}e^{-\tilde{c}t^2}$ of selecting $\Theta \in S^{n-1}$,

$$\#\left\{1 \leq i \leq N ; \langle x_i, \Theta \rangle \geq ct/\sqrt{n}\right\} \geq \frac{N}{3\lceil \exp(t^2) \rceil} \geq \frac{N}{6} \cdot e^{-t^2}.$$

We now let the parameter t vary. Let I be the collection of all integer powers of two that lie in the interval $[\sqrt{\log n}, \sqrt{\log N}]$. Then,

$$\mathbb{P}\left(\forall t \in I, \frac{\#\left\{1 \leq i \leq N ; \langle x_i, \Theta \rangle \geq ct/\sqrt{n}\right\}}{N} \geq \frac{e^{-t^2}}{6}\right) \geq 1 - \sum_{t \in I} 2\tilde{C}e^{-\tilde{c}t^2} \geq 1 - \frac{\hat{C}}{n^{\hat{c}}}.$$

The restriction $t \in I$ may be upgraded to the condition $t \in [\sqrt{\log n}, \sqrt{\log N}]$ by adjusting the constants. The lemma is thus proven. $\qquad\square$

Recall the construction of θ_1 and θ_2 from Sect. 2, and also the definition (8) of the parameter t_0. From the construction we see that for any $v \in S^{n-1}$ with $|\langle v, \theta_1 \rangle| \leq 1/10$,

$$\mathbb{P}\left(|X| \geq M \text{ and } \left|\frac{X}{|X|} - v\right| \leq \frac{1}{5}\right) \leq 2e^{-t_0^2}, \tag{28}$$

where $M > 0$ satisfies $\mathbb{P}(|X| \geq M) \geq 1/3$ and $\mathbb{P}(|X| \leq M) \geq 2/3$.

Lemma 4.4 *Assume that $t_0 \geq 5\sqrt{\log n}$ and set $N = \lfloor e^{t_0^2/4} \rfloor$. Let $X_1, \ldots, X_N$ be independent copies of X. Then with a probability of at least $1 - C/n$ of selecting $X_1, \ldots, X_N$, there exists $I \subseteq \{1, \ldots, N\}$ with the following three properties:*

(i) $\#(I) \geq N/10$.
(ii) For any $i, j \in I$ with $i \neq j$ we have $\langle X_i, X_j \rangle \leq (49/50) \cdot |X_i| \cdot |X_j|$.
(iii) For any $i \in I$,

$$|X_i| \geq M, \quad |\langle X_i, \theta_1 \rangle| \leq 10|X_i|/\sqrt{n} \quad \text{and} \quad |\langle X_i, \theta_2 \rangle| \leq 10|X_i|/\sqrt{n}.$$

Here, $C > 0$ is a universal constant.

Proof We may assume that $n \geq 10^4$, as otherwise for an appropriate choice of the constant C, all we claim is that a certain event holds with a non-negative probability. Write

$$\mathcal{A} = \{v \in \mathbb{R}^n ; |v| \geq M, \max_{j=1,2} |\langle v/|v|, \theta_j \rangle| \leq 10/\sqrt{n}\}.$$

According to Lemma 3.2, for $i = 1, \ldots, N$,

$$\mathbb{P}(X_i \in \mathcal{A}) > 1/5.$$

Denote $I = \{i = 1, \ldots, N; X_i \in \mathcal{A}\}$. By the Chernoff large deviation bound,

$$\mathbb{P}(\#(I) \geq N/10) \geq 1 - C\exp(-cN).$$

Note that $10/\sqrt{n} \leq 1/10$ and that if $v \in \mathcal{A}$ then $|\langle v/|v|, \theta_1 \rangle| \leq 1/10$. It thus follows from (28) that for any $i, j \in \{1, \ldots, N\}$ with $i \neq j$,

$$\mathbb{P}\left(i, j \in I \text{ and } \left| \frac{X_j}{|X_j|} - \frac{X_i}{|X_i|} \right| \leq \frac{1}{5}\right)$$

$$\leq \mathbb{P}\left(X_j \in \mathcal{A} \text{ and } \left| \frac{X_j}{|X_j|} - \frac{X_i}{|X_i|} \right| \leq \frac{1}{5} \;\middle|\; X_i \in \mathcal{A}\right) \leq 2e^{-t_0^2} \leq \frac{2}{N^4}.$$

Consequently,

$$\mathbb{P}\left(\exists i, j \in I \text{ with } i \neq j \text{ and } \left| \frac{X_i}{|X_i|} - \frac{X_j}{|X_j|} \right| \leq \frac{1}{5}\right) \leq \frac{N(N-1)}{2} \cdot \frac{2}{N^4} \leq \frac{1}{N^2}.$$

We conclude that with a probability of at least $1 - C\exp(-cN) - 1/N^2 \geq 1 - \tilde{C}/n$,

$$\#(I) \geq N/10 \quad \text{and} \quad \forall i, j \in I, i \neq j \implies \left| \frac{X_i}{|X_i|} - \frac{X_j}{|X_j|} \right| > \frac{1}{5}.$$

Note that $\langle X_i, X_j \rangle \leq (49/50) \cdot |X_i| \cdot |X_j|$ if and only if $|X_i/|X_i| - X_j/|X_j|| \geq 1/5$. Thus conclusions (i)–(iii) hold true with a probability of at least $1 - \tilde{C}/n$, thereby completing the proof. □

By combining Lemmas 4.3 and 4.4 we arrive at the following:

Lemma 4.5 *Assume that $t_0 \geq 5\sqrt{\log n}$ and set $N = \lfloor e^{t_0^2/4} \rfloor$. Then there exists a fixed subset $\mathcal{F} \subseteq S^{n-1}$ with $\sigma_{n-1}(\mathcal{F}) \geq 1 - C/n^c$ such that for any $\theta_3 \in \mathcal{F}$ the following holds: Define θ via (5). Let $X_1, \ldots, X_N$ be independent copies of the random vector X. Then with a probability of at least $1 - \tilde{C}/n^{\tilde{c}}$ of selecting $X_1, \ldots, X_N$,*

$$\frac{\#\left\{1 \leq i \leq N; \langle X_i, \theta \rangle \geq c_1 \frac{M}{\sqrt{n}} \cdot t\right\}}{N} \geq c_2 e^{-C_3 t^2}, \qquad \text{for all } t \in [\sqrt{\log n}, t_0], \tag{29}$$

and

$$\frac{\#\left\{1 \leq i \leq N; \langle X_i, \theta \rangle \leq -c_1 \frac{M}{\sqrt{n}} \cdot t\right\}}{N} \geq c_2 e^{-C_3 t^2}, \qquad \text{for all } t \in [\sqrt{\log n}, t_0]. \tag{30}$$

Here, $c_1, c_2, C_3, c, C, \tilde{c}, \tilde{C} > 0$ are universal constants.

Proof This proof is almost identical to the deduction of Lemma 3.5 from Lemmas 3.3 and 3.4. Let us spell out the details. Set $\vec{X} = (X_1, \ldots, X_N)$ and let Θ be a random vector, independent of $\vec{X}$, distributed uniformly on S^{n-1}. We say that $\vec{X} \in \mathcal{A}_1$ if the event described in Lemma 4.4 holds true. Thus,

$$\mathbb{P}(\vec{X} \in \mathcal{A}_1) \geq 1 - C/n.$$

Assuming that $\vec{X} \in \mathcal{A}_1$, we may apply Lemma 4.3 and obtain that with a probability of at least $1 - \tilde{C}/n^{\tilde{c}}$ of selecting $\Theta \in S^{n-1}$,

$$\#\left\{ 1 \leq i \leq N ; \left\langle \frac{X_i}{|X_i|}, \Theta \right\rangle \geq c_1 t / \sqrt{n} \right\} \geq c_2 e^{-C_3 t^2} \cdot (N/10) \quad \text{for all } t \in [\sqrt{\log n}, \sqrt{\log N}].$$

Assuming that $\vec{X} \in \mathcal{A}_1$, we may use Lemma 4.4(iii) in order to conclude that with a probability of at least $1 - \tilde{C}/n^{\tilde{c}}$ of selecting $\Theta \in S^{n-1}$, for $t \in [\sqrt{\log n}, 4\sqrt{\log N}]$,

$$\#\left\{ 1 \leq i \leq N ; \left\langle X_i, \frac{\theta_1 - \theta_2 + \Theta}{|\theta_1 - \theta_2 + \Theta|} \right\rangle \geq \tilde{c}_1 \frac{M}{\sqrt{n}} \cdot t \right\} \geq \tilde{c}_2 e^{-\tilde{C}_3 t^2} \cdot N. \tag{31}$$

Write $\mathcal{A}_2$ for the event that (31) holds true for all $t \in [\sqrt{\log n}, 4\sqrt{\log N}]$. Thus,

$$\mathbb{P}((\Theta, \vec{X}) \in \mathcal{A}_2) \geq 1 - C/n - \tilde{C}/n^{\tilde{c}} \geq 1 - \bar{C}/n^{\bar{c}}.$$

Consequently, there exists $\mathcal{F} \subseteq S^{n-1}$ with

$$\sigma_{n-1}(\mathcal{F}) \geq 1 - \hat{C}/n^{\hat{c}}$$

with the following property: For any $\theta_3 \in \mathcal{F}$, with a probability of at least $1 - \hat{C}/n^{\hat{c}}$ of selecting $X_1, \ldots, X_N$, for all $t \in [\sqrt{\log n}, 4\sqrt{\log N}]$,

$$\#\left\{ 1 \leq i \leq N ; \left\langle X_i, \frac{\theta_1 - \theta_2 + \theta_3}{|\theta_1 - \theta_2 + \theta_3|} \right\rangle \geq c_1 \frac{M}{\sqrt{n}} \cdot t \right\} \geq c_2 e^{-C_3 t^2} \cdot N.$$

Recalling that $4\sqrt{\log N} \geq t_0$, we have established (29). The proof of (30) is similar. $\square$

The short proof of the following corollary is analogous to that of Corollary 3.6.

Corollary 4.6 *There exists a fixed subset $\mathcal{F} \subseteq S^{n-1}$ with $\sigma_{n-1}(\mathcal{F}) \geq 1 - C/n^c$ such that for any $\theta_3 \in \mathcal{F}$, defining θ via (5),*

$$\forall t \in [\sqrt{\log n}, t_0], \quad \min\left\{ \mathbb{P}\left(\langle X, \theta \rangle \geq c_1 \frac{M}{\sqrt{n}} \cdot t \right), \mathbb{P}\left(\langle X, \theta \rangle \leq -c_1 \frac{M}{\sqrt{n}} \cdot t \right) \right\} \geq c_2 e^{-C_3 t^2},$$

where $c, C, c_1, c_2, C_3 > 0$ are universal constants.

Proof We may assume that n exceeds a certain fixed universal constant. Let $\mathcal{F}$ be the set from Lemma 4.5, denote $N = \lfloor \exp(t_0^2/4) \rfloor$, and let $X_1, \ldots, X_N$ be independent copies of X. Then for any $\theta_3 \in \mathcal{F}$, defining θ via (5) we have that for any $t \in [\sqrt{\log n}, t_0]$,

$$\mathbb{P}\left(\langle X, \theta \rangle \geq c_1 \frac{M}{\sqrt{n}} \cdot t\right) \geq c_2 e^{-C_3 t^2} \cdot \mathbb{P}\left(\frac{\#\left\{i; \langle X_i, \theta \rangle \geq c_1 \frac{M}{\sqrt{n}} \cdot t\right\}}{N} \geq c_2 e^{-C_3 t^2}\right)$$

$$\geq \frac{c_2}{2} e^{-C_3 t^2},$$

where the last passage is the content of Lemma 4.5. The bound for $\mathbb{P}(\langle X, \theta \rangle \leq -c_1 t M/\sqrt{n})$ is proven similarly. $\square$

5   Proof of the Main Proposition

In this section we complete the proof of Proposition 2.1. We begin with the following standard observation:

Lemma 5.1 *Suppose that X is a random vector in $\mathbb{R}^n$ with $\mathbb{P}(X = 0) = 0$. Then there exists a fixed subset $\mathcal{F} \subseteq S^{n-1}$ of full measure, such that $\mathbb{P}(\langle X, \theta \rangle = 0) = 0$ for all $\theta \in \mathcal{F}$.*

Proof For $a > 0$, we say that a subspace $E \subseteq \mathbb{R}^n$ is a-basic if $\mathbb{P}(X \in E) \geq a$ while $\mathbb{P}(X \in F) < a$ for all subspaces $F \subsetneq E$. Lemma 7.1 in [5] states that there are only finitely many subspaces that are a-basic for any fixed $a > 0$. Write $\mathcal{S}$ for the collection of all subspaces that are a-basic for some rational number $a > 0$. Then $\mathcal{S}$ is a countable family which does not contain the subspace $\{0\}$. Consequently, the set

$$\mathcal{F} = \{\theta \in S^{n-1} \, ; \, \forall E \in \mathcal{S}, \, E \not\subseteq \theta^\perp\}$$

is a set of full measure in S^{n-1}, as its complement is the countable union of spheres of lower dimension. Here, $\theta^\perp = \{x \in \mathbb{R}^n \, ; \, \langle x, \theta \rangle = 0\}$. Suppose that $\theta \in \mathcal{F}$, and let us prove that $\mathbb{P}(\langle X, \theta \rangle = 0) = 0$. Otherwise, there exists a rational number $a > 0$ such that

$$\mathbb{P}(\langle X, \theta \rangle = 0) \geq a.$$

Thus $\theta^\perp$ contains an a-basic subspace, contradicting the definition of $\mathcal{F}$. $\square$

Recall the definition of M, θ_1 and θ_2 from Sect. 2.

Lemma 5.2 Let $\mathcal{F}_3 \subseteq \{\theta_3 \in S^{n-1} ; |\langle \theta_3, \theta_1 \rangle| \le \frac{1}{10} \text{ and } |\langle \theta_3, \theta_2 \rangle| \le \frac{1}{10}\}$. Then for any $\theta_3 \in \mathcal{F}_3$ and $v \in S^{n-1}$,

$$|v - \theta_1| \le \frac{1}{5} \quad \Longrightarrow \quad \langle v, \theta_1 - \theta_2 + \theta_3 \rangle \ge \frac{1}{10}, \tag{32}$$

and

$$|v - \theta_2| \le \frac{1}{5} \quad \Longrightarrow \quad \langle v, \theta_1 - \theta_2 + \theta_3 \rangle \le -\frac{1}{10}. \tag{33}$$

Proof Recall that $|\langle \theta_1, \theta_2 \rangle| \le 1/10$. Note that for any $\theta_3 \in \mathcal{F}_3$ and $i, j \in \{1, 2, 3\}$ with $i \ne j$,

$$\sqrt{9/5} \le |\theta_i - \theta_j| \le \sqrt{11/5}.$$

Let $v \in S^{n-1}$ be any vector with $|v - \theta_1| \le 1/5$. Then for any $\theta_3 \in \mathcal{F}_3$ and $j = 2, 3$ we have that

$$\sqrt{\frac{9}{5}} - \frac{1}{5} \le |\theta_j - \theta_1| - |\theta_1 - v| \le |v - \theta_j| \le |\theta_j - \theta_1| + |\theta_1 - v| \le \sqrt{\frac{11}{5}} + \frac{1}{5},$$

and hence for $j = 2, 3$,

$$\langle v, \theta_j \rangle = 1 - \frac{1}{2} \cdot |v - \theta_j|^2 \in \left[1 - \frac{1}{2} \cdot \left(\sqrt{\frac{11}{5}} + \frac{1}{5} \right)^2 , 1 - \frac{1}{2} \cdot \left(\sqrt{\frac{9}{5}} - \frac{1}{5} \right)^2 \right]$$

$$\subseteq \left[-\frac{3}{7}, \frac{3}{7} \right]. \tag{34}$$

However, $\langle v, \theta_1 \rangle \ge 49/50$ for such v, and hence (32) follows from (34). By replacing the triplet $(\theta_1, \theta_2, \theta_3)$ by $(\theta_2, \theta_1, -\theta_3)$ and repeating the above argument, we obtain (33). $\qquad \square$

Proof of Proposition 2.1 From Corollaries 3.6 and 4.6 we learn that there exists $\mathcal{F} \subseteq S^{n-1}$ with $\sigma_{n-1}(\mathcal{F}_3) \ge 1 - C/n^c$ such that for any $\theta_3 \in \mathcal{F}$, defining θ via (5),

$$\forall t \in [0, t_0], \quad \min \left\{ \mathbb{P} \left(\langle X, \theta \rangle \ge c_1 \frac{M}{\sqrt{n}} \cdot t \right), \mathbb{P} \left(\langle X, \theta \rangle \le -c_1 \frac{M}{\sqrt{n}} \cdot t \right) \right\} \ge c_2 e^{-c_3 t^2}. \tag{35}$$

According to Lemma 5.1, we may remove a set of measure zero from $\mathcal{F}$ and additionally assume that $\mathbb{P}(\langle X, \theta \rangle = 0) = 0$. From Lemma 3.1 we learn that any median of $|\langle X, \theta \rangle|$ is at most $CM/\sqrt{n}$. Hence (35) shows that for any $\theta_3 \in \mathcal{F}$, defining θ via (5) we have that $\langle X, \theta \rangle$ is Super-Gaussian of length $c_1 t_0$, with parameters $c_2, c_3 > 0$. We still need to increase the length to $c_1 \sqrt{n}$. To this end, denote

$$\mathcal{F}_3 = \left\{ \theta_3 \in \mathcal{F} ; \ |\langle \theta_3, \theta_1 \rangle| \leq \frac{1}{10} \quad \text{and} \quad |\langle \theta_3, \theta_2 \rangle| \leq \frac{1}{10} \right\}.$$

Then $\sigma_{n-1}(\mathcal{F}_3) \geq \sigma_{n-1}(\mathcal{F}) - C \exp(-cn) \geq 1 - \tilde{C}/n^{\tilde{c}}$. Recall from Sect. 2 that for $j = 1, 2$,

$$\mathbb{P} \left(|X| \geq M \text{ and } \left| \frac{X}{|X|} - \theta_j \right| \leq \frac{1}{5} \right) \geq \frac{1}{2} \cdot e^{-t_0^2}. \tag{36}$$

Let us fix $t \in [t_0, \sqrt{n}]$, $\theta_3 \in \mathcal{F}_3$ and define θ via (5). Since $0 < |\theta_1 - \theta_2 + \theta_3| \leq 3$, by (36) and Lemma 5.2,

$$\mathbb{P} \left(\langle X, \theta \rangle \geq \frac{Mt}{30\sqrt{n}} \right) \geq \mathbb{P} \left(\langle X, \theta_1 - \theta_2 + \theta_3 \rangle \geq \frac{Mt}{10\sqrt{n}} \right)$$

$$\geq \mathbb{P} \left(\left\langle \frac{X}{|X|}, \theta_1 - \theta_2 + \theta_3 \right\rangle \geq \frac{M}{10|X|} \right)$$

$$\geq \mathbb{P} \left(|X| \geq M, \left| \frac{X}{|X|} - \theta_1 \right| \leq \frac{1}{5} \right) \geq \frac{1}{2} \cdot e^{-t_0^2} \geq \frac{1}{2} \cdot e^{-t^2}.$$

Similarly,

$$\mathbb{P} \left(\langle X, \theta \rangle \leq -\frac{Mt}{30\sqrt{n}} \right) \geq \mathbb{P} \left(\left\langle \frac{X}{|X|}, \theta_1 - \theta_2 + \theta_3 \right\rangle \leq -\frac{M}{10|X|} \right)$$

$$\geq \mathbb{P} \left(|X| \geq M, \left| \frac{X}{|X|} - \theta_2 \right| \leq \frac{1}{5} \right) = e^{-t_0^2} \geq e^{-t^2}.$$

Therefore, we may upgrade (35) to the following statement: For any $\theta_3 \in \mathcal{F}$ and $t \in [0, \sqrt{n}]$, defining θ via (5),

$$\min \left\{ \mathbb{P} \left(\langle X, \theta \rangle \geq c_1 \frac{M}{\sqrt{n}} \cdot t \right), \mathbb{P} \left(\langle X, \theta \rangle \leq -\hat{c}_1 \frac{M}{\sqrt{n}} \cdot t \right) \right\} \geq \hat{c}_2 e^{-\hat{C}_3 t^2}.$$

We have thus proven that $\langle X, \theta \rangle$ is Super-Gaussian of length $\bar{c}_1 \sqrt{n}$ with parameters $\bar{c}_2, \bar{c}_3 > 0$. $\qquad \square$

6 Angularly-Isotropic Position

In this section we deduce Theorem 1.3 from Proposition 2.1 by using the angularly-isotropic position which is discussed below. We begin with the following:

Lemma 6.1 *Let $d, X, \mathcal{B}$ be as in Theorem 1.3. Set $n = \lceil d \rceil$. Then there exists a fixed linear map $T : \mathcal{B} \to \mathbb{R}^n$ such that for any $\varepsilon > 0$, the random vector $T(X)$ is of class eff.rank$_{\geq d - \varepsilon}$.*

Proof We will show that a generic linear map T works. Denote $N = \dim(\mathcal{B})$ and identify $\mathcal{B} \cong \mathbb{R}^N$. Since the effective rank of X is at least d, necessarily $d \leq N$ and hence also $n = \lceil d \rceil \leq N$. Let $L \subseteq \mathbb{R}^N$ be a random n-dimensional subspace, distributed uniformly in the Grassmannian $G_{N,n}$. Denote $T = \mathrm{Proj}_L : \mathbb{R}^N \to L$, the orthogonal projection operator onto the subspace L.

For any fixed subspace $E \subseteq \mathbb{R}^N$, with probability one of selecting $L \in G_{N,n}$,

$$\dim(\ker(T) \cap E) = \max\{0, \dim(E) - n\},$$

or equivalently,

$$\dim(T(E)) = \dim(E) - \dim(\ker(T) \cap E) = \min\{n, \dim(E)\}. \tag{37}$$

Recall that for $a > 0$, a subspace $E \subseteq \mathbb{R}^N$ is a-basic if $\mathbb{P}(X \in E) \geq a$ while $\mathbb{P}(X \in F) < a$ for all subspaces $F \subsetneq E$. Lemma 7.1 in [5] states that there exist only countably many subspaces that are a-basic with a being a positive, rational number. Write $\mathcal{G}$ for the collection of all these basic subspaces. Then with probability one of selecting $L \in G_{N,n}$,

$$\forall E \in \mathcal{G}, \qquad \dim(T(E)) = \min\{n, \dim(E)\}. \tag{38}$$

We now fix a subspace $L \in G_{N,n}$ for which $T = \mathrm{Proj}_L$ satisfies (38). Let $S \subseteq L$ be any subspace and assume that $a \in \mathbb{Q} \cap (0, 1]$ satisfies

$$\mathbb{P}(T(X) \in S) \geq a.$$

Then $\mathbb{P}(X \in T^{-1}(S)) \geq a$. Therefore $T^{-1}(S)$ contains an a-basic subspace E. Thus $E \in \mathcal{G}$ while $E \subseteq T^{-1}(S)$ and $\mathbb{P}(X \in E) \geq a$. Since the effective rank of X is at least d, necessarily $\dim(E) \geq a \cdot d$. Since $T(E) \subseteq S$, from (38),

$$\dim(S) \geq \dim(T(E)) = \min\{n, \dim(E)\} \geq \min\{n, \lceil a \cdot d \rceil\} = \lceil a \cdot d \rceil.$$

We have thus proven that for any subspace $S \subseteq L$ and $a \in \mathbb{Q} \cap (0, 1]$,

$$\mathbb{P}(T(X) \in S) \geq a \quad \Longrightarrow \quad \dim(S) \geq \lceil a \cdot d \rceil. \tag{39}$$

It follows from (39) that for any subspace $S \subseteq L$,

$$\mathbb{P}(T(X) \in S) \leq \dim(S)/d.$$

This implies that for any $\varepsilon > 0$, the random vector $T(X)$ is of class eff.rank$_{\geq d - \varepsilon}$. $\quad\square$

Lemma 6.2 *Let $d, X, \mathcal{B}$ be as in Theorem 1.3. Assume that $d < \dim(\mathcal{B})$ and that for any subspace $\{0\} \neq E \subsetneq \mathcal{B}$,*

$$\mathbb{P}(X \in E) < \dim(E)/d. \tag{40}$$

Then there exists $\varepsilon > 0$ such that X is of class eff.rank$_{\geq d+\varepsilon}$.

Proof Since the effective rank of X is at least d, necessarily $\mathbb{P}(X = 0) = 0$. Assume by contradiction that for any $\varepsilon > 0$, the random vector X is not of class eff.rank$_{\geq d+\varepsilon}$. Then for any $\varepsilon > 0$ there exists a subspace $\{0\} \neq E \subseteq \mathcal{B}$ with

$$\mathbb{P}(X \in E) \geq -\varepsilon + \dim(E)/d.$$

The Grassmannian of all k-dimensional subspaces of $\mathcal{B}$ is compact. Hence there is a dimension $1 \leq k \leq \dim(\mathcal{B})$ and a converging sequence of k-dimensional subspaces $E_1, E_2, \ldots \subseteq \mathcal{B}$ with

$$\mathbb{P}(X \in E_\ell) \geq -1/\ell + \dim(E_\ell)/d = -1/\ell + k/d \qquad \text{for all } \ell \geq 1. \tag{41}$$

Denote $E_0 = \lim_\ell E_\ell$, which is a k-dimensional subspace in $\mathcal{B}$. Let $U \subseteq \mathcal{B}$ be an open neighborhood of E_0 with the property that $tx \in U$ for all $x \in U, t \in \mathbb{R}$. Then $E_\ell \subseteq U$ for a sufficiently large ℓ, and we learn from (41) that

$$\mathbb{P}(X \in U) \geq k/d. \tag{42}$$

Since E_0 is the intersection of a decreasing sequence of such neighborhoods U, it follows from (42) that

$$\mathbb{P}(X \in E_0) \geq k/d = \dim(E_0)/d. \tag{43}$$

Since $d < \dim(\mathcal{B})$, the inequality in (43) shows that $E_0 \neq \mathcal{B}$. Hence $1 \leq \dim(E_0) \leq \dim(\mathcal{B}) - 1$, and (43) contradicts (40). The lemma is thus proven. □

The following lemma is a variant of Lemma 5.4 from [5].

Lemma 6.3 *Let $d, X, \mathcal{B}$ be as in Theorem 1.3. Then there exists a fixed scalar product $\langle \cdot, \cdot \rangle$ on $\mathcal{B}$ such that denoting $|\theta| = \sqrt{\langle \theta, \theta \rangle}$, we have*

$$\mathbb{E}\left\langle \frac{X}{|X|}, \theta \right\rangle^2 \leq \frac{|\theta|^2}{d} \qquad \text{for all } \theta \in \mathcal{B}. \tag{44}$$

Proof By induction on the dimension $n = \dim(\mathcal{B})$. Assume first that there exists a subspace $\{0\} \neq E \subsetneq \mathcal{B}$, such that equality holds true in (3). In this case, there exists a subspace $F \subseteq \mathcal{B}$ with $E \oplus F = \mathcal{B}$ and $\mathbb{P}(X \in E \cup F) = 1$. We will construct a scalar product in $\mathcal{B}$ as follows: Declare that E and F are orthogonal subspaces, and use the induction hypothesis in order to find appropriate scalar products in the subspace E and in the subspace F. This induces a scalar product in $\mathcal{B}$ which satisfies

$$\mathbb{E}\left\langle \frac{X}{|X|}, \theta \right\rangle^2 \leq \frac{|\theta|^2}{d} \qquad \text{for all } \theta \in E \cup F.$$

For any $\theta \in \mathcal{B}$ we may decompose $\theta = \theta_E + \theta_F$ with $\theta_E \in E, \theta_F \in F$. Since $\mathbb{P}(X \in E \cup F) = 1$, we obtain

$$\mathbb{E}\left\langle \frac{X}{|X|}, \theta \right\rangle^2 = \mathbb{E}\left\langle \frac{X}{|X|}, \theta_E \right\rangle^2 + \mathbb{E}\left\langle \frac{X}{|X|}, \theta_F \right\rangle^2 \leq \frac{|\theta_E|^2 + |\theta_F|^2}{d} = \frac{|\theta|^2}{d},$$

proving (44).

Next, assume that for any subspace $\{0\} \neq E \subsetneq \mathcal{B}$, the inequality in (3) is strict. There are two distinct cases, either $d = n$ or $d < n$. Consider first the case where $d = n = \dim(\mathcal{B})$. Thus, for any subspace $E \subseteq \mathcal{B}$ with $E \neq \{0\}$ and $E \neq \mathcal{B}$,

$$\mathbb{P}(X \in E) < \dim(E)/n.$$

This is precisely the main assumption of Corollary 5.3 in [5]. By the conclusion of the corollary, there exists a scalar product in $\mathcal{B}$ such that (44) holds true. We move on to the case where $d < n$. Here, we apply Lemma 6.2 and conclude that X is of class eff.rank$_{\geq d+\varepsilon}$ for some $\varepsilon > 0$. Therefore, for some $\varepsilon > 0$,

$$\mathbb{P}(X \in E) < \dim(E)/(d + \varepsilon) \qquad \forall E \subseteq \mathcal{B}. \qquad (45)$$

Now we invoke Lemma 5.4 from [5]. Its assumptions are satisfies thanks to (45). From the conclusion of that lemma, there exists a scalar product in $\mathcal{B}$ for which (44) holds true. $\qquad \square$

The condition that the effective rank of X is at least d is not only sufficient but is also necessary for the validity of conclusion (44) from Lemma 6.3. Indeed, it follows from (44) that for any subspace $E \subseteq \mathcal{B}$,

$$\mathbb{P}(X \in E) \leq \mathbb{E}\left| \text{Proj}_E \frac{X}{|X|} \right|^2 = \sum_{i=1}^{\dim(E)} \mathbb{E}\left\langle \frac{X}{|X|}, u_i \right\rangle^2 \leq \frac{\dim(E)}{d}, \qquad (46)$$

where $u_1, \ldots, u_m$ is an orthonormal basis of the subspace E with $m = \dim(E)$. Equality in (46) holds true if and only if $\mathbb{P}(X \in E \cup E^\perp) = 1$, where $E^\perp$ is the orthogonal complement to E. Consequently, the effective rank of X is at least d.

Definition 6.4 Let X be a random vector in $\mathbb{R}^n$ with $\mathbb{P}(X = 0) = 0$. We say that X is angularly-isotropic if

$$\mathbb{E}\left\langle \frac{X}{|X|}, \theta \right\rangle^2 = \frac{1}{n} \qquad \text{for all } \theta \in S^{n-1}. \qquad (47)$$

For $0 < d \leq n$ we say that $X/|X|$ is sub-isotropic with parameter d if

$$\mathbb{E}\left\langle \frac{X}{|X|}, \theta \right\rangle^2 \leq \frac{1}{d} \qquad \text{for all } \theta \in S^{n-1}. \tag{48}$$

We observe that X is angularly-isotropic if and only if $X/|X|$ is sub-isotropic with parameter n. Indeed, suppose that (48) holds true with $d = n$. Given any $\theta \in S^{n-1}$ we may find an orthonormal basis $\theta_1, \ldots, \theta_n \in \mathbb{R}^n$ with $\theta_1 = \theta$. Hence

$$1 = \mathbb{E}\left| \frac{X}{|X|} \right|^2 = \mathbb{E} \sum_{i=1}^{n} \left\langle \frac{X}{|X|}, \theta_i \right\rangle^2 \leq \sum_{i=1}^{n} \frac{1}{n} = 1,$$

and (47) is proven.

Proof of Theorem 1.3 According to Lemma 6.1, we may project X to a lower-dimensional space, and assume that $\dim(\mathcal{B}) = n = \lceil d \rceil$ and that the effective rank of X is at least $n/2$. Lemma 6.3 now shows that there exists a scalar product in $\mathcal{B}$ with respect to which $X/|X|$ is sub-isotropic with parameter $n/2$. We may therefore identify $\mathcal{B}$ with $\mathbb{R}^n$ so that

$$\mathbb{E}\left\langle \frac{X}{|X|}, \theta \right\rangle^2 \leq \frac{2}{n} \qquad \text{for all } \theta \in S^{n-1}.$$

Thus condition (4) of Proposition 2.1 is verified. By the conclusion of Proposition 2.1, there exists a non-zero linear functional $\ell : \mathbb{R}^n \to \mathbb{R}$ such that $\ell(X)$ is Super-Gaussian of length $c_1 \sqrt{n} \geq c\sqrt{d}$ with parameters $c_2, c_3 > 0$. □

Proof of Corollary 1.4 By assumption, $\mathbb{P}(X \in E) \leq \dim(E)/d$ for any finite-dimensional subspace $E \subseteq \mathcal{B}$. Lemma 7.2 from [5] states that there exists a continuous, linear map $T : \mathcal{B} \to \mathbb{R}^N$ such that $T(X)$ has an effective rank of at least $d/2$. We may now invoke Theorem 1.3 for the random vector $T(X)$, and conclude that for some non-zero, fixed, linear functional $\ell : \mathbb{R}^N \to \mathbb{R}$, the random variable $(\ell \circ T)(X)$ is Super-Gaussian of length $c_1 \sqrt{d}$ with parameters $c_2, c_3 > 0$. □

Remark 6.5 We were asked by Yaron Oz about analogs of Theorem 1.1 in the hyperbolic space. We shall work with the standard hyperboloid model

$$\mathbb{H}^n = \left\{ (x_0, \ldots, x_n) \in \mathbb{R}^{n+1} \,;\, -x_0^2 + \sum_{i=1}^{n} x_i^2 = -1, x_0 > 0 \right\}$$

where the Riemannian metric tensor is $g = -dx_0^2 + \sum_{i=1}^{n} dx_i^2$. For any linear subspace $L \subseteq \mathbb{R}^{n+1}$, the intersection $L \cap \mathbb{H}^n$ is a totally-geodesic submanifold of $\mathbb{H}^n$ which is called a hyperbolic subspace. When we discuss the dimension of a hyperbolic subspace, we refer to its dimension as a smooth manifold. Note that an $(n-1)$-dimensional hyperbolic subspace $E \subseteq \mathbb{H}^n$ divides $\mathbb{H}^n$ into two sides.

A signed distance function $d_E : \mathbb{H}^n \to \mathbb{R}$ is a function that equals the hyperbolic distance to E on one of these sides, and minus the distance to E on the other side. Given a linear functional $\ell : \mathbb{R}^{n+1} \to \mathbb{R}$ such that $E = \mathbb{H}^n \cap \{x \in \mathbb{R}^{n+1} ; \ell(x) = 0\}$ we may write

$$d_E(x) = \operatorname{arcsinh}(\alpha \cdot \ell(x)) \qquad (x \in \mathbb{H}^n)$$

for some $0 \neq \alpha \in \mathbb{R}$. It follows from Theorem 1.3 that for any absolutely-continuous random vector X in $\mathbb{H}^n$, there exists an $(n-1)$-dimensional hyperbolic subspace $E \subseteq \mathbb{H}^n$ and an associated signed distance function d_E such that the random variable $\sinh(d_E(X))$ is Super-Gaussian of length $c_1 \sqrt{n}$ with parameters $c_2, c_3 > 0$. In general, we cannot replace the random variable $\sinh(d_E(X))$ in the preceding statement by $d_E(X)$ itself. This is witnessed by the example of the random vector

$$X = \left(\sqrt{1 + R^2 \sum_{i=1}^{n} Z_i^2}, RZ_1, \ldots, RZ_n \right) \in \mathbb{R}^{n+1}$$

which is supported in $\mathbb{H}^n$. Here, $Z_1, \ldots, Z_n$ are independent standard Gaussian random variables, and $R > 1$ is a fixed, large parameter.

Acknowledgements I would like to thank Bo Berndtsson and Emanuel Milman for interesting discussions and for encouraging me to write this paper. Supported by a grant from the European Research Council.

References

1. K. Böröczky, E. Lutwak, D. Yang, G. Zhang, The logarithmic Minkowski problem. J. Am. Math. Soc. **26**(3), 831–852 (2013)
2. K. Böröczky, E. Lutwak, D. Yang, G. Zhang, Affine images of isotropic measures. J. Differ. Geom. **99**(3), 407–442 (2015)
3. S. Boucheron, G. Lugosi, P. Massart, *Concentration Inequalities. A Nonasymptotic Theory of Independence* (Oxford University Press, Oxford, 2013)
4. M. Henk, E. Linke, Cone-volume measures of polytopes. Adv. Math. **253**, 50–62 (2014)
5. B. Klartag, On nearly radial marginals of high-dimensional probability measures. J. Eur. Math. Soc. **12**(3), 723–754 (2010)
6. M. Ledoux, M. Talagrand, *Probability in Banach Spaces. Isoperimetry and Processes*, vol. 23 (Springer, Berlin, 1991). Ergeb. Math. Grenzgeb. (3)
7. V.D. Milman, G. Schechtman, *Asymptotic Theory of Finite-Dimensional Normed Spaces*. Lecture Notes in Mathematics, vol. 1200 (Springer, Berlin, 1986)
8. G. Paouris, On the existence of supergaussian directions on convex bodies . Mathematika **58**(2), 389–408 (2012)
9. P. Pivovarov, On the volume of caps and bounding the mean-width of an isotropic convex body. Math. Proc. Camb. Philos. Soc. **149**(2), 317–331 (2010)

A Remark on Measures of Sections of L_p-balls

Alexander Koldobsky and Alain Pajor

Abstract We prove that there exists an absolute constant C so that

$$\mu(K) \leq C\sqrt{p} \max_{\xi \in S^{n-1}} \mu(K \cap \xi^{\perp}) |K|^{1/n}$$

for any $p > 2$, any $n \in \mathbb{N}$, any convex body K that is the unit ball of an n-dimensional subspace of L_p, and any measure μ with non-negative even continuous density in $\mathbb{R}^n$. Here $\xi^{\perp}$ is the central hyperplane perpendicular to a unit vector $\xi \in S^{n-1}$, and $|K|$ stands for volume.

1 Introduction

The slicing problem [1, 4, 5, 29], a major open question in convex geometry, asks whether there exists a constant C so that for any $n \in \mathbb{N}$ and any origin-symmetric convex body K in $\mathbb{R}^n$,

$$|K|^{\frac{n-1}{n}} \leq C \max_{\xi \in S^{n-1}} |K \cap \xi^{\perp}|,$$

where $|K|$ stands for volume of proper dimension, and $\xi^{\perp}$ is the central hyperplane in $\mathbb{R}^n$ perpendicular to a unit vector ξ. The best-to-date result $C \leq O(n^{1/4})$ is due to Klartag [15], who improved an earlier estimate of Bourgain [6]. The answer is affirmative for unconditional convex bodies (as initially observed by Bourgain; see also [3, 14, 29]), intersection bodies [10, Theorem 9.4.11], zonoids, duals of bodies with bounded volume ratio [29], the Schatten classes [23], k-intersection bodies [21, 22]; see [7] for more details.

A. Koldobsky (✉)
Department of Mathematics, University of Missouri, Columbia, MO 65211, USA
e-mail: koldobskiya@missouri.edu

A. Pajor
Université Paris-Est, Laboratoire d'Analyse et Mathématiques Appliquées (UMR 8050), UPEM, Marne-la-Vallée F-77454, France
e-mail: alain.pajor@u-pem.fr

© Springer International Publishing AG 2017
B. Klartag, E. Milman (eds.), *Geometric Aspects of Functional Analysis*,
Lecture Notes in Mathematics 2169, DOI 10.1007/978-3-319-45282-1_14

213

The case of unit balls of finite dimensional subspaces of L_p is of particular interest in this note. It was shown by Ball [2] that the slicing problem has an affirmative answer for the unit balls of finite dimensional subspaces of L_p, $1 \leq p \leq 2$. Junge [13] extended this result to every $p \in (1, \infty)$, with the constant C depending on p and going to infinity when $p \to \infty$. Milman [27] gave a different proof for subspaces of L_p, $2 < p < \infty$, with the constant $C \leq O(\sqrt{p})$. Another proof of this estimate can be found in [22].

A generalization of the slicing problem to arbitrary measures was considered in [18–21]. Does there exist a constant C so that for every $n \in N$, every origin-symmetric convex body K in $\mathbb{R}^n$, and every measure μ with non-negative even continuous density f in $\mathbb{R}^n$,

$$\mu(K) \leq C \max_{\xi \in S^{n-1}} \mu(K \cap \xi^{\perp}) |K|^{1/n} ? \qquad (1)$$

For every k-dimensional subspace of $\mathbb{R}^n$, $1 \leq k \leq n$ and any Borel set $A \subset E$,

$$\mu(A) = \int_A f(x)dx,$$

where the integration is with respect to the k-dimensional Lebesgue measure on E.

Inequality (1) was proved with an absolute constant C for intersection bodies [18] (see [16], this includes the unit balls of subspaces of L_p with $0 < p \leq 2$), unconditional bodies and duals of bodies with bounded volume ratio in [20], for k-intersection bodies in [21]. For arbitrary origin-symmetric convex bodies, (1) was proved in [19] with $C \leq O(\sqrt{n})$. A different proof of the latter estimate was recently given in [8], where the symmetry condition was removed.

For the unit balls of subspaces of L_p, $p > 2$, (1) was proved in [21] with $C \leq O(n^{1/2-1/p})$. In this note we improve the estimate to $C \leq O(\sqrt{p})$, extending Milman's result [27] to arbitrary measures in place of volume. In fact, we prove a more general inequality

$$\mu(K) \leq (C\sqrt{p})^k \max_{H \in Gr_{n-k}} \mu(K \cap H) |K|^{k/n}, \qquad (2)$$

where $1 \leq k < n$, Gr_{n-k} is the Grassmanian of $(n - k)$-dimensional subspaces of $\mathbb{R}^n$, K is the unit ball of any n-dimensional subspace of L_p, $p > 2$, μ is a measure on $\mathbb{R}^n$ with even continuous density, and C is a constant independent of p, n, k, K, μ.

The proof is a combination of two known results. Firstly, we use the reduction of the slicing problem for measures to computing the outer volume ratio distance from a body to the class of intersection bodies established in [20]; see Proposition 1. Note that outer volume ratio estimates have been applied to different cases of the original slicing problem by Ball [2], Junge [13], and Milman [27]. Secondly, we use an estimate for the outer volume ratio distance from the unit ball of a subspace of L_p, $p > 2$, to the class of origin-symmetric ellipsoids proved by Milman in [27].

This estimate also follows from results of Davis, Milman and Tomczak-Jaegermann [9]. We include a concentrated version of the proof in Proposition 2.

2 Slicing Inequalities

We need several definitions and facts. A closed bounded set K in $\mathbb{R}^n$ is called a *star body* if every straight line passing through the origin crosses the boundary of K at exactly two points, the origin is an interior point of K, and the *Minkowski functional* of K defined by

$$\|x\|_K = \min\{a \geq 0 : x \in aK\}$$

is a continuous function on $\mathbb{R}^n$.

The *radial function* of a star body K is defined by

$$\rho_K(x) = \|x\|_K^{-1}, \qquad x \in \mathbb{R}^n, \ x \neq 0.$$

If $x \in S^{n-1}$ then $\rho_K(x)$ is the radius of K in the direction of x.

We use the polar formula for volume of a star body

$$|K| = \frac{1}{n} \int_{S^{n-1}} \|\theta\|_K^{-n} d\theta. \tag{3}$$

The class of intersection bodies was introduced by Lutwak [25]. Let K, L be origin-symmetric star bodies in $\mathbb{R}^n$. We say that K is the intersection body of L if the radius of K in every direction is equal to the $(n - 1)$-dimensional volume of the section of L by the central hyperplane orthogonal to this direction, i.e. for every $\xi \in S^{n-1}$,

$$\rho_K(\xi) = \|\xi\|_K^{-1} = |L \cap \xi^\perp|$$

$$= \frac{1}{n-1} \int_{S^{n-1} \cap \xi^\perp} \|\theta\|_L^{-n+1} d\theta = \frac{1}{n-1} R\left(\| \cdot \|_L^{-n+1}\right)(\xi),$$

where $R : C(S^{n-1}) \to C(S^{n-1})$ is the *spherical Radon transform*

$$Rf(\xi) = \int_{S^{n-1} \cap \xi^\perp} f(x)dx, \qquad \forall f \in C(S^{n-1}).$$

All bodies K that appear as intersection bodies of different star bodies form *the class of intersection bodies of star bodies*. A more general class of *intersection bodies* is defined as follows. If μ is a finite Borel measure on S^{n-1}, then the spherical Radon transform $R\mu$ of μ is defined as a functional on $C(S^{n-1})$ acting by

$$(R\mu, f) = (\mu, Rf) = \int_{S^{n-1}} Rf(x) d\mu(x), \qquad \forall f \in C(S^{n-1}).$$

A star body K in $\mathbb{R}^n$ is called an *intersection body* if $\|\cdot\|_K^{-1} = R\mu$ for some measure μ, as functionals on $C(S^{n-1})$, i.e.

$$\int_{S^{n-1}} \|x\|_K^{-1} f(x) dx = \int_{S^{n-1}} Rf(x) d\mu(x), \qquad \forall f \in C(S^{n-1}).$$

Intersection bodies played a crucial role in the solution of the Busemann-Petty problem and its generalizations; see [17, Chap. 5].

A generalization of the concept of an intersection body was introduced by Zhang [30] in connection with the lower dimensional Busemann-Petty problem. For $1 \leq k \leq n - 1$, the $(n - k)$-*dimensional spherical Radon transform* $R_{n-k} : C(S^{n-1}) \rightarrow C(Gr_{n-k})$ is a linear operator defined by

$$R_{n-k}g(H) = \int_{S^{n-1} \cap H} g(x)\, dx, \quad \forall H \in Gr_{n-k}$$

for every function $g \in C(S^{n-1})$.

We say that an origin symmetric star body K in $\mathbb{R}^n$ is a *generalized k-intersection body*, and write $K \in \mathcal{BP}_k^n$, if there exists a finite Borel non-negative measure μ on Gr_{n-k} so that for every $g \in C(S^{n-1})$

$$\int_{S^{n-1}} \|x\|_K^{-k} g(x)\, dx = \int_{Gr_{n-k}} R_{n-k}g(H)\, d\mu(H). \qquad (4)$$

When $k = 1$ we get the class of intersection bodies. It was proved by Goodey and Weil [11] for $k = 1$ and by Grinberg and Zhang [12, Lemma 6.1] for arbitrary k (see also [28] for a different proof) that the class $\mathcal{BP}_k^n$ is the closure in the radial metric of k-radial sums of origin-symmetric ellipsoids. In particular, the classes $\mathcal{BP}_k^n$ contain all origin-symmetric ellipsoids in $\mathbb{R}^n$ and are invariant with respect to linear transformations. Recall that the k-radial sum $K +_k L$ of star bodies K and L is defined by

$$\rho_{K+_k L}^k = \rho_K^k + \rho_L^k.$$

For a convex body K in $\mathbb{R}^n$ and $1 \leq k < n$, denote by

$$\text{o.v.r.}(K, \mathcal{BP}_k^n) = \inf \left\{ \left(\frac{|C|}{|K|} \right)^{1/n} : K \subset C,\, C \in \mathcal{BP}_k^n \right\}$$

the outer volume ratio distance from a body K to the class $\mathcal{BP}_k^n$.

Let B_2^n be the unit Euclidean ball in $\mathbb{R}^n$, let $|\cdot|_2$ be the Euclidean norm in $\mathbb{R}^n$, and let σ be the uniform probability measure on the sphere S^{n-1} in $\mathbb{R}^n$. For every $x \in \mathbb{R}^n$, let x_1 be the first coordinate of x. We use the fact that for every $p > -1$

$$\int_{S^{n-1}} |x_1|^p d\sigma(x) = \frac{\Gamma(\frac{p+1}{2})\Gamma(\frac{n}{2})}{\sqrt{\pi}\Gamma(\frac{n+p}{2})}; \tag{5}$$

see for example [17, Lemma 3.12], where one has to divide by $|S^{n-1}| = 2\pi^{(n-1)/2}/\Gamma(\frac{n}{2})$, because the measure σ on the sphere is normalized.

In [20], the slicing problem for arbitrary measures was reduced to estimating the outer volume ratio distance from a convex body to the classes $\mathcal{BP}_k^n$, as follows.

Proposition 1 *For any* $n \in \mathbb{N}$, $1 \le k < n$, *any origin-symmetric star body* K *in* $\mathbb{R}^n$, *and any measure* μ *with even continuous density on* K,

$$\mu(K) \le \left(o.v.r.(K, \mathcal{BP}_k^n) \right)^k \frac{n}{n-k} c_{n,k} \max_{H \in Gr_{n-k}} \mu(K \cap H) |K|^{k/n},$$

where $c_{n,k} = |B_2^n|^{(n-k)/n}/|B_2^{n-k}| \in (e^{-k/2}, 1)$.

It appears that for the unit balls of subspaces of L_p, $p > 2$ the outer volume ration distance to the classes of intersection bodies does not depend on the dimension. As mentioned in the introduction, the following estimate was proved in [27] and also follows from results of [9]. We present a short version of the proof.

Proposition 2 *Let* $p > 2$, $n \in \mathbb{N}$, $1 \le k < n$, *and let* K *be the unit ball of an* n-*dimensional subspace of* L_p. *Then*

$$o.v.r.(K, \mathcal{BP}_k^n) \le C\sqrt{p},$$

where C *is an absolute constant.*

Proof Since the classes $\mathcal{BP}_k^n$ are invariant under linear transformations, we can assume that K is in the Lewis position. By a result of Lewis in the form of [26, Theorem 8.2], this means that there exists a measure ν on the sphere so that for every $x \in \mathbb{R}^n$

$$\|x\|_K^p = \int_{S^{n-1}} |(x, u)|^p d\nu(u),$$

and

$$|x|_2^2 = \int_{S^{n-1}} |(x, u)|^2 d\nu(u).$$

Also, by the same result of Lewis [24], $K \subset n^{1/2-1/p} B_2^n$.

Let us estimate the volume of K from below. By the Fubini theorem, formula (5) and Stirling's formula, we get

$$\int_{S^{n-1}} \|x\|_K^p d\sigma(x) = \int_{S^{n-1}} \int_{S^{n-1}} |(x, u)|^p d\sigma(x) dv(u)$$

$$= \int_{S^{n-1}} |x_1|^p d\sigma(x) \int_{S^{n-1}} dv(u) \le \left(\frac{Cp}{n+p}\right)^{p/2} \int_{S^{n-1}} dv(u).$$

Now

$$\frac{Cp}{n+p} \left(\int_{S^{n-1}} dv(u)\right)^{2/p} \ge \left(\int_{S^{n-1}} \|x\|_K^p d\sigma(x)\right)^{2/p}$$

$$\ge \left(\int_{S^{n-1}} \|x\|_K^{-n} d\sigma(x)\right)^{-2/n} = \left(\frac{|K|}{|B_2^n|}\right)^{-2/n} \sim \frac{1}{n} |K|^{-2/n},$$

because $|B_2^n|^{1/n} \sim n^{-1/2}$. On the other hand,

$$1 = \int_{S^{n-1}} |x|_2^2 d\sigma(x) = \int_{S^{n-1}} \int_{S^{n-1}} (x, u)^2 dv(u) d\sigma(x)$$

$$= \int_{S^{n-1}} \int_{S^{n-1}} |x_1|^2 d\sigma(x) dv(u) = \frac{1}{n} \int_{S^{n-1}} dv(u),$$

so

$$\frac{Cp}{n+p} n^{2/p} \ge \frac{1}{n} |K|^{-2/n},$$

and

$$|K|^{1/n} \ge cn^{-1/p} \sqrt{\frac{n+p}{np}} \ge \frac{cn^{1/2-1/p}}{\sqrt{p}} |B_2^n|^{1/n}.$$

Finally, since $K \subset n^{1/2-1/p} B_2^n$, and $B_2^n \in \mathcal{BP}_k^n$ for every k, we have

$$\text{o.v.r.}(K, \mathcal{BP}_k^n) \le \left(\frac{|n^{1/2-1/p} B_2^n|}{|K|}\right)^{1/n} \le C\sqrt{p},$$

where C is an absolute constant.

We now formulate the main result of this note.

Corollary 1 *There exists a constant C so that for any $p > 2$, $n \in \mathbb{N}$, $1 \le k < n$, any convex body K that is the unit ball of an n-dimensional subspace of L_p, and any* *measure μ with non-negative even continuous density in $\mathbb{R}^n$,*

$$\mu(K) \leq (C\sqrt{p})^k \max_{H \in Gr_{n-k}} \mu(K \cap H) \, |K|^{k/n}.$$

Proof Combine Proposition 1 with Proposition 2. Note that $\frac{n}{n-k} \in (1, e^k)$, and $c_{n,k} \in (e^{-k/2}, 1)$, so these constants can be incorporated in the constant C. $\qquad\square$

Acknowledgements The first named author was partially supported by the US National Science Foundation, grant DMS-1265155.

References

1. K. Ball, Isometric problems in ℓ_p and sections of convex sets, Ph.D. dissertation, Trinity College, Cambridge, 1986
2. K. Ball, *Normed Spaces with a Weak Gordon-Lewis Property*. Lecture Notes in Mathematics, vol. 1470 (Springer, Berlin 1991), pp. 36–47
3. S. Bobkov, F. Nazarov, On convex bodies and log-concave probability measures with unconditional basis, in *Geometric Aspects of Functional Analysis*, ed. by V.D. Milman, G. Schechtman. Lecture Notes in Mathematics, vol. 1807 (Springer, Berlin, 2003), pp. 53–69
4. J. Bourgain, On high-dimensional maximal functions associated to convex bodies. Am. J. Math. **108**, 1467–1476 (1986)
5. J. Bourgain, Geometry of Banach spaces and harmonic analysis, in *Proceedings of the International Congress of Mathematicians (Berkeley, Calif., 1986)* (American Mathematical Society, Providence, RI, 1987), pp. 871–878
6. J. Bourgain, On the distribution of polynomials on high-dimensional convex sets, in *Geometric Aspects of Functional Analysis, Israel Seminar (1989–90)*. Lecture Notes in Mathematics, vol. 1469 (Springer, Berlin, 1991), pp. 127–137
7. S. Brazitikos, A. Giannopoulos, P. Valettas, B. Vritsiou, *Geometry of Isotropic Convex Bodies* (American Mathematical Society, Providence RI, 2014)
8. G. Chasapis, A. Giannopoulos, D.-M. Liakopoulos, Estimates for measures of lower dimensional sections of convex bodies. Adv. Math. **306**, 880–904 (2017)
9. W.J. Davis, V.D. Milman, N. Tomczak-Jaegermann, The distance between certain n-dimensional Banach spaces. Isr. J. Math. **39**, 1–15 (1981)
10. R.J. Gardner, *Geometric Tomography*, 2nd edn. (Cambridge University Press, Cambridge, 2006)
11. P. Goodey, W. Weil, Intersection bodies and ellipsoids. Mathematika **42**, 295–304 (1995)
12. E. Grinberg, G. Zhang, Convolutions, transforms and convex bodies. Proc. Lond. Math. Soc. **78**, 77–115 (1999)
13. M. Junge, On the hyperplane conjecture for quotient spaces of L_p. Forum Math. **6**, 617–635 (1994)
14. M. Junge, Proportional subspaces of spaces with unconditional basis have good volume properties, in *Geometric Aspects of Functional Analysis (Israel Seminar, 1992–1994)*. Operator Theory Advances and Applications, vol. 77 (Birkhauser, Basel, 1995), pp. 121–129
15. B. Klartag, On convex perturbations with a bounded isotropic constant. Geom. Funct. Anal. **16**, 1274–1290 (2006)
16. A. Koldobsky, Intersection bodies, positive definite distributions and the Busemann-Petty problem. Am. J. Math. **120**, 827–840 (1998)

17. A. Koldobsky, *Fourier Analysis in Convex Geometry* (American Mathematical Society, Providence RI, 2005)
18. A. Koldobsky, A hyperplane inequality for measures of convex bodies in $\mathbb{R}^n$, $n \leq 4$, Discret. Comput. Geom. **47**, 538–547 (2012)
19. A. Koldobsky, A $\sqrt{n}$ estimate for measures of hyperplane sections of convex bodies. Adv. Math. **254**, 33–40 (2014)
20. A. Koldobsky, Slicing inequalities for measures of convex bodies. Adv. Math. **283**, 473–488 (2015)
21. A. Koldobsky, Slicing inequalities for subspaces of L_p. Proc. Am. Math. Soc. **144**, 787–795 (2016)
22. A. Koldobsky, A. Pajor, V. Yaskin, Inequalities of the Kahane-Khinchin type and sections of L_p-balls. Stud. Math. **184**, 217–231 (2008)
23. H. König, M. Meyer, A. Pajor, The isotropy constants of the Schatten classes are bounded. Math. Ann. **312**, 773–783 (1998)
24. D.R. Lewis, Finite dimensional subspaces of L_p. Stud. Math. **63**, 207–212 (1978)
25. E. Lutwak, Intersection bodies and dual mixed volumes. Adv. Math. **71**, 232–261 (1988)
26. E. Lutwak, D. Yang, G. Zhang, L_p John ellipsoids. Proc. Lond. Math. Soc. **90**, 497–520 (2005)
27. E. Milman, Dual mixed volumes and the slicing problem. Adv. Math. **207**, 566–598 (2006)
28. E. Milman, Generalized intersection bodies. J. Funct. Anal. **240**(2), 530–567 (2006)
29. V. Milman, A. Pajor, Isotropic position and inertia ellipsoids and zonoids of the unit ball of a normed n-dimensional space, in *Geometric Aspects of Functional Analysis*, ed. by J. Lindenstrauss, V. Milman. Lecture Notes in Mathematics, vol. 1376 (Springer, Heidelberg, 1989), pp. 64–104
30. G. Zhang, Section of convex bodies. Am. J. Math. **118**, 319–340 (1996)

Sharp Poincaré-Type Inequality for the Gaussian Measure on the Boundary of Convex Sets

Alexander V. Kolesnikov and Emanuel Milman

Abstract A sharp Poincaré-type inequality is derived for the restriction of the Gaussian measure on the boundary of a convex set. In particular, it implies a Gaussian mean-curvature inequality and a Gaussian iso-second-variation inequality. The new inequality is nothing but an infinitesimal equivalent form of Ehrhard's inequality for the Gaussian measure. While Ehrhard's inequality does not extend to general $CD(1, \infty)$ measures, we formulate a sufficient condition for the validity of Ehrhard-type inequalities for general measures on $\mathbb{R}^n$ via a certain property of an associated Neumann-to-Dirichlet operator.

1 Introduction

We consider Euclidean space $(\mathbb{R}^n, \langle \cdot, \cdot \rangle)$ equipped with the standard Gaussian measure $\gamma = \Psi_\gamma dx$, $\Psi_\gamma(x) = (2\pi)^{-n/2} \exp(-|x|^2/2)$. Let $K \subset \mathbb{R}^n$ denote a convex domain with C^2 smooth boundary and outer unit-normal field $\nu = \nu_{\partial K}$. The second fundamental form $\mathrm{II} = \mathrm{II}_{\partial K}$ of ∂K at $x \in \partial K$ is as usual (up to sign) defined by $\mathrm{II}_x(X, Y) = \langle \nabla_X \nu, Y \rangle$, $X, Y \in T_x \partial K$. The quantities:

$$H(x) := tr(\mathrm{II}_x) \, , \ H_\gamma(x) := H(x) - \langle x, \nu(x) \rangle \, ,$$

are called the mean-curvature and *Gaussian* mean-curvature of ∂K at $x \in \partial K$, respectively. It is well-known that H governs the first variation of the (Lebesgue) boundary-measure $\mathrm{Vol}_{\partial K}$ under the normal-map $t \mapsto \exp(t\nu)$, and similarly H_γ governs the first variation of the Gaussian boundary-measure $\gamma_{\partial K} := \Psi_\gamma \mathrm{Vol}_{\partial K}$, see e.g. [15] or Sect. 2.

Recall that the Gaussian isoperimetric inequality of Borell [4] and Sudakov–Tsirelson [20] asserts that if E is a half-plane with $\gamma(E) = \gamma(K)$, then

A.V. Kolesnikov
Faculty of Mathematics, Higher School of Economics, Moscow, Russia
e-mail: akolesnikov@hse.ru; sascha77@mail.ru

E. Milman (✉)
Department of Mathematics, Technion - Israel Institute of Technology, Haifa 32000, Israel
e-mail: emilman@tx.technion.ac.il

© Springer International Publishing AG 2017
B. Klartag, E. Milman (eds.), *Geometric Aspects of Functional Analysis*,
Lecture Notes in Mathematics 2169, DOI 10.1007/978-3-319-45282-1_15

$\gamma_{\partial K}(\partial K) \geq \gamma_{\partial E}(\partial E)$ (in fact, this applies not just to convex sets but to all Borel sets, with an appropriate interpretation of Gaussian boundary measure). In other words:

$$\gamma_{\partial K}(\partial K) \geq I_\gamma(\gamma(K))$$

with equality for half-planes, where $I_\gamma : [0, 1] \rightarrow \mathbb{R}_+$ denotes the Gaussian isoperimetric profile, given by $I_\gamma := \varphi \circ \Phi^{-1}$ with $\varphi(t) = \frac{1}{\sqrt{2\pi}} \exp(-t^2/2)$ and $\Phi(t) = \int_{-\infty}^t \varphi(s)ds$. Note that I_γ is concave and symmetric around $1/2$, hence it is increasing on $[0, 1/2]$ and decreasing on $[1/2, 1]$.

Our main result is the following new Poincaré-type inequality for the Gaussian boundary-measure on ∂K:

Theorem 1.1 *For all convex K and $f \in C^1(\partial K)$ for which the following expressions make sense, we have:*

$$\int_{\partial K} H_\gamma f^2 d\gamma_{\partial K} - (\log I_\gamma)'(\gamma(K)) \left(\int_{\partial K} f d\gamma_{\partial K} \right)^2 \leq \int_{\partial K} \langle II_{\partial K}^{-1} \nabla_{\partial K} f, \nabla_{\partial K} f \rangle d\gamma_{\partial K}.$$

$$(1)$$

Here $\nabla_{\partial K} f$ denotes the gradient of f on ∂K with its induced metric, and $(\log I_\gamma)'(v) = -\Phi^{-1}(v)/I_\gamma(v)$.

This inequality is already interesting for the constant function $f \equiv 1$:

Corollary 1.2 (Gaussian Mean-Curvature Inequality)

$$\int_{\partial K} H_\gamma d\gamma_{\partial K} \leq (\log I_\gamma)'(\gamma(K))\gamma_{\partial K}(\partial K)^2.$$

$$(2)$$

In particular, if $\gamma(K) \geq 1/2$ then necessarily $\int_{\partial K} H_\gamma d\gamma_{\partial K} \leq 0$.

The latter inequality is sharp, yielding an equality when K is any half-plane E. Indeed, since $I_\gamma(\gamma(E)) = \gamma_{\partial E}(\partial E)$, it is enough to note that $E = (-\infty, t] \times \mathbb{R}^{n-1}$ has constant Gaussian mean-curvature $H_\gamma = -t = (\log \varphi)'(t) = I_\gamma'(\gamma(E))$.

More surprisingly, we will see in Sect. 4 that Corollary 1.2 in fact implies the Gaussian isoperimetric inequality (albeit only for convex sets). Furthermore, we have:

Corollary 1.3 (Gaussian Iso-Curvature Inequality) *If E is a half-plane with $\gamma(E) = \gamma(K) \geq 1/2$, then the following iso-curvature inequality holds:*

$$\int_{\partial K} H_\gamma d\gamma_{\partial K} \leq \int_{\partial E} H_\gamma d\gamma_{\partial E} \ (\leq 0).$$

Proof This is immediate from (2), the Gaussian isoperimetric inequality $\gamma_{\partial K}(\partial K) \geq \gamma_{\partial E}(\partial E)$, the assumption that $(\log I_\gamma)'(\gamma(K)) \leq 0$, and the equality in (2) for half-planes.

Clearly, by passing to complements, the latter corollary yields a reverse inequality when applied to K, the complement to a convex set C satisfying $\gamma(K) \leq 1/2$ (since $\partial K = \partial C$ with reverse orientation and thus their generalized mean-curvature simply changes sign). It is also easy to check that a reverse inequality holds when K is a small Euclidean (convex) ball centered at the origin. It is probably unreasonable to expect that a reverse inequality holds for all convex K with $\gamma(K) \leq 1/2$, but we have not seriously searched for a counterexample.

We proceed to give the following interpretation of the latter two corollaries. Denoting:

$$\delta^0_\gamma(K) = \gamma(K)\,, \ \delta^1_\gamma(K) = \gamma_{\partial K}(\partial K)\,, \ \delta^2_\gamma(K) = \int_{\partial K} H_\gamma d\gamma_{\partial K}\,,$$

we note that $\delta^i_\gamma(K)$ is precisely the i-th variation of the function $t \mapsto \gamma(K_t)$, where $K_t := \{x \in \mathbb{R}^n \,;\, d(x, K) \leq t\}$ and d denotes Euclidean distance. Consequently, Corollary 1.3 may be rewritten as:

Corollary 1.4 (Gaussian Iso-Second-Variation Inequality) *If E is a half-plane with $\gamma(E) = \gamma(K) \geq 1/2$, then the following iso-second-variation inequality holds:*

$$\delta^2_\gamma(K) \leq \delta^2_\gamma(E) \ (\leq 0).$$

It is interesting to note that we are not aware of an analogous statement on any other metric-measure space, and in particular, for the Lebesgue measure in Euclidean space, as all known isoperimetric inequalities only pertain (by definition) to the first-variation (and with reversed direction of the inequality). Furthermore, in contrast to the isoperimetric inequality, it is easy to see that the second-variation inequality above is false without the assumption that K is convex, as witnessed for instance by taking the complement of any non-degenerate slab $\{x \in \mathbb{R}^n \,;\, a \leq x_1 \leq b\}$ of measure $1/2$. As for Corollary 1.2, we see that it may be rewritten as:

Corollary 1.5 (Minkowski's Second Inequality for Euclidean Gaussian Extensions)

$$\delta^2_\gamma(K) \leq (\log I_\gamma)'(\delta^0_\gamma(K))(\delta^1_\gamma(K))^2. \tag{3}$$

This already hints at the proof of Theorem 1.1. To describe the proof, and put the latter interpretation in the appropriate context, let us recall some classical facts from the Brunn-Minkowski theory (for the Lebesgue measure).

1.1   Brunn–Minkowski Inequality

The Brunn–Minkowski inequality [13, 19] asserts that:

$$Vol((1-t)K + tL)^{1/n} \geq (1-t)Vol(K)^{1/n} + tVol(L)^{1/n} , \ \forall t \in [0,1] , \tag{4}$$

for all convex $K, L \subset \mathbb{R}^n$; it was extended to arbitrary Borel sets by Lyusternik. Here Vol denotes Lebesgue measure and $A + B := \{a + b \ ; \ a \in A, b \in B\}$ denotes Minkowski addition. We refer to the excellent survey by Gardner [13] for additional details and references.

For convex sets, (4) is equivalent to the concavity of the function $t \mapsto Vol(K + tL)^{1/n}$. By Minkowski's theorem, extending Steiner's observation for the case that L is the Euclidean ball, $Vol(K + tL)$ is an n-degree polynomial $\sum_{i=0}^{n} \binom{n}{i} W_{n-i}(K, L)t^i$, whose coefficients

$$W_{n-i}(K, L) := \frac{(n-i)!}{n!} \left(\frac{d}{dt} \right)^i \bigg|_{t=0} Vol(K + tL) , \tag{5}$$

are called mixed-volumes. The above concavity thus amounts to the following "Minkowski's second inequality", which is a particular case of the Alexandrov–Fenchel inequalities:

$$W_{n-1}(K, L)^2 \geq W_{n-2}(K, L)W_n(K, L) . \tag{6}$$

Specializing to the case that L is the Euclidean unit-ball D, noting that $K_t = K + tD$, and denoting by $\delta^i(K)$ the i-th variation of $t \mapsto Vol(K_t)$, we have as before:

$$\delta^0(K) = Vol(K) , \ \delta^1(K) = Vol_{\partial K}(\partial K) , \ \delta^2(K) = \int_{\partial K} H dVol_{\partial K}.$$

The corresponding distinguished mixed-volumes $W_{n-i}(K) = W_{n-i}(K, D)$, which are called intrinsic-volumes or quermassintegrals, are related to $\delta^i(K)$ via (5). Consequently, when $L = D$, Minkowski's second inequality amounts to the inequality:

$$\delta^2(K) \leq \frac{n-1}{n} \frac{1}{\delta^0(K)} (\delta^1(K))^2.$$

The analogy with (3) becomes apparent, in view of the fact that $(\log I)'(v) = \frac{n-1}{n} \frac{1}{v}$, where $I(v) = c_n v^{\frac{n-1}{n}}$ is the standard isoperimetric profile of Euclidean space $(\mathbb{R}^n, \langle \cdot, \cdot \rangle)$ endowed with the Lebesgue measure.

An important difference to note with respect to the classical theory, is that in the Gaussian theory, $\delta_\gamma^2(K)$ may actually be negative, in contrast to the non-negativity of all mixed-volumes, and in particular of $\delta^2(K)$. One reason for this is that the Gaussian measure is finite whereas the Lebesgue measure is not, so that I is monotone increasing whereas I_y is not. This feature seems to also be responsible for the peculiar iso-second-variation corollary.

1.2   Ehrhard Inequality

A remarkable extension of the Brunn-Minkowski inequality to the Gaussian setting was obtained by Ehrhard [11], who showed that:

$$\Phi^{-1}(\gamma((1-t)K + tL)) \geq (1-t)\Phi^{-1}(\gamma(K)) + t\Phi^{-1}(\gamma(L)) \quad \forall t \in [0,1],$$

for all convex sets $K, L \subset \mathbb{R}^n$, with equality when K and L are parallel half-planes (pointing in the same direction). This was later extended by Latała [16] to the case that only one of the sets is assumed convex, and finally by Borell [6, 7] to arbitrary Borel sets. As before, for K, L convex sets, Ehrhard's inequality is equivalent to the concavity of the function $t \mapsto F_\gamma(t) := \Phi^{-1}(\gamma((1-t)K + tL))$.

To prove Theorem 1.1, we repeat an idea of A. Colesanti. In [9] (see also [10]), Colesanti showed that the Brunn-Minkowski concavity of $t \mapsto F(t) := Vol((1-t)K + tL)^{1/n}$ is equivalent to a certain Poincaré-type inequality on ∂K, by parametrizing K, L via their support functions and calculating the second variation of $F(t)$. Repeating the calculation for $F_\gamma(t)$, Theorem 1.1 turns out to be an equivalent infinitesimal reformulation of Ehrhard's inequality for convex sets.

1.3   Comparison with Previous Results

Going in the other direction, we have recently shown in our previous work [15] how to directly derive a Poincaré-type inequality on the boundary of a locally-convex subset of a weighted Riemannian manifold, which may then be used to infer a Brunn-Minkowski inequality in the weighted Riemannian setting via an appropriate geometric flow. In particular, in the Euclidean setting, our results apply to Borell's class of $1/N$-concave measures [5] ($\frac{1}{N} \in [-\infty, \frac{1}{n}]$), defined as those measures μ on $\mathbb{R}^n$ satisfying the following generalized Brunn-Minkowski inequality:

$$\mu((1-t)A + tB) \geq \left((1-t)\mu(A)^{1/N} + t\mu(B)^{1/N}\right)^N,$$

for all $t \in [0,1]$ and Borel sets $A, B \subset \mathbb{R}^n$ with $\mu(A), \mu(B) > 0$. It was shown by Brascamp–Lieb [8] and Borell [5] that the absolutely continuous members of this class are precisely characterized by having density Ψ so that $(N-n)\Psi^{1/(N-n)}$ is concave on its convex support Ω (interpreted as $\log \Psi$ being concave when $N = \infty$), amounting to the Bakry–Émery $CD(0, N)$ condition [1, 2, 14]. Our results from

[15] then imply that for any (say) compact convex K in the interior of Ω with C^2 boundary, and any $f \in C^1(\partial K)$, one has:

$$\int_{\partial K} H_\mu f^2 d\mu_{\partial K} - \frac{N-1}{N} \frac{1}{\mu(K)} \left(\int_{\partial K} f d\mu_{\partial K} \right)^2 \leq \int_{\partial K} \langle \mathrm{II}_{\partial K}^{-1} \nabla_{\partial K} f, \nabla_{\partial K} f \rangle d\mu_{\partial K} , \tag{7}$$

with $\mu_{\partial K} = \Psi \mathrm{Vol}_{\partial K}$ denoting the boundary measure and $H_\mu = H + \langle \log \Psi, \nu \rangle$ the μ-weighted mean-curvature. Note that the Gaussian measure γ satisfies $CD(1, \infty)$ and in particular $CD(0, \infty)$, as $\log \Psi_\gamma$ is concave on $\mathbb{R}^n$. Consequently, applying (7) with $N = \infty$, we have:

$$\int_{\partial K} H_\mu f^2 d\gamma_{\partial K} - \frac{1}{\gamma(K)} \left(\int_{\partial K} f d\gamma_{\partial K} \right)^2 \leq \int_{\partial K} \langle \mathrm{II}_{\partial K}^{-1} \nabla_{\partial K} f, \nabla_{\partial K} f \rangle d\gamma_{\partial K} . \tag{8}$$

It is easy to verify that $(\log I_\gamma)'(v) < \frac{1}{v}$ for all $v \in (0, 1)$, and hence Theorem 1.1 constitutes an improvement over (8).

A very important point is that the latter improvement is strict only for test functions f with non-zero mean, $\int_{\partial K} f d\gamma_{\partial K} \neq 0$. Put differently, the entire significance of Theorem 1.1 lies in the coefficient in front of the $\left(\int_{\partial K} f d\gamma_{\partial K} \right)^2$ term, since by (7), for zero-mean test functions, the inequality asserted in Theorem 1.1 holds not only for the Gaussian measure, but in fact for Borell's entire class of concave (or $CD(0, 0)$) measures (using our convention from [14, 15] that $\frac{N-1}{N} = -\infty$ when $N = 0$ and that $\infty \cdot 0 = 0$).

Unfortunately, our method from [15], involving L^2-duality and the Reilly formula from Riemannian geometry, cannot be used in the Gaussian setting without some additional ingredients, like information on an associated Neumann-to-Dirichlet operator, see Sect. 3. In particular, we observe in Sect. 4 that Theorem 1.1 (or equivalently, Ehrhard's inequality for convex sets) and even Corollary 1.2, are simply false for a general $CD(1, \infty)$ probability measure in Euclidean space, having density $\Psi = \exp(-V)$ with $\nabla^2 V \geq Id$.

2 Proof of Theorem 1.1

The general formulation of Theorem 1.1 is reduced to the case that K is compact with strictly-convex C^3 smooth boundary ($\mathrm{II}_{\partial K} > 0$) by a standard (Euclidean) approximation argument—this class of convex sets is denoted by $\mathscr{C}_+^3$ (and analogously we define the class $\mathscr{C}_+^2$). As explained in the Introduction, the proof of Theorem 1.1 boils down to a direct calculation of the second variation of the function:

$$t \mapsto \Phi^{-1}(\gamma((1-t)K + tL))$$

for an appropriately chosen L. Ehrhard's inequality ensures that this function is concave when K, L are convex.

The second variation will be conveniently expressed using support functions. Recall that the support function of a convex body (convex compact set with non-empty interior) C is defined as the following function on the Euclidean unit sphere S^{n-1}:

$$h_C(\theta) := \sup \{ \langle \theta, x \rangle ; x \in C \} \, , \, \theta \in S^{n-1}.$$

It is easy to see that the correspondence $C \mapsto h_C$ between convex bodies and functions on S^{n-1} is injective and positively linear: $h_{aC_1 + bC_2} = ah_{C_1} + bh_{C_2}$ for all $a, b \geq 0$. As $K \in \mathscr{C}_+^3$ we know that h_K is C^3 smooth [19, p. 106].

Now let $f \in C^2(\partial K)$, and consider the function $h_\Delta := f \circ v_{\partial K}^{-1} : S^{n-1} \to \mathbb{R}$. Since $K \in \mathscr{C}_+^3$ this function is well-defined and C^2 smooth. Moreover, it is not hard to show (e.g. [19, pp. 38, 111], [9]) that for $\epsilon > 0$ small enough, $h_K + th_\Delta$ is the support function of a convex body $K_t \in \mathscr{C}_+^2$ for all $t \in [0, \epsilon]$. It follows by linearity of the support functions that $K_{\epsilon t} = (1 - t)K + tK_\epsilon$ for all $t \in [0, 1]$, and so Ehrhard's inequality implies that:

$$t \mapsto F_\gamma(t) := \Phi^{-1}(\gamma(K_t))$$

is concave on $[0, \epsilon]$.

The first and second variations of $t \mapsto \mu(K_t)$ were calculated by Colesanti in [9] for the case that μ is the Lebesgue measure, and for general measures μ with positive density Ψ by the authors in [15] (in fact in a general weighted Riemannian setting, with an appropriate interpretation of K_t avoiding support functions):

$$\delta^0 := (d/dt)^0|_{t=0} \, \mu(K_t) = \mu(K) \, ,$$

$$\delta^1 := (d/dt)^1|_{t=0} \, \mu(K_t) = \int_{\partial K} f d\mu_{\partial K} \, ,$$

$$\delta^2 := (d/dt)^2|_{t=0} \, \mu(K_t) = \int_{\partial K} H_\mu f^2 d\mu_{\partial K} - \int_{\partial K} \langle \mathrm{II}_{\partial K}^{-1} \nabla_{\partial K} f, \nabla_{\partial K} f \rangle d\mu_{\partial K}.$$

Applying the above formulae for $\mu = \gamma$, calculating:

$$0 \geq F_\gamma''(0) = (\Phi^{-1})''(\delta^0)(\delta^1)^2 + (\Phi^{-1})'(\delta^0)\delta^2,$$

dividing by $(\Phi^{-1})'(\delta^0) > 0$ and using that:

$$\frac{(\Phi^{-1})''(v)}{(\Phi^{-1})'(v)} = -(\log I_\gamma)'(v),$$

Theorem 1.1 readily follows for $f \in C^2(\partial K)$. The general case for $f \in C^1(\partial K)$ is obtained by a standard approximation argument.

Going in the other direction, it should already be clear that Theorem 1.1 implies back Ehrhard's inequality. Indeed, given $K, L \in \mathscr{C}_+^2$, consider $K_t = (1 - t)K + tL$ for $t \in [0, 1]$, and note that $h_{K_t} = (1 - t)h_K + th_L$. Fixing $t_0 \in (0, 1)$, it follows that $h_{K_{t_0+\epsilon}} = h_{K_{t_0}} + \epsilon(h_L - h_K)$. Inspecting the proof above, we see that the statement of Theorem 1.1 for K_{t_0} and $f = (h_L - h_K) \circ \nu \in C^1(\partial K)$, is precisely equivalent to the concavity of the function $\epsilon \mapsto \Phi^{-1}(\gamma(K_{t_0+\epsilon}))$ at $\epsilon = 0$. Since the point $t_0 \in (0, 1)$ was arbitrary, we see that Theorem 1.1 implies the concavity of $[0, 1] \ni t \mapsto \Phi^{-1}(\gamma((1-t)K+tL))$ for $K, L \in \mathscr{C}_+^2$. The case of general convex K, L follows by approximation.

3   Neumann-to-Dirichlet Operator

In this section, we mention how a certain property of a Neumann-to-Dirichlet operator can be used to directly obtain an Ehrhard-type inequality for general measures $\mu = \exp(-V(x))dx$ on $\mathbb{R}^n$ (say with C^2 positive density). Define the associated weighted Laplacian $L = L_\mu$ as:

$$L = L_\mu := \exp(V)\nabla \cdot (\exp(-V)\nabla) = \Delta - \langle \nabla V, \nabla \rangle .$$

Given a compact set $\Omega \subset \mathbb{R}^n$ with C^1 smooth boundary, note that the usual integration by parts formula is satisfied for $f, g \in C^2(\Omega)$:

$$\int_\Omega L(f)g d\mu = \int_{\partial M} f_\nu g d\mu_{\partial M} - \int_\Omega \langle \nabla f, \nabla g \rangle d\mu = \int_{\partial \Omega} (f_\nu g - g_\nu f)d\mu_{\partial \Omega} + \int_\Omega L(g)f d\mu ,$$

where $u_\nu = \nu \cdot u$.

Given a compact convex body $K \in \mathscr{C}_+^2$ and $f \in C^{1,\alpha}(\partial K)$, let us now solve the following Neumann Laplace equation:

$$Lu \equiv \frac{1}{\mu(K)} \int_{\partial K} f d\mu_{\partial K} \text{ on } K , \ u_\nu = f \text{ on } \partial K. \tag{9}$$

Since the compatibility condition $\int_K Lu d\mu = \int_{\partial K} f d\mu_{\partial K}$ is satisfied, it is known (e.g. [15]) that a solution $u \in C^{2,\alpha}(K)$ exists (and is unique up to an additive constant). The operator mapping $f \mapsto u$ is called the Neumann-to-Dirichlet operator.

Theorem 3.1 *Assume that there exists a function $F : \mathbb{R}_+ \to \mathbb{R}$ so that for all K, f and u as above:*

$$F(\mu(K))(\int_{\partial K} f d\mu_{\partial K})^2 \leq \int_{\partial K} (\langle \nabla_{\partial K} f, \nabla_{\partial K} u \rangle + u_{\nu,\nu}f) d\mu_{\partial K}. \tag{10}$$

Denote $G(v) := \frac{1}{v} - F(v)$ *and* $\Phi_\mu^{-1}(v) := \int_{1/2}^v \exp(-\int_{1/2}^t G(s)ds)dt$. *Then for all* $f \in C^1(\partial K)$:

$$\int_{\partial K} H_\mu f^2 d\mu_{\partial K} - G(\mu(K)) \left(\int_{\partial K} f d\mu_{\partial K}\right)^2 \le \int_{\partial K} \langle II_{\partial K}^{-1} \nabla_{\partial K} f, \nabla_{\partial K} f\rangle d\mu_{\partial K}, \quad (11)$$

and for all convex $K, L \subset \mathbb{R}^n$ *and* $t \in [0, 1]$:

$$\Phi_\mu^{-1}((1 - t)K + tL) \ge (1 - t)\Phi_\mu^{-1}(K) + t\Phi_\mu^{-1}(L). \quad (12)$$

For the proof, we require the following lemma. We denote by $\|\nabla^2 u\|$ the Hilbert-Schmidt norm of the Hessian $\nabla^2 u$.

Lemma 3.2 *With the above notation:*

$$\int_K \left(\langle \nabla^2 V \nabla u, \nabla u\rangle + \|\nabla^2 u\|^2\right) d\mu = \int_{\partial K} (\langle \nabla_{\partial K} f, \nabla_{\partial K} u\rangle + u_{v,v}f) d\mu_{\partial K}.$$

Remark 3.3 The integrand on the left-hand-side above is the celebrated Bakry–Émery iterated carré-du-champ $\Gamma_2(u)$, associated to $(K, \langle \cdot, \cdot \rangle, \mu)$ [2].

Proof Denoting $\nabla u = (u_1, \dots, u_n)$, we calculate:

$$\int_K \|\nabla^2 u\|^2 d\mu = \int_K \sum_{i=1}^n |\nabla u_i|^2 d\mu = \int_{\partial K} \sum_{i=1}^n u_{i,v} u_i d\mu_{\partial K} - \int_K \sum_{i=1}^n L(u_i)u_i d\mu.$$

To handle the $L(u_i)$ terms, we take the i-th partial derivative in the Laplace equation (9), yielding:

$$0 = (Lu)_i = L(u_i) - \langle \nabla u, \nabla V_i \rangle.$$

Consequently, we have:

$$\sum_{i=1}^n L(u_i)u_i = \langle \nabla^2 V \nabla u, \nabla u\rangle,$$

and therefore:

$$\int_K \left(\langle \nabla^2 V \nabla u, \nabla u\rangle + \|\nabla^2 u\|^2\right) d\mu = \int_{\partial K} \sum_{i=1}^n u_{i,v} u_i d\mu_{\partial K}.$$

Recalling that $f = u_v$, the assertion follows.

Proof (Proof of Theorem 3.1) As in [15], our starting point is the generalized Reilly formula [14], which is an integrated form of Bochner's formula in the presence of a boundary. In the Euclidean setting, it states that for any $u \in C^2(K)$ (see [14] for less restrictions on u):

$$\int_K (Lu)^2 d\mu = \int_K \|\nabla^2 u\|^2 d\mu + \int_K \langle \nabla^2 V \, \nabla u, \nabla u \rangle d\mu$$
$$+ \int_{\partial K} H_\mu (u_\nu)^2 d\mu_{\partial K} + \int_{\partial K} \langle \mathrm{II}_{\partial K} \, \nabla_{\partial K} u, \nabla_{\partial K} u \rangle d\mu_{\partial K} - 2 \int_{\partial K} \langle \nabla_{\partial K} u_\nu, \nabla_{\partial K} u \rangle d\mu_{\partial K}.$$

$$(13)$$

As we assume that $\mathrm{II}_{\partial K} > 0$, we may apply the Cauchy–Schwarz inequality to the last-term above:

$$2 \langle \nabla_{\partial K} u_\nu, \nabla_{\partial K} u \rangle \leq \langle \mathrm{II}_{\partial K} \, \nabla_{\partial K} u, \nabla_{\partial K} u \rangle + \langle \mathrm{II}_{\partial K}^{-1} \nabla_{\partial K} u_\nu, \nabla_{\partial K} u_\nu \rangle , \qquad (14)$$

yielding:

$$\int_K (Lu)^2 d\mu \geq \int_K \|\nabla^2 u\|^2 d\mu + \int_K \langle \nabla^2 V \, \nabla u, \nabla u \rangle d\mu$$
$$+ \int_{\partial K} H_\mu (u_\nu)^2 d\mu_{\partial K} - \int_{\partial K} \langle \mathrm{II}_{\partial K}^{-1} \, \nabla_{\partial K} u_\nu, \nabla_{\partial K} u_\nu \rangle d\mu_{\partial K} .$$

Given $f \in C^{1,\alpha}(\partial K)$, we now apply the above inequality to the solution u of the Neumann Laplace equation (9). Together with Lemma 3.2, this yields:

$$\int_{\partial K} H_\mu (u_\nu)^2 d\mu_{\partial K} - \frac{1}{\mu(K)} \left(\int_K f d\mu \right)^2 + \int_{\partial K} (\langle \nabla_{\partial K} f, \nabla_{\partial K} u \rangle + u_{\nu,\nu} f) \, d\mu_{\partial K}$$
$$\leq \int_{\partial K} \langle \mathrm{II}_{\partial K}^{-1} \, \nabla_{\partial K} f, \nabla_{\partial K} f \rangle d\mu_{\partial K}.$$

Invoking our assumption (10), the asserted inequality (11) follows for $f \in C^{1,\alpha}(\partial K)$. The case of a general $f \in C^1(\partial K)$ follows by a standard approximation argument.

Lastly, (12) is an equivalent version of (11). Indeed, the proof provided in Sect. 2 demonstrates how to pass from (12) to (11), with:

$$G(v) = \frac{(\Phi_\mu^{-1})''(v)}{(\Phi_\mu^{-1})'(v)} = (\log((\Phi_\mu^{-1})'))'(v).$$

To see the other direction, repeat the argument described in the previous section. After establishing (12) for $K, L \in \mathscr{C}_+^2$, the general case follows by a standard approximation argument.

Unfortunately, we cannot claim that condition (10) is equivalent to the Ehrhard-type inequality (12), since the proof of Theorem 3.1 involved an application of the Cauchy-Schwarz inequality (14). Consequently, we pose this as a question:

Question 1 (Gaussian Neumann-to-Dirichlet Operator on Convex Domains)
Does (10) hold for $\mu = \gamma$ the Gaussian measure with $F(v) = \frac{1}{v} - (\log I_\gamma)'(v)$?

Note that the analogous question for $\frac{1}{N}$-concave measures μ, $\frac{1}{N} \in (-\infty, \frac{1}{n}]$, has a positive answer: (10) holds for any $K \in \mathscr{C}_+^2$ in the support of μ with $F(v) = \frac{1}{N} \frac{1}{v}$. Indeed, if $\mu = \Psi(x)dx = \exp(-V(x))dx$ satisfies on its support:

$$-(N-n)\frac{\nabla^2 \Psi^{\frac{1}{N-n}}}{\Psi^{\frac{1}{N-n}}} = \nabla^2 V - \frac{1}{N-n}\nabla V \otimes \nabla V \geq 0,$$

then by several applications of the Cauchy–Schwarz inequality (see [14]):

$$\int_{\partial K} (\langle \nabla_{\partial K} f, \nabla_{\partial K} u \rangle + u_{\nu,\nu} f)\, d\mu_{\partial K} = \int_K \left(\langle \nabla^2 V \, \nabla u, \nabla u \rangle + \|\nabla^2 u\|^2 \right) d\mu$$

$$\geq \int_K \left(\frac{1}{N-n}\langle \nabla u, \nabla V \rangle^2 + \frac{1}{n}(\Delta u)^2 \right) d\mu \geq \int_K \frac{1}{N}(\Delta u - \langle \nabla u, \nabla V \rangle)^2 d\mu$$

$$= \frac{1}{N}\int_K (Lu)^2 d\mu = \frac{1}{N}\frac{1}{\mu(K)}\left(\int_{\partial K} f\, d\mu_{\partial K} \right)^2.$$

4   Concluding Remarks

4.1   Refined Version

Peculiarly, as in [15], it is possible to strengthen Theorem 1.1 by applying it to $f + z$ and optimizing over $z \in \mathbb{R}$. This results in the following stronger inequality:

$$\int_{\partial M} H_\gamma f^2 d\mu_{\partial M} - (\log I_\gamma)'(\gamma(K))\left(\int_{\partial K} f d\gamma_{\partial K} \right)^2 + \frac{\left(\int_{\partial K} f\beta d\gamma_{\partial K} \right)^2}{\int_{\partial K} \beta d\gamma_{\partial K}}$$

$$\leq \int_{\partial K} \langle \mathrm{II}_{\partial K}^{-1} \nabla_{\partial K} f, \nabla_{\partial K} f \rangle d\gamma_{\partial K},$$

where:

$$\beta(x) := (\log I_\gamma)'(\gamma(K))\gamma_{\partial K}(\partial K) - H_\gamma(x).$$

Note that indeed $\int_{\partial K} \beta d\gamma_{\partial K} \geq 0$ by Corollary 1.2, so the additional third term appearing above is always non-negative.

Recall that our original weaker inequality (1) is an equivalent infinitesimal form of Ehrhard's inequality, and so one cannot hope to obtain a strict improvement in the cases when Ehrhard's inequality is sharp (and indeed when K is a half-plane we see that $\beta \equiv 0$). On the other hand, it would be interesting to integrate back the stronger inequality above and obtain a refined version of Ehrhard's inequality, which would perhaps be better suited for obtaining delicate stability results (cf. [12, 18] and the references therein). We leave this for another occasion.

4.2 Mean-Curvature Inequality Implies Isoperimetric Inequality

As explained in Sect. 2, Theorem 1.1 is an equivalent infinitesimal form of Ehrhard's inequality (for convex domains K, L), i.e. equivalent to the concavity of $[0, 1] \ni t \mapsto \Phi^{-1}(\gamma((1-t)K + tL))$. Similarly, Corollary 1.2, which is obtained by setting $f \equiv 1$ in Theorem 1.1, is an equivalent infinitesimal form of the concavity of:

$$\mathbb{R}_+ \ni t \mapsto F(t) := \Phi^{-1}(\gamma(K + tB_2^n)),$$

where B_2^n denotes the Euclidean unit-ball; indeed, Corollary 1.2 expresses precisely that $F''(0) \leq 0$.

It is worthwhile to note that the latter concavity may be used to recover the Gaussian isoperimetric inequality (albeit only for convex sets). The following is a variant on an argument due to Ledoux (private communication), who showed how Ehrhard's inequality with L being a multiple of B_2^n, may be used to recover the Gaussian isoperimetric inequality (for general Borel sets). Indeed, the concavity of F implies that:

$$F'(0) \geq \lim_{t \to \infty} \frac{F(t) - F(0)}{t} = \lim_{t \to \infty} \frac{F(t)}{t} \geq \lim_{t \to \infty} \frac{\Phi^{-1}(\gamma(tB_2^n))}{t}.$$

A straightforward calculation (e.g. [3]) verifies that the right-hand-side is equal to 1, and hence:

$$1 \leq F'(0) = (\Phi^{-1})'(\gamma(K))\gamma_{\partial K}(\partial K),$$

or equivalently:

$$\gamma_{\partial K}(\partial K) \geq I_\gamma(\gamma(K)),$$

as asserted.

4.3   Ehrhard's Inequality is False for $CD(1, \infty)$ Measures

It is well known (e.g. [17]) that various isoperimetric, functional and concentration inequalities which are valid for the Gaussian measure are also valid for any measure $\mu = \exp(-V)dx$ on $\mathbb{R}^n$ with $\nabla^2 V \geq Id$, the so-called class of $CD(1, \infty)$ measures in Euclidean space.

However, we remark that it is not possible to extend Ehrhard's inequality (and hence Theorem 1.1) to this more general class, providing in particular a negative answer to Question 1 for several natural members of this class. Indeed, this is witnessed already by considering the probability measure μ obtained by conditioning the one-dimensional Gaussian measure onto a half-line $(-\infty, b]$ (which may clearly be approximated in total-variation by probability measures $\exp(-V)dx$ with $V'' \geq 1$). It is not true that:

$$\Phi^{-1}(\mu((1 - t)K + tL)) \geq (1 - t)\Phi^{-1}(\mu(K)) + t\Phi^{-1}(\mu(L)),$$

even for half-lines K, L. If that were the case, it would mean that the function $(-\infty, b] \ni t \mapsto \Phi^{-1}(\Phi(t)/\Phi(b))$ is concave, but it is easy to see that this is not the case as $t \to b$. The same argument shows that $\mathbb{R}_+ \ni t \mapsto \Phi^{-1}(\mu(K + t[-1, 1]))$ is not concave even for a half-line K, and so we see that even Corollary 1.2 cannot be extended to the $CD(1, \infty)$ setting.

4.4   Dual Inequality for Mean-Convex Domains

Lastly, for completeness, we specialize a dual Poincaré-type inequality obtained in [15], for the case of the Gaussian measure:

Theorem 4.1 (Dual Inequality for Mean-Convex Domains) *Let $K \subset \mathbb{R}^n$ denote a compact set with C^2 smooth boundary which is strictly Gaussian mean-convex, i.e. $H_\gamma > 0$ on ∂K. Then for any $f \in C^2(\partial K)$ and $C \in \mathbb{R}$:*

$$\int_{\partial K} \langle II_{\partial K} \, \nabla_{\partial K} f, \nabla_{\partial K} f \rangle \, d\gamma_{\partial K} \leq \int_{\partial K} \frac{1}{H_\gamma}\left(L_{\partial K} f + \frac{(f - C)}{2}\right)^2 d\gamma_{\partial K} .$$

Here $L_{\partial K} = \Delta_{\partial K} - \langle x, \nabla_{\partial K} \rangle$ denotes the induced Ornstein–Uhlenbeck generator on ∂K.

Acknowledgements The first named author was supported by the RFBR project 17-01-00662, and the DFG project RO 1195/12-1. The second named author is supported by BSF (grant no. 2010288) and Marie-Curie Actions (grant no. PCIG10-GA-2011-304066). The article was prepared within the framework of the Academic Fund Program at the National Research University Higher School of Economics (HSE) in 2017–2018 (grant no. 17-01-0102) and by the Russian Academic Excellence Project "5-100". The research leading to these results is part of a project that has received funding from the European Research Council (ERC) under the European Union's Horizon 2020 research and innovation programme (grant agreement no. 637851).

References

1. D. Bakry, L'hypercontractivité et son utilisation en théorie des semigroupes, in *Lectures on Probability Theory (Saint-Flour, 1992)*. Lecture Notes in Mathematics, vol. 1581 (Springer, Berlin, 1994), pp. 1–114
2. D. Bakry, M. Émery, Diffusions hypercontractives, in *Séminaire de probabilités, XIX, 1983/84*. Lecture Notes in Mathematics, vol. 1123 (Springer, Berlin, 1985), pp. 177–206
3. V.I. Bogachev, *Gaussian Measures*. Mathematical Surveys and Monographs, vol. 62 (American Mathematical Society, Providence, RI, 1998)
4. Ch. Borell, The Brunn–Minkowski inequality in Gauss spaces. Invent. Math. **30**, 207–216 (1975)
5. Ch. Borell, Convex set functions in d-space. Period. Math. Hung. **6**(2), 111–136 (1975)
6. C. Borell, The Ehrhard inequality. C. R. Math. Acad. Sci. Paris **337**(10), 663–666 (2003)
7. C. Borell, Minkowski sums and Brownian exit times. Ann. Fac. Sci. Toulouse Math. (6) **16**(1), 37–47 (2007)
8. H.J. Brascamp, E.H. Lieb, On extensions of the Brunn-Minkowski and Prékopa-Leindler theorems, including inequalities for log concave functions, and with an application to the diffusion equation. J. Funct. Anal. **22**(4), 366–389 (1976)
9. A. Colesanti, From the Brunn-Minkowski inequality to a class of Poincaré-type inequalities. Commun. Contemp. Math. **10**(5), 765–772 (2008)
10. A. Colesanti, E. Saorín Gómez, Functional inequalities derived from the Brunn-Minkowski inequalities for quermassintegrals. J. Convex Anal. **17**(1), 35–49 (2010)
11. A. Ehrhard, Symétrisation dans l'espace de Gauss. Math. Scand. **53**(2), 281–301 (1983)
12. R. Eldan, A two-sided estimate for the Gaussian noise stability deficit. Invent. Math. **201**(2), 561–624 (2015)
13. R.J. Gardner, The Brunn–Minkowski inequality. Bull. Am. Math. Soc. (N.S.) **39**(3), 355–405 (2002)
14. A.V. Kolesnikov, E. Milman, Brascamp–Lieb type inequalities on weighted Riemannian manifolds with boundary. J. Geom. Anal. (2017, to appear). arxiv.org/abs/1310.2526
15. A.V. Kolesnikov, E. Milman, Poincaré and Brunn–Minkowski inequalities on the boundary of weighted Riemannian manifolds. Submitted, arxiv.org/abs/1310.2526
16. R. Latała, A note on the Ehrhard inequality. Stud. Math. **118**(2), 169–174 (1996)
17. M. Ledoux, *The Concentration of Measure Phenomenon*. Mathematical Surveys and Monographs, vol. 89 (American Mathematical Society, Providence, RI, 2001)
18. E. Mossel, J. Neeman, Robust dimension free isoperimetry in Gaussian space. Ann. Probab. **43**(3), 971–991 (2015)
19. R. Schneider, *Convex Bodies: The Brunn-Minkowski Theory*. Encyclopedia of Mathematics and Its Applications, vol. 44 (Cambridge University Press, Cambridge, 1993)
20. V.N. Sudakov, B.S. Cirel'son [Tsirelson], Extremal properties of half-spaces for spherically invariant measures. Zap. Naučn. Sem. Leningrad. Otdel. Mat. Inst. Steklov. (LOMI) **41**, 14–24, 165 (1974). Problems in the theory of probability distributions, II

Rigidity of the Chain Rule and Nearly Submultiplicative Functions

Hermann König and Vitali Milman

Abstract Assume that $T : C^1(\mathbb{R}) \to C(\mathbb{R})$ nearly satisfies the chain rule in the sense that

$$|T(f \circ g)(x) - (Tf)(g(x))(Tg)(x)| \leq S(x, (f \circ g)(x), g(x))$$

holds for all $f, g \in C^1(\mathbb{R})$ and $x \in \mathbb{R}$, where $S : \mathbb{R}^3 \to \mathbb{R}$ is a suitable fixed function. We show under a weak non-degeneracy and a weak continuity assumption on T that S may be chosen to be 0, i.e. that T satisfies the chain rule operator equation, the solutions of which are explicitly known. We also determine the solutions of one-sided chain rule inequalities like

$$T(f \circ g)(x) \leq (Tf)(g(x))(Tg)(x) + S(x, (f \circ g)(x), g(x))$$

under a further localization assumption. To prove the above results, we investigate the solutions of nearly submultiplicative inequalities on $\mathbb{R}$

$$\phi(\alpha\beta) \leq \phi(\alpha)\phi(\beta) + d$$

and characterize the nearly multiplicative functions on $\mathbb{R}$

$$|\phi(\alpha\beta) - \phi(\alpha)\phi(\beta)| \leq d$$

under weak restrictions on ϕ.

H. König (✉)
Mathematisches Seminar, Universität Kiel, 24098 Kiel, Germany
e-mail: hkoenig@math.uni-kiel.de

V. Milman
School of Mathematical Sciences, Tel Aviv University, Ramat Aviv, Tel Aviv 69978, Israel

© Springer International Publishing AG 2017
B. Klartag, E. Milman (eds.), *Geometric Aspects of Functional Analysis,*
Lecture Notes in Mathematics 2169, DOI 10.1007/978-3-319-45282-1_16

1 Introduction and Results

Several fundamental operations in analysis and geometry such as derivatives, the Fourier transform, the Legendre transform, multiplicative maps or the duality of convex bodies may be characterized, essentially, by elementary properties or as solutions of simple operator functional equations on classical function spaces, cf. [2–4]. The latter may be abstract versions of the Leibniz or the chain rule. In this paper, we concentrate on the question to what extent the chain rule and perturbations of the chain rule determine the derivative. It turns out that the chain rule shows a remarkable rigidity and stability which we will study in this paper. This involves the investigation of nearly multiplicative functions on the real line, i.e. functions which are multiplicative up to some fixed error. We start with known results on the solutions of chain rule equation before considering perturbations of it.

Let $T : C^1(\mathbb{R}) \to C(\mathbb{R})$ be an operator satisfying the chain rule equation

$$T(f \circ g) = ((Tf) \circ g) \cdot Tg \quad , \quad f, g \in C^1(\mathbb{R}) \,.$$

By Artstein-Avidan et al. [3], if T is not identically zero on the bounded functions and $T(-\mathrm{Id})(0) < 0$, T has the form

$$Tf = \frac{H \circ f}{H} \, \mathrm{sgn} f' \, |f'|^p$$

for a suitable $p > 0$ and a continuous function $H : \mathbb{R} \to \mathbb{R}_{>0}$. The equation is very stable: if it is replaced by

$$V(f \circ g) = ((T_1 f) \circ g) \cdot (T_2 g) \quad , \quad f, g \in C^1(\mathbb{R})$$

for operators $V, T_1, T_2 : C^1(\mathbb{R}) \to C(\mathbb{R})$, its solutions under a mild condition of non-degeneracy of V are of a very similar type:

$$Vf = (c_1 \circ f) \cdot c_2 \cdot Tf \,, \ T_1 f = (c_1 \circ f) \cdot Tf \,, \ T_2 f = c_2 \cdot Tf$$

where $c_1, c_2 \in C(\mathbb{R})$ and T has the above form, cf. [5]. It is also stable in another sense: if $S : \mathbb{R}^3 \to \mathbb{R}$ is a function such that

$$T(f \circ g) = ((Tf) \circ g) \cdot Tg + S(\cdot, f \circ g(\cdot), g(\cdot)) \quad , \quad f, g \in C^1(\mathbb{R})$$

holds, under weak conditions one may show that the only possible choice is $S = 0$, i.e. that T satisfies the chain rule equation properly, cf. [5]. The stability even extends to the chain rule inequality

$$T(f \circ g) \le ((Tf) \circ g) \cdot Tg \quad , \quad f, g \in C^1(\mathbb{R})$$

which under weak assumptions on T has only solutions of the form

$$Tf = \left\{ \begin{array}{ll} \frac{H \circ f}{H} f'^p & f' \geq 0 \\ -A \frac{H \circ f}{H} |f'|^p & f' < 0 \end{array} \right\}$$

with p and H as before and a constant $A \geq 1$; $A = |T(-\mathrm{Id})(0)|$, cf. [6]. In this paper, we study a joint extension of the last two problems. We consider the one-sided operator inequality

$$T(f \circ g) \leq ((Tf) \circ g) \cdot Tg + S(\cdot, f \circ g(\cdot), g(\cdot)) \quad , \quad f, g \in C^1(\mathbb{R})$$

and the two-sided operator inequality

$$|T(f \circ g) - ((Tf) \circ g) \cdot Tg| \leq S(\cdot, f \circ g(\cdot), g(\cdot)) \quad , \quad f, g \in C^1(\mathbb{R})$$

and determine the general form of their solutions under reasonable assumptions on T. In the case of the last operator inequality, S may be chosen to be 0, i.e. T actually again satisfies the chain rule operator equation. Hence these equations and inequalities are very rigid and stable under perturbations.

In our previous papers the difference $T(f \circ g) - ((Tf) \circ g) \cdot Tg$ was assumed to be a function of $(x, (f \circ g)(x), g(x))$ which is much stronger than assuming that it is only bounded by a function of these three parameters as done in this paper.

After localizing the problem of the two-sided operator inequality, i.e. showing that there is a function $F : \mathbb{R}^3 \to \mathbb{R}$ such that $Tf(x) = F(x, f(x), f'(x)), f \in C^1(\mathbb{R})$, $x \in \mathbb{R}$, the two operator inequalities for T turn into functional inequalities for F. To solve these, we have to characterize continuous functions $\phi : \mathbb{R} \to \mathbb{R}$ which are submultiplicative up to constants. We start with a result for these functions which has some independent interest.

Assumption 1 Let $\phi : \mathbb{R} \to \mathbb{R}$ be continuous with $\overline{\lim}_{\alpha \to \infty} \phi(\alpha) = \infty$. Suppose also that there is $\alpha_0 > 0$ such that $\phi(-\alpha_0) < 0$.

Theorem 1 *Let $\phi : \mathbb{R} \to \mathbb{R}$ satisfy Assumption 1 and suppose that there is $d \in \mathbb{R}$ such that for all $\alpha, \beta \in \mathbb{R}$*

$$\phi(\alpha\beta) \leq \phi(\alpha)\phi(\beta) + d . \tag{1}$$

Then $d \geq 0$ and there are $p > 0$ and $A \geq 1$ such that for all $\alpha > 0$

$$\phi(\alpha) = \alpha^p , \quad -A\alpha^p \leq \phi(-\alpha) \leq \min(-\frac{1}{A}\alpha^p, -A\alpha^p + d) . \tag{2}$$

Moreover the limit $\lim_{\alpha \to \infty} \frac{\phi(-\alpha)}{-\alpha^p}$ exists and $A = \lim_{\alpha \to \infty} \frac{\phi(-\alpha)}{-\alpha^p}$.

The case $d = 0$ was considered in Theorem 1.1 of [6]. For $d \neq 0$, in general $\phi|_{\mathbb{R}_{<0}}$ is not power type, although it is a bounded perturbation of the power type function $-A\alpha^p$, and the estimates in (2) are the best possible, as the following example shows for $p = 1, A = 2$ and $d = \frac{3}{2}$:

Example 1 Define the continuous, piecewise affine function $\phi : \mathbb{R} \to \mathbb{R}$ by

$$
\phi(\alpha) := \left\{
\begin{array}{ll}
\alpha & \alpha \geq 0 \\[2mm]
\frac{1}{2}\alpha & \alpha \in [-1, 0) \\[2mm]
3 + \frac{7}{2}\alpha & \alpha \in [-2, -1) \\[2mm]
2\alpha & \alpha \in (-\infty, -2)
\end{array}
\right\}.
$$

Then $\phi(\alpha\beta) \leq \phi(\alpha)\phi(\beta) + d$ for all $\alpha, \beta \in \mathbb{R}$, where $d = \frac{3}{2}$. Obviously, $\phi(\alpha\beta) = \phi(\alpha)\phi(\beta)$ for $\alpha, \beta \geq 0$.

For $\alpha, \beta \leq 0$, $\phi(\alpha\beta) = \alpha\beta$. Clearly $\phi(\alpha) \leq \frac{1}{2}\alpha$. Hence if $\beta \leq -2$, $\phi(\alpha)\phi(\beta) \geq \frac{1}{2}\alpha 2\beta = \alpha\beta = \phi(\alpha\beta)$. If $\alpha, \beta \in [-2, 0]$, $\phi(\alpha)\phi(\beta) \geq \frac{1}{2}\alpha\phi(\beta) \geq \alpha\beta - \frac{3}{2}$. The last inequality is easily checked for $\beta \in [-1, 0]$ and $\beta \in [-2, -1]$ separately.

For $\alpha < 0 < \beta$, $\phi(\beta) = \beta$ and $\phi(\alpha) \geq 2\alpha$, hence $\phi(\alpha)\phi(\beta) \geq 2\alpha\beta$. If $\alpha\beta \in (-\infty, -2]$, $\phi(\alpha\beta) = 2\alpha\beta \leq \phi(\alpha)\phi(\beta)$. If $\alpha\beta \in [-1, 0)$, $\phi(\alpha\beta) = \frac{1}{2}\alpha\beta \leq 2\alpha\beta + \frac{3}{2} \leq \phi(\alpha)\phi(\beta) + \frac{3}{2}$. If $\alpha\beta \in [-2, -1)$, $\phi(\alpha\beta) - \phi(\alpha)\phi(\beta) \leq (3 + \frac{7}{2}\alpha\beta) - 2\alpha\beta = 3 + \frac{3}{2}\alpha\beta \leq \frac{3}{2}$.

This shows that $\phi(\alpha\beta) \leq \phi(\alpha)\phi(\beta) + \frac{3}{2}$ holds for all $\alpha, \beta \in \mathbb{R}$, and that the estimate in (2) with $A = 2, p = 1$ cannot be improved, in general.

For the analogue of Theorem 1 for nearly supermultiplicative functions we need a modified assumption.

Assumption 2 Let $\phi : \mathbb{R} \to \mathbb{R}$ be continuous with $\underline{\lim}_{\alpha \to \infty} \phi(-\alpha) = -\infty$. Suppose also that there is $\alpha_0 > 0$ such that $\phi(\alpha_0) > 0$.

Theorem 2 *Let $\phi : \mathbb{R} \to \mathbb{R}$ satisfy Assumption 2 and suppose that there is $d \in \mathbb{R}$ such that for all $\alpha, \beta \in \mathbb{R}$*

$$
\phi(\alpha\beta) \geq \phi(\alpha)\phi(\beta) - d .
$$

Then $d \geq 0$ and there are $p > 0$ and $0 < B \leq 1$ such that for all $\alpha > 0$

$$
\phi(\alpha) = \alpha^p , \quad \max(-\frac{1}{B}\alpha^p, -B\alpha^p - d) \leq \phi(-\alpha) \leq -B\alpha^p .
$$

Moreover the limit $\lim_{\alpha \to \infty} \frac{\phi(-\alpha)}{-\alpha^p}$ exists and $B = \lim_{\alpha \to \infty} \frac{\phi(-\alpha)}{-\alpha^p}$.

As an immediate consequence of both theorems we get

Corollary 3 *Suppose that $\phi : \mathbb{R} \to \mathbb{R}$ satisfies Assumptions 1 and 2 and that there is $d \in \mathbb{R}$ such that for all $\alpha, \beta \in \mathbb{R}$*

$$|\phi(\alpha\beta) - \phi(\alpha)\phi(\beta)| \le d.$$

Then there is $p > 0$ such that for all $\alpha \in \mathbb{R}$

$$\phi(\alpha) = \text{sgn}\alpha \, |\alpha|^p.$$

We will use the previous results to study the rigidity and the stability of the chain rule operator equation. To formulate our result, we need the following assumptions.

Definition An operator $T : C^1(\mathbb{R}) \to C(\mathbb{R})$ is called *pointwise continuous* provided that for any functions $f, f_n \in C^1(\mathbb{R})$, $n \in \mathbb{N}$ such that $f_n \to f$ and $f_n' \to f'$ converge uniformly on compact subsets of $\mathbb{R}$, we have that $(Tf_n)(x) \to (Tf)(x)$ converges pointwise for all $x \in \mathbb{R}$.

Definition An operator $T : C^1(\mathbb{R}) \to C(\mathbb{R})$ is called *non-degenerate* provided that

(a) for all open intervals $I \subset \mathbb{R}$, all $x \in I$ and all $t > 0$ there are functions $f_1, f_2 \in C^1(\mathbb{R})$ with $f_1(x) = f_2(x) = x$, $\text{Im} f_1 \subset I$, $\text{Im} f_2 \subset I$ and $(Tf_1)(x) \ge t$, $(Tf_2)(x) \le -t$,
(b) for some $x_0 \in \mathbb{R}$, $T(-\text{Id})(x_0) < 0$.

We then have the following rigidity result for the chain rule operator inequality.

Theorem 4 *Assume that $T : C^1(\mathbb{R}) \to C(\mathbb{R})$ is pointwise continuous and non-degenerate. Suppose further that there is a function $S : \mathbb{R}^3 \to \mathbb{R}$ such that the perturbed chain operator inequality*

$$|T(f \circ g)(x) - (Tf)(g(x)) \cdot (Tg)(x)| \le S(x, (f \circ g)(x), g(x)) \tag{3}$$

holds for all $f, g \in C^1(\mathbb{R})$ and all $x \in \mathbb{R}$. Then there are $p > 0$ and a positive continuous function $H : \mathbb{R} \to \mathbb{R}_{>0}$ such that for all $f \in C^1(\mathbb{R})$ and all $x \in \mathbb{R}$

$$Tf(x) = \frac{H(f(x))}{H(x)} \text{sgn} f'(x) \, |f'(x)|^p.$$

This implies, in particular, that we may choose $S = 0$, i.e. that we have equality

$$T(f \circ g)(x) = (Tf)(g(x)) \cdot (Tg)(x), \quad f, g \in C^1(\mathbb{R}), x \in \mathbb{R}.$$

The proof of Theorem 4 relies on the following *localization result*:

Proposition 5 *Assume that* $T : C^1(\mathbb{R}) \to C(\mathbb{R})$ *is non-degenerate and pointwise continuous. Suppose further that there is a function* $S : \mathbb{R}^3 \to \mathbb{R}$ *such that the perturbed chain rule inequality*

$$|T(f \circ g)(x) - (Tf)(g(x))(Tg)(x)| \le S(x, (f \circ g)(x), g(x))$$

holds for all $f, g \in C^1(\mathbb{R})$ *and all* $x \in \mathbb{R}$. *Then there is a function* $F : \mathbb{R}^3 \to \mathbb{R}$ *such that for all* $f \in C^1(\mathbb{R})$ *and all* $x \in \mathbb{R}$

$$(Tf)(x) = F(x, f(x), f'(x)) .$$

This means that $Tf(x)$ depends only on $x, f(x)$ and $f'(x)$, i.e. the germ of f at x, and does not depend on values or derivatives of f on values y different from x. The two-sided chain rule operator inequality then turns into two functional inequalities for F which we then will solve using Theorems 1 and 2. In the case of the one-sided operator inequality

$$T(f \circ g)(x) \le (Tf)(g(x))(Tg)(x) + S(x, (f \circ g)(x), g(x)) ,$$

localization is not true, in general, as the following example shows, even though it satisfies the non-degeneracy and the pointwise continuity assumption.

Example 2 For $f \in C^1(\mathbb{R})$, $x \in \mathbb{R}$ with $f'(x) \in (-1, 0)$, let $I_{f,x}$ denote the interval $I_{f,x} := [x + f'(x)(1 + f'(x)), x]$. Then $0 < |I_{f,x}| \le \frac{1}{4}$. Let $Jf(x) := \frac{1}{|I_{f,x}|} \int_{I_{f,x}} f(y)\, dy$. Choose any non-constant function $H \in C(\mathbb{R})$ with $4 \le H \le 5$. For $f \in C^1(\mathbb{R})$, $x \in \mathbb{R}$ put

$$Tf(x) := \begin{cases} \dfrac{H(f(x))}{H(x)} f'(x) & f'(x) \ge 0 \\[2mm] \dfrac{H(f(x))}{H(x)} 4f'(x) & f'(x) \le -2 \\[2mm] \dfrac{H(f(x))}{H(x)} (7 + \frac{15}{2} f'(x)) & -2 < f'(x) \le -1 \\[2mm] \dfrac{H(Jf(x))}{H(x)} \frac{1}{2} f'(x) & -1 < f'(x) < 0 \end{cases} .$$

Then T maps $C^1(\mathbb{R})$ into $C(\mathbb{R})$ and satisfies

$$T(f \circ g)(x) \le (Tf)(g(x)) \cdot (Tg)(x) + 5 \quad ; \quad f, g \in C^1(\mathbb{R}), \; x \in \mathbb{R} .$$

We will prove this statement in Sect. 4. Obviously, T is not localized.

However, assuming that T is defined locally, we can determine the general form of solutions of the one-sided chain rule operator inequality:

Theorem 6 *Assume that* $T : C^1(\mathbb{R}) \to C(\mathbb{R})$ *is pointwise continuous and non-degenerate. Suppose further that there is a function* $S : \mathbb{R}^3 \to \mathbb{R}$ *such that the perturbed chain operator inequality*

$$T(f \circ g)(x) \leq (Tf)(g(x)) \cdot (Tg)(x) + S(x, (f \circ g)(x), g(x)) \qquad (4)$$

holds for all $f, g \in C^1(\mathbb{R})$ *and all* $x \in \mathbb{R}$. *Assume also that there is a function* $F : \mathbb{R}^3 \to \mathbb{R}$ *such that*

$$Tf(x) = F(x, f(x), f'(x)) \quad ; \quad f \in C^1(\mathbb{R}), \, x \in \mathbb{R}.$$

Then there are $p > 0$, $A \geq 1$, *a positive continuous function* $H : \mathbb{R} \to \mathbb{R}_{>0}$ *and a function* $K : \mathbb{R}^2 \times \mathbb{R}_{<0} \to \mathbb{R}_{<0}$ *which is continuous in the second and third variable with*

$$-A\alpha^p \leq K(x, z, -\alpha) \leq \min(-\frac{1}{A}\alpha^p, -A\alpha^p + \frac{H(x)}{H(z)} \min[S(x, z, x), S(x, z, z)]) \,,$$

for all $x, z \in \mathbb{R}$, $\alpha > 0$, *where the limits* $\lim_{\beta \to \infty} \frac{K(x,z,-\beta)}{-\beta^p}$ *exist for all* $x, z \in \mathbb{R}$ *with* $A = \lim_{\beta \to \infty} \frac{K(x,z,-\beta)}{-\beta^p}$ *such that for all* $f \in C^1(\mathbb{R})$ *and* $x \in \mathbb{R}$

$$Tf(x) := \left\{ \begin{array}{ll} \frac{H(f(x))}{H(x)} f'(x)^p & f'(x) \geq 0 \\ \frac{H(f(x))}{H(x)} K(x, f(x), f'(x)) & f'(x) < 0 \end{array} \right\}.$$

The property of K means that for negative values of $f'(x)$, $Tf(x)$ is reasonably close to $-A \frac{H(f(x))}{H(x)} |f'(x)|^p$, deviating from this value by at most $\max[S(x, f(x), x), S(x, f(x), f(x))]$. The inequality case with $S = 0$ has been considered in Theorem 1.2 of [6] and the equality case that

$$T(f \circ g)(x) = (Tf)(g(x)) \cdot (Tg)(x) + S(x, (f \circ g)(x), g(x))$$

with general S has been solved in Theorem 8 of [5].

A similar result holds for the perturbed supermultiplicative operator inequality

$$T(f \circ g)(x) \geq (Tf)(g(x)) \cdot (Tg)(x) - S(x, (f \circ g)(x), g(x))$$

with the property of K being replaced by

$$-\frac{1}{B}\alpha^p \leq K(x, y, -\alpha) \leq -B\alpha^p \quad , \quad 0 < \lim_{\beta \to \infty} \frac{K(x, y, -\beta)}{-\beta^p} = B \leq 1$$

for all $x, y \in \mathbb{R}$ and $\alpha > 0$.

Of course, Theorem 4 is a consequence of Proposition 5, Theorem 6 and its supermultiplicative analogue.

2 Proof of Theorems 1 and 2

To prove Theorem 1, we need a lemma.

Lemma 7 *Under the assumptions of Theorem 1, $\phi(1) = 1$, $\phi(0) = 0$ and $\phi|_{\mathbb{R}_{<0}} < 0 < \phi|_{\mathbb{R}_{>0}}$. Moreover*

$$\lim_{\alpha \to \infty} \phi(\alpha) = \infty \quad , \quad \lim_{\alpha \to \infty} \phi(-\alpha) = -\infty \, ,$$

where both limits exist. The same is true under the assumptions of Theorem 2.

Proof

(i) If there would be $0 \neq \bar{\alpha} \in \mathbb{R}$ with $\phi(\bar{\alpha}) = 0$, by (1)

$$\phi(\alpha) \leq \phi(\bar{\alpha})\phi(\frac{\alpha}{\bar{\alpha}}) + d \, , \, \alpha \in \mathbb{R}$$

so that ϕ would be bounded from above, contradicting Assumption 1. Since ϕ is continuous and $\overline{\lim}_{\alpha \to \infty} \phi(\alpha) = \infty$, we have that $\phi(\alpha) > 0$ for all $\alpha > 0$ since otherwise there would be a zero of ϕ. Hence $\phi|_{\mathbb{R}_{>0}} > 0$.

(ii) By assumption, $\overline{\lim}_{\alpha \to \infty} \phi(\alpha) = \infty$. Choose $\alpha_n \to \infty$ with $\lim_{n \to \infty} \phi(\alpha_n) = \infty$. Then $\phi(\alpha_n) \leq \phi(1)\phi(\alpha_n) + d$ implies $\phi(1) \geq 1$: if $0 < \phi(1) < 1$, $\sup_{n \in \mathbb{N}} \phi(\alpha_n) \leq d/(1 - \phi(1))$ would be bounded. Since ϕ is continuous, $M := \sup \phi|_{[0,1]}$ is finite. Choose $n_0 \in \mathbb{N}$ such that for all $n \geq n_0$, $\phi(\alpha_n) \geq 2d$. Then for all $\alpha \geq \alpha_n$, $n \geq n_0$

$$\phi(\alpha_n) \leq \phi(\frac{\alpha_n}{\alpha})\phi(\alpha) + d \leq M\phi(\alpha) + \frac{1}{2}\phi(\alpha_n) \, ,$$

$\phi(\alpha) \geq \frac{1}{2M}\phi(\alpha_n)$. Hence $\underline{\lim}_{\alpha \to \infty} \phi(\alpha) = \infty$ and therefore the limit $\lim_{\alpha \to \infty} \phi(\alpha)$ exists and is ∞.

(iii) By Assumption 1 there is $\alpha_0 > 0$ with $\phi(-\alpha_0) < 0$. For all $\alpha > 0$

$$\phi(-\alpha) \leq \phi(-\alpha_0)\phi(\frac{\alpha}{\alpha_0}) + d \, .$$

Hence $\lim_{\alpha \to \infty} \phi(\frac{\alpha}{\alpha_0}) = \infty$ implies that $\lim_{\alpha \to \infty} \phi(-\alpha) = -\infty$ and

$$\phi(-\alpha) \leq \phi(1)\phi(-\alpha) + d$$

yields $\phi(1) \leq 1$. Hence $\phi(1) = 1$. Again by continuity, $\phi|_{\mathbb{R}_{<0}} < 0$, since otherwise ϕ would have a zero in $\mathbb{R}_{<0}$. Since ϕ is continuous and $\phi|_{\mathbb{R}_{<0}} < 0 < \phi|_{\mathbb{R}_{>0}}$, $\phi(0) = 0$ follows.

(iv) In the case of Theorem 2, under Assumption 2, again there is no non-zero $\bar{\alpha}$ with $\phi(\bar{\alpha}) = 0$, since otherwise in view of

$$\phi(-\alpha) \geq \phi(\bar{\alpha})\phi(-\frac{\alpha}{\bar{\alpha}}) - d = -d$$

ϕ would be bounded from below, contradicting $\underline{\lim}_{\alpha \to \infty} \phi(-\alpha) = -\infty$. This again yields that $\phi|_{\mathbb{R}_{<0}} < 0 < \phi|_{\mathbb{R}_{>0}}$. We claim that $\lim_{\alpha \to \infty} \phi(-\alpha) = -\infty$: Choose $\alpha_n \to \infty$ with $\phi(-\alpha_n) \to -\infty$. There is $n_0 \in \mathbb{N}$ such that for all $n \geq n_0$, $|\phi(-\alpha_n)| = -\phi(-\alpha_n) \geq 2d$. Also $M := \sup \phi|_{[0,1]}$ is finite. Therefore for all $\alpha \geq \alpha_n$, $n \geq n_0$

$$M\phi(-\alpha) \leq \phi(\frac{\alpha_n}{\alpha})\phi(-\alpha) \leq \phi(-\alpha_n) + d \leq \frac{1}{2}\phi(-\alpha_n),$$

$\phi(-\alpha) \leq \frac{1}{2M}\phi(-\alpha_n)$ proving $\lim_{\alpha \to \infty} \phi(-\alpha) = -\infty$. Since $\phi(\alpha) \geq \phi(-1)\phi(-\alpha) - d$, also $\lim_{\alpha \to \infty} \phi(\alpha) = \infty$. □

Proof of Theorem 1

(a) For any $b > 1$, by Lemma 7 there is $\gamma_0 = \gamma_0(b) \geq 1$ such that for any $\alpha, \beta \in \mathbb{R}$ with $\alpha\beta \geq \gamma_0$ we have $\phi(\alpha\beta) \geq \frac{b}{b-1}d$ and for all $\alpha, \beta \in \mathbb{R}$ with $\alpha\beta \leq -\gamma_0$ we have $\phi(\alpha\beta) \leq -\frac{1}{b-1}d$. Then by (1) for $\alpha\beta \geq \gamma_0$

$$\frac{1}{b}\phi(\alpha\beta) + d \leq \frac{1}{b}\phi(\alpha\beta) + \frac{b-1}{b}\phi(\alpha\beta)$$

$$= \phi(\alpha\beta) \leq \phi(\alpha)\phi(\beta) + d, \phi(\alpha\beta) \leq b\phi(\alpha)\phi(\beta), \alpha\beta \geq \gamma_0.$$

Also by (1) for $\alpha\beta \leq -\gamma_0$

$$b\phi(\alpha\beta) = \phi(\alpha\beta) + (b-1)\phi(\alpha\beta) \leq \phi(\alpha\beta) - d \leq \phi(\alpha)\phi(\beta),$$

$$b\phi(\alpha\beta) \leq \phi(\alpha)\phi(\beta), \alpha\beta \leq -\gamma_0.$$

Let $\phi_1 := b\phi$, $\phi_2 := \frac{1}{b}\phi$. Then $\phi_1 = b^2\phi_2$ and

$$\begin{cases} \phi_1(\alpha\beta) \leq \phi_1(\alpha)\phi_1(\beta), & \alpha\beta \geq \gamma_0 \\ \phi_2(\alpha\beta) \leq \phi_2(\alpha)\phi_2(\beta), & \alpha\beta \leq -\gamma_0, \end{cases} \tag{5}$$

i.e. we have submultiplicativity for large $|\alpha\beta|$.

(b) Define $f : \mathbb{R} \to \mathbb{R}$ by $f(t) := \ln \phi_1(\exp(t))$, $t \in \mathbb{R}$ and put $t_0 := t_0(b) :=$ $\ln \gamma_0(b) \geq 0$. For all $\alpha > 0$

$$\phi_1(\alpha) = \exp(f(\ln \alpha)) \, , \quad \phi(\alpha) = \exp(f(\ln \alpha) - \ln b) \, .$$

For all $s, t \in \mathbb{R}$ with $t + s \geq t_0$, $\exp(t) \exp(s) \geq \gamma_0$ and hence by (5)

$$f(t + s) = \ln \phi_1(\exp(t) \exp(s))$$
$$\leq \ln \phi_1(\exp(t)) + \ln \phi_1(\exp(s)) = f(t) + f(s) \qquad (6)$$

for $t + s \geq t_0$. Since $\lim_{\alpha \to \infty} \phi_1(\alpha) = \infty$, $\phi_1(0) = 0$ and ϕ_1 is continuous, we have

$$\lim_{t \to \infty} f(t) = \infty \, , \quad \lim_{t \to -\infty} f(t) = -\infty \, .$$

By (6), $f(t_0) \leq f(t_0 - t) + f(t)$, $t \in \mathbb{R}$ which yields for $t \to \infty$ with

$$\frac{f(t_0)}{t} + \frac{f(t_0 - t)}{-t} \leq \frac{f(t)}{t}$$

that $\underline{\lim}_{t \to \infty} \frac{f(t)}{t} \geq 0$. Note here that for large $t > 0$, $f(t_0 - t) < 0$ and $\frac{f(t_0 - t)}{-t} \geq 0$. Hence $p := \inf_{t \geq t_0} \frac{f(t)}{t} \in \mathbb{R}$ is finite, using that f is continuous since ϕ_1 is. We claim that

$$p := \inf_{t \geq t_0} \frac{f(t)}{t} = \lim_{t \to \infty} \frac{f(t)}{t} \, , \qquad (7)$$

and that the limit exists. To see this, let $\epsilon > 0$. Choose $c > t_0 \geq 0$ with $\frac{f(c)}{c} \leq p + \epsilon$. For any $t \geq 2c$, choose $n \in \mathbb{N}$ with $t \in [(n+1)c, (n+2)c]$. Then $t - nc \in [c, 2c]$. Let $M := \sup f|_{[c, 2c]}$. By (6)

$$p \leq \frac{f(t)}{t} = \frac{f(nc + (t - nc))}{t} \leq \frac{nf(c)}{t} + \frac{f(t - nc)}{t} \leq \frac{nc}{t} \frac{f(c)}{c} + \frac{M}{t} \, .$$

Since $\frac{nc}{t} \to 1$ for $t \to \infty$, we find for any $\epsilon > 0$

$$p \leq \overline{\lim_{t \to \infty}} \frac{f(t)}{t} \leq p + \epsilon = \inf_{t \geq t_0} \frac{f(t)}{t} + \epsilon \, .$$

Hence $\lim_{t \to \infty} \frac{f(t)}{t}$ exists and is equal to $p = \inf_{t \geq t_0} \frac{f(t)}{t}$. Since $\lim_{t \to \infty} f(t) = \infty$, it follows that $p \geq 0$.

(c) By (6), $f(t) \leq f(0) + f(t)$ for any $t \geq t_0$. Therefore $f(0) \geq 0$. For any $t \in \mathbb{R}$, choose $s \in \mathbb{R}$ such that $s \geq t_0$ and $s + t \geq t_0$. Then by (6)

$$f(s) \leq f(s+t) + f(-t) \leq f(s) + f(t) + f(-t) .$$

Hence $f(t) + f(-t) \geq 0$ for any $t \in \mathbb{R}$. Define $a : \mathbb{R} \to \mathbb{R}$ by $f(t) = pt + a(t)$, $t \in \mathbb{R}$. Then by (7), $a(t) \geq 0$ for any $t \geq t_0$ and $\lim_{t \to \infty} \frac{a(t)}{t} = 0$ as well as $a(s+t) \leq a(s) + a(t)$ for $s + t \geq t_0$. Note that $f(t) \geq -f(-t)$ implies

$$\phi_1(\alpha) = \exp(f(\ln \alpha)) \geq \exp(-f(-\ln \alpha)) = \frac{1}{\phi_1(\frac{1}{\alpha})} .$$

(d) We now turn to $\phi|_{\mathbb{R}_{<0}}$. Define $g : \mathbb{R} \to \mathbb{R}$ by $g(t) := \ln |\phi_2(-\exp(t))|$, $t \in \mathbb{R}$. Recall that $\phi_2 = \frac{1}{b}\phi$, $\phi_1 = b^2\phi_2$ and $\phi_2|_{\mathbb{R}_{<0}} < 0$ so that $\phi_2(-\alpha) = -\exp(g(\ln \alpha))$ for any $\alpha > 0$. For any $t, s \in \mathbb{R}$ with $t + s \geq t_0 = \ln \gamma_0$, $-\exp(t)\exp(s) \leq -\gamma_0$ and by (5)

$$\phi_2(-\exp(t)\exp(s)) \leq \phi_2(-\exp(t))\phi_2(\exp(s)) ,$$

$$|\phi_2(-\exp(t)\exp(s))| \geq |\phi_2(-\exp(t))|\phi_2(\exp(s)) = |\phi_2(-\exp(t))|\frac{1}{b^2}\phi_1(\exp(s)) .$$

This yields

$$g(t+s) \geq g(t) + f(s) - 2\ln b , \quad t + s \geq t_0 . \tag{8}$$

We find for any $t \in \mathbb{R}$ using (8) and $f(s) \geq -f(-s)$

$$g(t_0) \geq g(t) + f(t_0 - t) - 2\ln b \geq g(t) - f(t - t_0) - 2\ln b$$
$$= g(t) - p(t - t_0) - a(t - t_0) - 2\ln b .$$

Define $\zeta := g(t_0) - pt_0$. Then

$$g(t) \leq \zeta + pt + a(t - t_0) + 2\ln b , \quad t \in \mathbb{R} . \tag{9}$$

For $t \geq t_0$ there is a reverse type inequality since by (8)

$$g(t) \geq g(t_0) + f(t - t_0) - 2\ln b$$
$$= g(t_0) + p(t - t_0) + a(t - t_0) - 2\ln b ,$$
$$g(t) \geq \zeta + pt + a(t - t_0) - 2\ln b , \quad t \geq t_0 . \tag{10}$$

Note that for $t \geq 2t_0$, $t - t_0 \geq t_0$, $a(t - t_0) \geq 0$ in (9) and (10). Since $\lim_{t \to \infty} \frac{a(t)}{t} = 0$, (9) and (10) imply, in particular, that $\lim_{t \to \infty} \frac{g(t)}{t} = p$ and for $t \geq t_0$

$$|g(t) - (\zeta + pt + a(t - t_0))| \leq 2 \ln b . \tag{11}$$

(e) We know that a is submultiplicative for large arguments. We now want to show that a satisfies a weak form of supermultiplicativity for large arguments and that a is bounded. Let $A := \exp(\zeta)$. Then for any $\alpha \geq \gamma_0 = \exp(t_0)$ by (11)

$$\exp(g(\ln(\alpha))) = A\alpha^p \exp(a(\ln \frac{\alpha}{\gamma_0})) \, \theta_\alpha ,$$

where $\frac{1}{b^2} \leq \theta_\alpha \leq b^2$. Therefore we have by definition of f and g

$$\left\{ \begin{array}{l} \phi(\alpha) = \frac{1}{b}\phi_1(\alpha) = \frac{1}{b}\alpha^p \exp(a(\ln \alpha)) , \; \alpha > 0 \\ \phi(-\alpha) = b\phi_2(-\alpha) = -b\exp(g(\ln \alpha)) = -A\alpha^p \exp(a(\ln \frac{\alpha}{\gamma_0}))b\theta_\alpha , \; \alpha \geq \gamma_0 \end{array} \right\} , \tag{12}$$

where $\frac{1}{b^2} \leq \theta_\alpha \leq b^2$. Let $\alpha \geq \gamma_0$, $\beta > 0$. By (1)

$$\phi(-\alpha\beta) \leq \phi(-\alpha)\phi(\beta) + d$$

where $\phi(-\alpha) < 0 < \phi(\beta)$, $\phi(-\alpha\beta) < 0$. Hence by (12)

$$-A(\alpha\beta)^p \exp(a(\ln(\frac{\alpha}{\gamma_0}\beta)))b^3 \leq \phi(-\alpha\beta) \leq \phi(-\alpha)\phi(\beta) + d$$

$$\leq -A\alpha^p \exp(a(\ln(\frac{\alpha}{\gamma_0})))\frac{1}{b}\beta^p \exp(a(\ln(\beta)))\frac{1}{b} + d$$

which yields

$$b^5 \exp[a(\ln(\frac{\alpha}{\gamma_0}) + \ln(\beta))] \geq \exp[a(\ln(\frac{\alpha}{\gamma_0})) + a(\ln(\beta))] - \frac{db^2}{A(\alpha\beta)^p} .$$

This yields for any $t = \ln(\frac{\alpha}{\gamma_0}) \geq 0$ and $s = \ln(\beta) \in \mathbb{R}$

$$a(t) + a(s) \leq \ln[b^5 \exp(a(t + s)) + \frac{db^2}{A\gamma_0^p \exp(p(t + s))}] .$$

Since for any $x \geq 1$ and $\epsilon > 0$

$$\ln(x + \epsilon) = \ln x + \ln(1 + \frac{\epsilon}{x}) \leq \ln x + \ln(1 + \epsilon) \leq \ln x + \epsilon ,$$

we find for $t + s \geq t_0$ when $a(t + s) \geq 0$ and $t \geq 0$

$$a(t) + a(s) \leq a(t + s) + 5\ln b + \frac{db^2}{A\gamma_0^p \exp(p(t + s))} .$$

In particular, for $t = s \geq \frac{1}{2}t_0 = \frac{1}{2}t_0(b)$

$$2a(t) \leq a(2t) + 5\ln b + \frac{db^2}{A\gamma_0^p \exp(2pt)} .$$

We know that $p \geq 0$. Assume first that $p > 0$. Then there is $t_1 = t_1(b) \geq \frac{1}{2}t_0(b)$ such that for all $t \geq t_1$

$$2a(t) \leq a(2t) + 6\ln b . \tag{13}$$

We claim that this implies $a(t) \leq \delta := 6\ln b$ for all $t \geq t_1$. If this would be false, there would be $B > 1$ and $\bar{t} \geq t_1$ such that $a(\bar{t}) \geq B\delta$. Then by (13)

$$2B\delta \leq 2a(\bar{t}) \leq a(2\bar{t}) + \delta , \quad (2B - 1)\delta \leq a(2\bar{t}) .$$

Iterating this, we get

$$2(2B - 1)\delta \leq 2a(2\bar{t}) \leq a(4\bar{t}) + \delta , \quad (4B - 3)\delta \leq a(4\bar{t})$$

and by induction

$$2^n(B - 1)\delta \leq (2^n B - 2^n + 1)\delta \leq a(2^n \bar{t}) ,$$

$$0 < (B - 1)\frac{\delta}{\bar{t}} \leq \frac{a(2^n \bar{t})}{2^n \bar{t}} .$$

However, $\lim_{n \to \infty} \frac{a(2^n \bar{t})}{2^n \bar{t}} = 0$. This yields a contradiction.

We now show that $p = 0$ is impossible. If $p = 0$, instead of (13) we would have for all $t \geq \frac{1}{2}t_0$

$$2a(t) \leq a(2t) + 5\ln b + \frac{db^2}{A} .$$

The same proof as above then shows that $a(t) \leq 5\ln b + \frac{db^2}{A} =: \delta$ for all $t \geq \frac{1}{2}t_0$. Hence a is bounded and for $\alpha \to \infty$

$$\phi(\alpha) = \alpha^p \exp(a(\ln \alpha)) = \exp(a(\ln \alpha))$$

would be bounded, contradicting Assumption 1. Hence $p > 0$ and $0 \leq a(t) \leq 6 \ln b$ holds for all $t \geq t_1$.

(f) Let $t_2 := \max(t_0, t_1)$ and $\gamma_2 := \exp(t_2)$. The previous bound for a and (12) together imply for all $\alpha \geq \gamma_2 \geq 1$

$$\frac{1}{b}\alpha^p \leq \phi(\alpha) = \frac{1}{b}\alpha^p \exp(a(\ln \alpha)) \leq b^5 \alpha^p$$

and for all $\alpha \geq \gamma_0 \gamma_2 =: \gamma_1$

$$-Ab^9\alpha^p \leq \phi(-\alpha) = -A\alpha^p \exp(a(\ln \frac{\alpha}{\gamma_0}))b\theta_\alpha \leq -A\frac{1}{b}\alpha^p .$$

Hence for all $\alpha \geq \gamma_1$

$$\frac{1}{b} \leq \frac{\phi(\alpha)}{\alpha^p} \leq b^5 \quad , \quad \frac{1}{b} \leq \frac{\phi(-\alpha)}{-A\alpha^p} \leq b^9 . \tag{14}$$

Since $b > 1$ was arbitrary, we find

$$\lim_{\alpha \to \infty} \frac{\phi(\alpha)}{\alpha^p} = 1 \quad , \quad \lim_{\alpha \to \infty} \frac{\phi(-\alpha)}{-A\alpha^p} = 1 .$$

Note here that by definition of A, A formally depends on b. However, $\frac{1}{b} \leq \frac{\phi(-\alpha)}{-A\alpha^p} \leq b^9$ for all $\alpha \geq \gamma_1(b)$ implies for $b \to 1$ and $\alpha \to \infty$ that $\lim_{b \to 1} A(b) =: A$ exists.

(g) We now claim that for all $\alpha > 0$, $\phi(\alpha) = \alpha^p$. Let $\alpha > 0$. Choose $\beta > 0$ so large that $\alpha\beta \geq \gamma_1(\geq \gamma_0)$ and $\beta \geq \gamma_1$. Then by (5) and (14)

$$\frac{1}{b}(\alpha\beta)^p \leq \phi(\alpha\beta) \leq b\phi(\beta)\phi(\alpha) \leq b^6\beta^p\phi(\alpha) , \quad \frac{1}{b^7}\alpha^p \leq \phi(\alpha) .$$

Since this holds for all $b > 1$, $\phi(\alpha) \geq \alpha^p$. For the same choice of $\beta > 0$, again by (5) and (14)

$$-Ab^9(\alpha\beta)^p \leq \phi(-\alpha\beta) \leq \frac{1}{b}\phi(-\beta)\phi(\alpha) \leq -A\frac{1}{b^2}\beta^p\phi(\alpha) , \quad \phi(\alpha) \leq b^{11}\alpha^p ,$$

i.e. $\phi(\alpha) \leq \alpha^p$ for $b \to 1$. This proves that $\phi(\alpha) = \alpha^p$ for all $\alpha > 0$.

(h) We now show $\phi(-\alpha) \geq -A\alpha^p$ for all $\alpha > 0$. Let $\alpha > 0$ and choose $\beta \geq \gamma_1$ such that $\alpha\beta \geq \gamma_1$. Then by (5), (14) and part (g)

$$-A(\alpha\beta)^pb^9 \leq \phi(-\alpha\beta) \leq \frac{1}{b}\phi(-\alpha)\phi(\beta) = \frac{1}{b}\phi(-\alpha)\beta^p , \quad -A\alpha^pb^{10} \leq \phi(-\alpha) .$$

Since $b > 1$ was arbitrary, $\phi(-\alpha) \geq -A\alpha^p$ for all $\alpha > 0$.

Next we prove $\phi(-\alpha) \leq -\frac{1}{A}\alpha^p$ for all $\alpha > 0$. Let $\alpha > 0$ and choose $\beta \geq \gamma_1$ such that $\alpha\beta \geq \gamma_1$. We get similarly as before, noting that $\phi(-\alpha) < 0$, $\phi(-\beta) < 0$,

$$(\alpha\beta)^p = \phi(\alpha\beta) \leq b\phi(-\alpha)\phi(-\beta) \leq b\phi(-\alpha)(-A\beta^p b^9) , \quad \phi(-\alpha) \leq -\frac{1}{A}\alpha^p\frac{1}{b^{10}}$$

and for $b \to 1$, $\phi(-\alpha) \leq -\frac{1}{A}\alpha^p$, $\alpha > 0$. Clearly, $A \geq 1$.

Let $\alpha > 0$ and choose $\beta > 0$ such that $\alpha\beta \geq \gamma_1$. Then by (1) and (14)

$$\phi(-\alpha) = \phi(-\alpha\beta\frac{1}{\beta}) \leq \phi(-\alpha\beta)\frac{1}{\beta^p} + d \leq -\frac{1}{b}A(\alpha\beta)^p\frac{1}{\beta^p} + d = -\frac{1}{b}A\alpha^p + d .$$

For $b \to 1$ we get a second estimate for $\phi(-\alpha)$: $\phi(-\alpha) \leq -A\alpha^p + d$. □

Proof of Theorem 2 The proof of Theorem 2 for nearly supermultiplicative functions is similar to the one of Theorem 1, reversing inequalities. We indicate some changes. In particular, by Lemma 7, for any $b > 1$ there is $\gamma_0 = \gamma_0(b) \geq 1$ such that for all $\alpha, \beta \in \mathbb{R}$

$$\phi(\alpha\beta) \geq \frac{1}{b-1}d , \quad \alpha\beta \geq \gamma_0 ,$$

$$\phi(\alpha\beta) \leq -\frac{b}{b-1}d , \quad \alpha\beta \leq -\gamma_0 .$$

Then for $\phi_1 := \frac{1}{b}\phi$, $\phi_2 := b\phi$, $\phi_2 = b^2\phi_1$

$$\phi_1(\alpha\beta) \geq \phi_1(\alpha)\phi_1(\beta) , \quad \alpha\beta \geq \gamma_0 ,$$

$$\phi_2(\alpha\beta) \geq \phi_2(\alpha)\phi_2(\beta) , \quad \alpha\beta \leq -\gamma_0 .$$

Let $f(t) := \ln\phi_1(\exp(t))$, $t \in \mathbb{R}$ and $t_0 := \ln\gamma_0 \geq 0$. Defining $p := \sup_{t \geq t_0}\frac{f(t)}{t}$, one shows that the limit $\lim_{t\to\infty}\frac{f(t)}{t}$ exists and is equal to p. Then $p \geq 0$. Define $a : \mathbb{R} \to \mathbb{R}$ by $f(t) = pt + a(t)$. Then $a(t) \leq 0$ for all $t \geq t_0$ and hence $\phi_1(\alpha) \leq \alpha^p$ for all $\alpha \geq \gamma_0$. Since $\lim_{\alpha\to\infty}f(t) = \infty$, $p > 0$ follows immediately. Also $a(0) \leq 0$ and $a(t) \leq -a(-t)$ for all $t \in \mathbb{R}$ as well as $a(t+s) \geq a(t) + a(s)$ for $t+s \geq t_0$. Again, let $g(t) := \ln|\phi_2(-\exp(t))|$, $t \in \mathbb{R}$. Then (9) and (10) are replaced by

$$g(t) \geq \zeta + pt + a(t - t_0) - 2\ln b , \quad t \in \mathbb{R} ,$$

$$g(t) \leq \zeta + pt + a(t - t_0) + 2\ln b , \quad t \geq t_0 ,$$

with $\zeta := g(t_0) - pt_0$. Again $\lim_{t\to\infty}\frac{g(t)}{t} = p$ follows. Parts (e) to (h) of the proof of Theorem 1 are easily adapted by reversing signs, with $0 < B := \exp(\zeta) \leq 1$. □

Corollary 3 is a combination of Theorems 1 and 2. However, there is a much simpler direct proof of Corollary 3 which we now give

Second Proof of Corollary 3 By Lemma 7, $\lim_{|\alpha| \to \infty} |\phi(\alpha)| = \infty$. Hence by assumption for all $\alpha \in \mathbb{R}$ and $\beta \to \infty$

$$\left| \frac{\phi(\alpha\beta)}{\phi(\beta)} - \phi(\alpha) \right| \leq \frac{d}{|\phi(\beta)|} \to 0 .$$

This yields that

$$\phi(\alpha) = \lim_{\beta \to \infty} \frac{\phi(\alpha\beta)}{\phi(\beta)} ,$$

where the limit exists. Therefore

$$\phi(\alpha)\phi(\beta) = \lim_{\gamma,\delta \to \infty} \frac{\phi(\alpha\gamma)}{\phi(\gamma)} \frac{\phi(\beta\delta)}{\phi(\delta)} .$$

Now $\phi(\alpha\gamma)\phi(\beta\delta) \leq \phi(\alpha\beta\gamma\delta) + d$ and $\phi(\gamma)\phi(\delta) \geq \phi(\gamma\delta) - d$. Hence

$$\phi(\alpha)\phi(\beta) \leq \lim_{\gamma,\delta \to \infty} \frac{\phi(\alpha\beta\gamma\delta) + d}{\phi(\gamma\delta) - d} = \lim_{\gamma,\delta \to \infty} \frac{\phi(\alpha\beta\gamma\delta)}{\phi(\gamma\delta)} = \phi(\alpha\beta) .$$

Similarly, $\phi(\alpha)\phi(\beta) \geq \phi(\alpha\beta)$. Hence $\phi : \mathbb{R} \to \mathbb{R}$ is multiplicative, $\phi(\alpha)\phi(\beta) = \phi(\alpha\beta)$ for all $\alpha, \beta \in \mathbb{R}$ which implies by Lemma 13 of [3] that $\phi(\alpha) = \text{sgn}\alpha\, |\alpha|^p$ for a suitable $p \in \mathbb{R}$ and all $\alpha \in \mathbb{R}$. Since ϕ is continuous in 0 and $\lim_{\alpha \to \infty} \phi(\alpha) = \infty$, we have $p > 0$. $\qquad\square$

3   Further Results on Submultiplicativity

To prove the rigidity and stability results for the chain operator inequalities, i.e. Theorems 6 and 4, we need a proposition on two nearly submultiplicative functions.

Proposition 8 *Let $\phi, \psi : \mathbb{R} \to \mathbb{R}$ be continuous functions and $d, e, f, g \in \mathbb{R}$ with $f > 0$ be such that for all $\alpha, \beta \in \mathbb{R}$*

$$\phi(\alpha\beta) \leq \phi(\alpha)\phi(\beta) + d \tag{15}$$

$$\psi(\alpha\beta) \leq \psi(\alpha)\phi(\beta) + e \tag{16}$$

$$\phi(\alpha) \leq f\psi(\alpha) + g . \tag{17}$$

Suppose also that ϕ satisfies Assumption 1. Then there exist $p > 0$, $A \geq 1$ and $C > 0$ such that for all $\alpha > 0$

$$\phi(\alpha) = \alpha^p \quad , \quad -A\alpha^p \leq \phi(-\alpha) \leq \min(-\frac{1}{A}\alpha^p, -A\alpha^p + d),$$

$$\psi(\alpha) = C\alpha^p \quad , \quad -AC\alpha^p \leq \psi(-\alpha) \leq \min(-\frac{1}{A}C\alpha^p, -AC\alpha^p + e).$$

Also, $\lim_{\alpha \to \infty} \frac{\phi(-\alpha)}{-\alpha^p} = A$ *and* $\lim_{\alpha \to \infty} \frac{\psi(-\alpha)}{-\alpha^p} = AC$, *where both limits exist.*

Hence $\phi|_{\mathbb{R}_{<0}}$ and $\psi|_{\mathbb{R}_{<0}}$ are bounded perturbations of the power type functions $-A\alpha^p$ and $-AC\alpha^p$.

Proof

(a) Since $\lim_{\alpha \to \infty} \phi(\alpha) = \infty$, (17) implies that $\lim_{\alpha \to \infty} \psi(\alpha) = \infty$. If there would be $\alpha_0 \neq 0$ with $\psi(\alpha_0) = 0$, (16) would yield

$$\psi(\alpha) \leq \psi(\alpha_0)\phi(\frac{\alpha}{\alpha_0}) + e = e,$$

i.e. ψ would be bounded above, a contradiction. Since ψ is continuous, this implies that $\psi|_{\mathbb{R}_{>0}} > 0$. By Lemma 7, $\phi(-1) < 0$, and hence (16) yields

$$\psi(-\alpha) \leq \psi(\alpha)\phi(-1) + e,$$

i.e. $\lim_{\alpha \to \infty} \psi(-\alpha) = -\infty$. Therefore by the continuity of ψ, $\psi|_{\mathbb{R}_{<0}} < 0$ and $\psi(0) = 0$.

(b) We claim that $C := \lim_{\alpha \to \infty} \frac{\psi(\alpha)}{\alpha^p}$ exists with $C \leq \psi(1)$. By (15) and Theorem 1, there are $p > 0$ and $A = \lim_{\alpha \to \infty} \frac{\phi(-\alpha)}{-\alpha^p} \geq 1$ such that for all $\alpha > 0$

$$\phi(\alpha) = \alpha^p \quad , \quad -A\alpha^p \leq \phi(-\alpha) \leq \min(-\frac{1}{A}\alpha^p, -A\alpha^p + d).$$

By (16), $\psi(\alpha) \leq \psi(1)\phi(\alpha) + e = \psi(1)\alpha^p + e$. Hence

$$0 \leq \varliminf_{\alpha \to \infty} \frac{\psi(\alpha)}{\alpha^p} \leq \varlimsup_{\alpha \to \infty} \frac{\psi(\alpha)}{\alpha^p} \leq \psi(1).$$

For any $\epsilon > 0$, there is $\alpha_0 > 0$ such that

$$\frac{\psi(\alpha_0)}{\alpha_0^p} \leq \varliminf_{\alpha \to \infty} \frac{\psi(\alpha)}{\alpha^p} + \epsilon.$$

Choose $\beta_n \to \infty$ such that

$$\lim_{n \to \infty} \frac{\psi(\alpha_0 \beta_n)}{(\alpha_0 \beta_n)^p} = \varlimsup_{\alpha \to \infty} \frac{\psi(\alpha)}{\alpha^p} \ .$$

Since $\psi(\alpha_0 \beta_n) \le \psi(\alpha_0)\phi(\beta_n) + e = \psi(\alpha_0)\beta^p + e$,

$$\varlimsup_{\alpha \to \infty} \frac{\psi(\alpha)}{\alpha^p} = \lim_{n \to \infty} \frac{\psi(\alpha_0 \beta_n)}{(\alpha_0 \beta_n)^p} \le \frac{\psi(\alpha_0)}{\alpha_0^p} \le \varliminf_{\alpha \to \infty} \frac{\psi(\alpha)}{\alpha^p} + \epsilon \ .$$

Hence $C := \lim_{\alpha \to \infty} \frac{\psi(\alpha)}{\alpha^p}$ exists and $C \le \psi(1)$.

(c) We claim that $D := \lim_{\alpha \to \infty} \frac{\psi(-\alpha)}{-\alpha^p}$ exists with $D \le |\psi(-1)| + e$. We use (16) for $\alpha > 0$ in the form

$$\psi(-1) = \psi(-\alpha \frac{1}{\alpha}) \le \psi(-\alpha)\phi(\frac{1}{\alpha}) + e = \psi(-\alpha)\frac{1}{\alpha^p} + e \ ,$$

$0 \le \frac{\psi(-\alpha)}{-\alpha^p} \le |\psi(-1)| + e$ and $e \ge 0$. Therefore

$$0 \le \varliminf_{\alpha \to \infty} \frac{\psi(-\alpha)}{-\alpha^p} \le \varlimsup_{\alpha \to \infty} \frac{\psi(-\alpha)}{-\alpha^p} < \infty \ .$$

For any $\epsilon > 0$, choose $\alpha_1 > 0$ such that $\alpha_1^p > 1/\epsilon$ and

$$\frac{\psi(-\alpha_1)}{-\alpha_1^p} \le \varliminf_{\alpha \to \infty} \frac{\psi(-\alpha)}{-\alpha^p} + \epsilon \ .$$

Choose $\beta_n \to \infty$ such that

$$\lim_{n \to \infty} \frac{\psi(-\alpha_1 \beta_n)}{-(\alpha_1 \beta_n)^p} = \varlimsup_{\alpha \to \infty} \frac{\psi(-\alpha)}{-\alpha^p} \ .$$

By (16)

$$\psi(-\alpha_1) = \psi(-\alpha_1 \beta_n \frac{1}{\beta_n}) \le \psi(-\alpha_1 \beta_n)\phi(\frac{1}{\beta_n}) + e = \psi(-\alpha_1 \beta_n)\frac{1}{\beta_n^p} + e \ ,$$

hence

$$\varlimsup_{\alpha \to \infty} \frac{\psi(-\alpha)}{-\alpha^p} = \lim_{n \to \infty} \frac{\psi(-\alpha_1 \beta_n)}{-(\alpha_1 \beta_n)^p} \le \frac{\psi(-\alpha_1)}{-\alpha_1^p} + \frac{e}{\alpha_1^p} \le \varliminf_{\alpha \to \infty} \frac{\psi(\alpha)}{\alpha^p} + (1+e)\epsilon \ .$$

Therefore $D := \lim_{\alpha \to \infty} \frac{\psi(-\alpha)}{-\alpha^p}$ exists and $D \le |\psi(-1)| + e$.

(d) We now show that $D = AC$. By (16), for $\alpha, \beta \to \infty$,

$$D = \lim_{\alpha,\beta\to\infty} \frac{\psi(-\alpha\beta)}{-(\alpha\beta)^p} \geq \lim_{\alpha\to\infty} \frac{\psi(\alpha)}{\alpha^p} \lim_{\beta\to\infty} \frac{\phi(-\beta)}{-\beta^p} = CA .$$

On the other hand, by (17), $\phi(\alpha) \leq f\psi(\alpha) + g$, yielding for $\alpha > 0$ and then $\alpha \to \infty$

$$1 = \frac{\phi(\alpha)}{\alpha^p} \leq f\frac{\psi(\alpha)}{\alpha^p} + \frac{g}{\alpha^p} , \quad 1 \leq fC .$$

Also by (17), $\phi(-\alpha) \leq f\psi(-\alpha) + g$, i.e. with $f \geq \frac{1}{C}$

$$A = \lim_{\alpha\to\infty} \frac{\phi(-\alpha)}{-\alpha^p} \geq f \lim_{\alpha\to\infty} \frac{\psi(-\alpha)}{-\alpha^p} = fD \geq \frac{D}{C} .$$

Therefore $D \geq CA \geq C\frac{D}{C} = D$, $D = CA$.

(e) We now claim that $\psi(\alpha) = C\alpha^p$ for all $\alpha > 0$. Let $\alpha > 0$. For any $\epsilon > 0$, there is a large $\beta > 0$ such that, using (16),

$$(C - \epsilon)(\alpha\beta)^p \leq \psi(\alpha\beta) \leq \psi(\alpha)\beta^p + e ,$$

$(C - \epsilon)\alpha^p \leq \psi(\alpha) + \frac{e}{\beta^p}$. This yields for $\epsilon \to 0, \beta \to \infty$ that $C\alpha^p \leq \psi(\alpha)$.
Similarly, for any $\epsilon > 0$, there is a large $\beta > 0$ such that by (16)

$$-(D + \epsilon)(\alpha\beta)^p \leq \psi(-\alpha\beta) \leq \psi(\alpha)\phi(-\beta) + e ,$$

$$(D + \epsilon)\alpha^p \geq \psi(\alpha)\frac{\phi(-\beta)}{-\beta^p} - \frac{e}{\beta^p} .$$

This together with (d) implies for $\epsilon \to 0, \beta \to \infty$ that

$$D\alpha^p \geq \psi(\alpha)A , \quad \psi(\alpha) \leq \frac{D}{A}\alpha^p = C\alpha^p .$$

Therefore $\psi(\alpha) = C\alpha^p$ for all $\alpha > 0$.

(f) We now show $-AC\alpha^p \leq \psi(-\alpha) \leq -\frac{1}{A}C\alpha^p$ for any $\alpha > 0$. Let $\alpha > 0$. For any $\epsilon > 0$, there is a large $\beta > 0$ such that

$$-(D + \epsilon)(\alpha\beta)^p \leq \psi(-\alpha\beta) \leq \psi(-\alpha)\phi(\beta) + e = \psi(-\alpha)\beta^p + e ,$$

i.e.

$$-AC\alpha^p = -D\alpha^p \leq \psi(-\alpha) .$$

Also, there is a large $\beta > 0$ such that

$$(C - \epsilon)(\alpha\beta)^p \leq \psi(\alpha\beta) \leq \psi(-\alpha)\phi(-\beta) + e \,,$$

$$C\alpha^p \leq -\psi(-\alpha)A \quad , \quad \psi(-\alpha) \leq -\frac{1}{A}C\alpha^p \,.$$

Finally we claim that $\psi(-\alpha) \leq -AC\alpha^p + e$ for any $\alpha > 0$. Given any $\alpha > 0$ and $\epsilon > 0$, there is a large $\beta > 0$ such that, using (16),

$$\psi(-\alpha) = \psi(-\alpha\beta\frac{1}{\beta}) \leq \psi(-\alpha\beta)\frac{1}{\beta^p} + e \leq -(D-\epsilon)(\alpha\beta)^p\frac{1}{\beta^p} + e = -(D-\epsilon)\alpha^p + e$$

which implies for $\epsilon \to 0$ that $\psi(-\alpha) \leq -AC\alpha^p + e$.   □

The supermultiplicative analogue of Proposition 8 is

Proposition 9 *Let* $\phi, \psi : \mathbb{R} \to \mathbb{R}$ *be continuous functions and* $d, e, f, g \in \mathbb{R}$ *with* $f > 0$ *be such that for all* $\alpha, \beta \in \mathbb{R}$

$$\phi(\alpha\beta) \geq \phi(\alpha)\phi(\beta) - d$$
$$\psi(\alpha\beta) \geq \psi(\alpha)\phi(\beta) - e$$
$$\phi(\alpha) \geq f\psi(\alpha) - g \,.$$

Suppose also that ϕ *satisfies Assumption 2. Then there exist* $p > 0, 0 < B \leq 1$ *and* $C > 0$ *such that for all* $\alpha > 0$

$$\phi(\alpha) = \alpha^p \quad , \quad \max(-\frac{1}{B}\alpha^p, -B\alpha^p - d) \leq \phi(-\alpha) \leq -B\alpha^p \,,$$

$$\psi(\alpha) = C\alpha^p \quad , \quad \max(-\frac{1}{B}C\alpha^p, -B\alpha^p - e) \leq \psi(-\alpha) \leq -BC\alpha^p \,.$$

Also, $\lim_{\alpha\to\infty} \frac{\phi(-\alpha)}{-\alpha^p} = B$ *and* $\lim_{\alpha\to\infty} \frac{\psi(-\alpha)}{-\alpha^p} = BC$, *where both limits exist.*

Both results together imply

Corollary 10 *Let* $\phi, \psi : \mathbb{R} \to \mathbb{R}$ *be continuous functions and* $d, e, f, g > 0$ *be such that for all* $\alpha, \beta \in \mathbb{R}$

$$|\phi(\alpha\beta) - \phi(\alpha)\phi(\beta)| \leq d$$
$$|\psi(\alpha\beta) - \psi(\alpha)\phi(\beta)| \leq e$$
$$|\phi(\alpha) - f\psi(\alpha)| \leq g \,.$$

Assume also that ϕ satisfies Assumptions 1 and 2. Then there exist $p > 0$ and $C > 0$ such that for all $\alpha \in \mathbb{R}$

$$\phi(\alpha) = \text{sgn}\alpha \, |\alpha|^p \quad , \quad \psi(\alpha) = C \, \text{sgn}\alpha \, |\alpha|^p = C\phi(\alpha) \, .$$

Proof Just note that by Propositions 8 and 9

$$1 \leq A = \lim_{\alpha \to \infty} \frac{\phi(-\alpha)}{-\alpha^p} = B \leq 1 \, ,$$

hence $A = B = 1$. Therefore we have with $AC = C = \lim_{\alpha \to \infty} \frac{\psi(-\alpha)}{-\alpha^p} > 0$ that $\phi(-\alpha) = -\alpha^p$, $\psi(-\alpha) = -C\alpha^p$ holds for all $\alpha > 0$. We remark that $C = \frac{1}{f}$. $\qquad\square$

4 Proof of the Rigidity and the Stability of the Chain Rule

We now show that in the case of the two-sided chain rule operator inequality T is determined locally by function and derivative evaluations. The proof of Proposition 5 relies on the following lemma.

Lemma 11 *Let $T : C^1(\mathbb{R}) \to C(\mathbb{R})$ be non-degenerate, pointwise continuous and satisfy the perturbed chain rule inequality*

$$T(f \circ g)(x) \leq ((Tf) \circ g)(x) \cdot (Tg)(x) + S(x, (f \circ g)(x), g(x)) \quad ; \quad f, g \in C^1(\mathbb{R}), x \in \mathbb{R} \, .$$

Then we have for any open interval $I \subset \mathbb{R}$:

(a) Let $c \in \mathbb{R}$, $f \in C^1(\mathbb{R})$ with $f|_I = c$. Then $Tf|_I = 0$.
(b) Let $f \in C^1(\mathbb{R})$ with $f|_I = \text{Id}|_I$. Then $Tf|_I = 1$.
(c) Assuming that

$$|T(f \circ g)(x) - ((Tf) \circ g)(x) \cdot Tg(x)| \leq S(x, f \circ g(x), g(x)) \quad ; \quad f, g \in C^1(\mathbb{R}), x \in \mathbb{R} \, ,$$

we have for any $f_1, f_2 \in C^1(\mathbb{R})$ with $f_1|_I = f_2|_I$ that $Tf_1|_I = Tf_2|_I$.

Proof

(a) For the constant function c, $c \circ g = c$ for any $g \in C^1(\mathbb{R})$, hence
   $Tc(x) = T(c \circ g)(x) \leq Tc(g(x)) \, Tg(x) + S(x, c, g(x))$ for any $x \in I$. By non-degeneration of T, we find $g_{1,i}, g_{2,i} \in C^1(\mathbb{R})$ with $g_{j,i}(x) = x$, $\text{Im}(g_{j,i}) \subset I$, $j \in \{1, 2\}$, $i \in \mathbb{N}$ and $\lim_{i \to \infty} Tg_{1,i}(x) = \infty$, $\lim_{i \to \infty} Tg_{2,i}(x) = -\infty$. Assuming $Tc(x) > 0$, we get $Tc(x) < 0$ by applying $Tc(x) \leq Tc(g(x)) \, Tg(x) + S(x, c, x)$ for $g = g_{2,i}$ and letting $i \to \infty$, a contradiction. Assuming $Tc(x) < 0$, and applying $g_{1,i}$, we find for $i \to \infty$ that $Tc(x) = -\infty$, again a contradiction. Therefore $Tc(x) = 0$ for all $x \in I$.

Now assume that $f \in C^1(\mathbb{R})$ satisfies $f|_I = c$. Choose $g_{1,i}, g_{2,i}$ for all $i \in \mathbb{N}$ as before. Then $f \circ g_{j,i} = c$, $j \in \{1,2\}, i \in \mathbb{N}$ and hence, by what we just showed, for $x \in I, 0 = Tc(x) \leq Tf(x)\, Tg_{j,i}(x) + S(x,c,x)$, implying $Tf(x) = 0$.

(b) Assume that $f \in C^1(\mathbb{R})$ satisfies $f|_I = \mathrm{Id}|_I$. Let $x \in I$ and choose again $g_{1,i}, g_{2,i} \in C^1(\mathbb{R})$ with $g_{j,i}(x) = x$, $\mathrm{Im}(g_{j,i}) \subset I, j \in \{1,2\}, i \in \mathbb{N}$ with $\lim_{i\to\infty} Tg_{1,i}(x) = \infty$ and $\lim_{i\to\infty} Tg_{2,i}(x) = -\infty$. Then $f \circ g_{j,i} = g_{j,i}$ for $j \in \{1,2\}, i \in \mathbb{N}$ and

$$Tg_{j,i}(x) = T(f \circ g_{j,i})(x) \leq Tf(x)\, Tg_{j,i}(x)$$
$$+ S(x,x,x) , (1 - Tf(x))Tg_{j,i}(x) \leq B(x,x,x) .$$

If $Tf(x) < 1$ would hold, $Tg_{j,i}(x)$ would be bounded from above, a contradiction when choosing $j = 1$. If $Tf(x) > 1$ would hold, $Tg_{j,i}(x)$ would be bounded from below, a contradiction when choosing $j = 2$. Therefore $Tf(x) = 1$.

(c) Now assume that the two-sided inequality holds

$$|T(f \circ g)(x) - ((Tf) \circ g)(x) \cdot Tg(x)|$$
$$\leq S(x, f \circ g(x), g(x)) , -S(x, f \circ g(x), g(x))$$
$$\leq T(f \circ g)(x) - ((Tf) \circ g)(x) \cdot Tg(x)$$
$$\leq S(x, f \circ g(x), g(x)) .$$

Let $I \subset \mathbb{R}$ be open and $f_1, f_2 \in C^1(\mathbb{R})$ be such that $f_1|_I = f_2|_I$. We claim that $Tf_1|_I = Tf_2|_I$ holds. Let $x \in I$. Choose g_i for all $i \in \mathbb{N}$ with $g_i(x) = x$, $\mathrm{Im}(g_i) \subset I$ and $\lim_{i\to\infty} Tg_i(x) = \infty$. Then for $g = g_i$ and $f = f_1$ by the above inequalities

$$-S(x, f_1(x), x) \leq T(f_1 \circ g_i)(x) - (Tf_1)(x) \cdot Tg_i(x) \leq S(x, f_1(x), x) .$$

Since $\lim_{i\to\infty} \frac{S(x, f_1(x), x)}{Tg_i(x)} = 0$, we get by dividing the previous inequality by $Tg_i(x)$ that

$$Tf_1(x) = \lim_{i\to\infty} \frac{T(f_1 \circ g_i)(x)}{Tg_i(x)} ,$$

where the limit exists. Now note that $f_1 \circ g_i = f_2 \circ g_i$. Therefore the quotient on the right side stays the same by exchanging f_1 with f_2 and hence $Tf_1(x) = Tf_2(x)$ for all $x \in I$. $\qquad \square$

Proof of Proposition 5 Fix $x_0 \in \mathbb{R}$ and consider $f \in C^1(\mathbb{R})$. Let $J_1 := (x_0, \infty)$ and $J_2 := (-\infty, x_0)$. Consider the tangent of f at x_0, $g(x) := f(x_0) + (x - x_0)f'(x_0)$, $x \in \mathbb{R}$. It suffices to prove that $(Tf)(x_0) = (Tg)(x_0)$. Define $h \in C^1(\mathbb{R})$ by

$$h(x) := \begin{cases} g(x) & x \in J_1 \\ f(x) & x \in \bar{J}_2 \end{cases} .$$

Then $h|_{J_1} = g|_{J_1}$ and $h|_{J_2} = f|_{J_2}$. Hence by Lemma 11(c)

$$(Tg)|_{J_1} = (Th)|_{J_1} \quad \text{and} \quad (Th)|_{J_2} = (Tf)|_{J_2} .$$

These equalities extend by continuity to $x_0 \in \bar{J}_1 \cap \bar{J}_2$. We conclude that $(Tg)(x_0) = (Th)(x_0) = (Tf)(x_0)$. Therefore the value $(Tf)(x_0)$ depends only on the two parameters $f(x_0)$ and $f'(x_0)$, for any fixed $x_0 \in \mathbb{R}$. We encode this information by letting $(Tf)(x_0) = F_{x_0}(f(x_0), f'(x_0))$, where $F_{x_0} : \mathbb{R}^2 \to \mathbb{R}$ is a fixed function for any $x_0 \in \mathbb{R}$. Finally denoting $F(x, y, z) := F_x(y, z)$, we have that for any $x \in \mathbb{R}$ and $f \in C^1(\mathbb{R})$

$$Tf(x) = F(x, f(x), f'(x)) .$$

$\square$

Proof of the Claim in Example 2 We show that the operator T defined by

$$Tf(x) := \begin{cases} \frac{H(f(x))}{H(x)} f'(x) & f'(x) \geq 0 \\[2mm] \frac{H(f(x))}{H(x)} 4f'(x) & f'(x) \leq -2 \\[2mm] \frac{H(f(x))}{H(x)} (7 + \frac{15}{2}f'(x)) & -2 < f'(x) \leq -1 \\[2mm] \frac{H(Jf(x))}{H(x)} \frac{1}{2}f'(x) & -1 < f'(x) < 0 \end{cases},$$

with $H \in C(\mathbb{R})$, $4 \leq H \leq 5$ non-constant, maps $C^1(\mathbb{R})$ into $C(\mathbb{R})$ and satisfies

$$T(f \circ g)(x) \leq (Tf)(g(x)) \cdot (Tg)(x) + 5 \quad ; \quad f, g \in C^1(\mathbb{R}), \ x \in \mathbb{R} .$$

This operator is not localized since the term containing $H(Jf(x))$ involves the integral of f over some interval. Note that $Tf \in C(\mathbb{R})$: If $x_n \in \mathbb{R}$ are such that $f'(x_n) \in (-1, 0)$ and $x_n \to x, f'(x_n) \to -1$ or $f'(x_n) \to 0$, then $Jf(x_n) \to f(x)$ since $|I_{f, x_n}| \to 0$. Further, $7 + \frac{15}{2}\alpha = 4\alpha$ for $\alpha = -2$ and $7 + \frac{15}{2}\alpha = \frac{1}{2}\alpha$ for $\alpha = -1$. Therefore T maps $C^1(\mathbb{R})$ into $C(\mathbb{R})$. The operator T is not localized since $Jf(x)$ is not locally defined. Clearly we have $\frac{4}{5} \leq \frac{H(y)}{H(z)} \leq \frac{5}{4}$ for all $y, z \in \mathbb{R}$. To prove the claimed inequality, we distinguish several cases. For $f, g \in C^1(\mathbb{R})$, $x \in \mathbb{R}$, denote $\alpha := f'(g(x))$ and $\beta := g'(x)$. Then $\alpha\beta = (f \circ g)'(x)$.

(i) If $\alpha \geq 0$ and $\beta \geq 0$, $T(f \circ g)(x) = (Tf)(g(x))(Tg)(x)$.

(ii) If $\alpha < 0$ and $\beta < 0$, $(f \circ g)'(x) > 0$ and $0 < T(f \circ g)(x) = \frac{H(f \circ g(x))}{H(x)}\alpha\beta \leq \frac{5}{4}\alpha\beta$.

   Note that $(Tg)(x) \leq \frac{1}{2}\frac{4}{5}\beta < 0$; indeed, if $\beta \in (-1, 0)$, this follows directly from the definition of Tg, if $\beta \in [-2, -1]$, we get $7 + \frac{15}{2}\beta \leq \frac{1}{2}\beta$, and if $\beta \leq -2$, we find $4\beta \leq \frac{1}{2}\beta$.

If $\alpha \leq -2$, $(Tf)(g(x)) = \frac{H(f \circ g(x))}{H(g(x))} 4\alpha \leq 4\frac{4}{5}\alpha$. Therefore

$$(Tf)(g(x))(Tg)(x) \geq (4\frac{4}{5}\alpha)(\frac{1}{2}\frac{4}{5}\beta) = \frac{32}{25}\alpha\beta ,$$

$$T(f \circ g)(x) - (Tf)(g(x))(Tg)(x) \leq (\frac{5}{4} - \frac{32}{25})\alpha\beta = -\frac{3}{100}\alpha\beta < 0 .$$

If $\beta \leq -2$ and $\alpha < 0$, the same argument holds with α and β exchanged.
  If both $\alpha, \beta \in [-2, 0)$, $0 < \alpha\beta \leq 4$ and

$$(Tf)(g(x))(Tg)(x) \geq (\frac{1}{2}\frac{4}{5})^2\alpha\beta = \frac{4}{25}\alpha\beta ,$$

$$T(f \circ g)(x) - (Tf)(g(x))(Tg)(x) \leq (\frac{5}{4} - \frac{4}{25})\alpha\beta < 5 .$$

(iii) If $\alpha < 0 < \beta$, $T(f \circ g)(x) < 0$, $(Tf)(g(x)) < 0$ and $0 < (Tg)(x) = \frac{H(g(x))}{H(x)}\beta \leq \frac{5}{4}\beta$.

  (a) Assume first that $\alpha \leq -1$. Then $(Tf)(g(x)) \geq 4\frac{H(f \circ g(x))}{H(g(x))}\alpha$, since for $\alpha \in [-2, -1]$, we have $7 + \frac{15}{2}\alpha \geq 4\alpha$. Hence

$$(Tf)(g(x))(Tg)(x) \geq (4\frac{H(f \circ g(x))}{H(g(x))}\alpha)(\frac{H(g(x))}{H(x)}\beta) = 4\frac{H(f \circ g(x))}{H(x)}\alpha\beta .$$

If $\alpha\beta \leq -2$, the right side is just $T(f \circ g)(x)$. If $\alpha\beta \in (-2, -1]$,

$$T(f \circ g)(x) - (Tf)(g(x))(Tg)(x) \leq \frac{H(f \circ g(x))}{H(x)}(7 + \frac{15}{2}\alpha\beta - 4\alpha\beta)$$

$$\leq \frac{5}{4}(7 + \frac{7}{2}\alpha\beta) \leq \frac{35}{8} < 5 .$$

If $\alpha\beta \in (-1, 0)$, $T(f \circ g)(x) \leq \frac{1}{2}\frac{4}{5}\alpha\beta$ and

$$T(f \circ g)(x) - (Tf)(g(x))(Tg)(x) \leq (\frac{1}{2}\frac{4}{5} - 4\frac{5}{4})\alpha\beta \leq \frac{23}{5} < 5 .$$

  b) Now assume that $-1 < \alpha < 0$. Then $(Tf)(g(x)) = \frac{1}{2}\frac{H(Jf(g(x)))}{H(g(x))}\alpha \geq \frac{1}{2}\frac{5}{4}\alpha$.

$$(Tf)(g(x))(Tg)(x) \geq (\frac{1}{2}\frac{5}{4}\alpha)(\frac{5}{4}\beta) = \frac{1}{2}(\frac{5}{4})^2\alpha\beta .$$

If $\alpha\beta \leq -2$, $T(f \circ g)(x) = 4\frac{H(f \circ g(x))}{H(x)}\alpha\beta \leq 4\frac{4}{5}\alpha\beta$ and

$$T(f \circ g)(x) - (Tf)(g(x))(Tg)(x) \leq (4\frac{4}{5} - \frac{1}{2}(\frac{5}{4})^2)\alpha\beta \leq 0 \ .$$

If $-2 < \alpha\beta < 0$, $T(f \circ g)(x) \leq \frac{1}{2}\frac{4}{5}\alpha\beta$ and therefore

$$T(f \circ g)(x) - (Tf)(g(x))(Tg)(x) \leq (\frac{1}{2}\frac{4}{5} - \frac{1}{2}(\frac{5}{4})^2)\alpha\beta \leq 1 \ .$$

(iv) For $\beta < 0 < \alpha$, the argument is essentially the same as in (iii), with α and β being exchanged. $\qquad\square$

In the situation of Theorems 4 and 6, we have localization: There exists a function $F : \mathbb{R}^3 \to \mathbb{R}$ such that

$$Tf(x) = F(x, f(x), f'(x)) \tag{18}$$

holds for all $f \in C^1(\mathbb{R})$ and all $x \in \mathbb{R}$. In the case of Theorem 4, this is true by Proposition 5, and in the case of Theorem 6, by assumption. Further by Lemma 11 (b) $T(\mathrm{Id}) = 1$, i.e. $F(x, x, 1) = 1$ for all $x \in \mathbb{R}$. The chain rule operator inequalities then translate into functional inequalities for the function F the structure of which we have to determine. Whereas there is extensive knowledge on functional equations, cf. [1], less is known about *functional inequalities*.

Proof of Theorem 6

(a) For any $x, y, z \in \mathbb{R}$, $\alpha, \beta \in \mathbb{R}$, choose $f, g \in C^1(\mathbb{R})$ with $g(x) = y$, $f(y) = z$ and $g'(x) = \beta$, $f'(y) = \alpha$. Using (18), the operator inequality (4) for T is equivalent to the functional inequality for F,

$$F(x, z, \alpha\beta) \leq F(y, z, \alpha)F(x, y, \beta) + S(x, z, y) \ , \tag{19}$$

$x, y, z, \alpha, \beta \in \mathbb{R}$. For $x, z \in \mathbb{R}$, define $\phi_x, \psi_{x,z} : \mathbb{R} \to \mathbb{R}$ by $\phi_x(\alpha) := F(x, x, \alpha)$, $\psi_{x,z}(\alpha) := F(x, z, \alpha)$. Let $d_x := S(x, x, x)$. By (19) for $x = y = z$

$$\phi_x(\alpha\beta) \leq \phi_x(\alpha)\phi_x(\beta) + d_x \ , \quad \alpha, \beta \in \mathbb{R} \ . \tag{20}$$

For $x, z, \alpha \in \mathbb{R}$, define $f_{x,z,\alpha} \in C^1(\mathbb{R})$ by $f_{x,z,\alpha}(y) := \alpha y + (z - \alpha x)$. Then $f_{x,z,\alpha}(x) = z$ and $f'_{x,z,\alpha}(x) = \alpha$. Hence $(Tf_{x,z,\alpha})(x) = F(x, z, \alpha)$. For sequences $z_n \to z$ and $\alpha_n \to \alpha$, $f_{x,z_n,\alpha_n} \to f_{x,z,\alpha}$ and $f'_{x,z_n,\alpha_n} \to f'_{x,z,\alpha}$ uniformly on compact subsets of $\mathbb{R}$. By the assumption of pointwise continuity of T,

$$F(x, z_n, \alpha_n) = (Tf_{x,z_n,\alpha_n})(x) \to (Tf_{x,z,\alpha})(x) = F(x, z, \alpha) \ .$$

Therefore F is continuous in the second and third variable. In particular, in the constant case $z_n = z$, all $\psi_{x,z}$ and for $x = z$ all ϕ_x are continuous functions.

Since for $f \in C^1(\mathbb{R})$ with $f(x) = x$ we have $Tf(x) = F(x, x, f'(x))$ by the assumption of non-degeneration of T, we conclude that for all $x \in \mathbb{R}$

$$\sup_{\alpha \in \mathbb{R}} \phi_x(\alpha) = \sup_{\alpha \in \mathbb{R}} F(x, x, \alpha) = \infty .$$

By the above, $F(x, \cdot, 1)$ and $F(x, \cdot, -1)$ are continuous functions. Assume there would be $y \in \mathbb{R}$ with $F(x, y, 1) \leq 0$. Then, since $F(x, x, 1) = T(\mathrm{Id})(x) = 1 > 0$, by continuity there would be also $z \in \mathbb{R}$ with $F(x, z, 1) = 0$. But then by (19)

$$F(x, x, \alpha) \leq F(z, x, \alpha)F(x, z, 1) + S(x, x, z) = S(x, x, z)$$

would imply $\sup_{\alpha \in \mathbb{R}} \phi_x(\alpha) \leq S(x, z, z)$, i.e. ϕ_x would be bounded from above on $\mathbb{R}$, a contradiction. Hence $F(x, z, 1) > 0$ for all $x, z \in \mathbb{R}$.

By assumption, there is $x_0 \in \mathbb{R}$ with $T(-\mathrm{Id})(x_0) = F(x_0, -x_0, -1) < 0$. Since $T(-\mathrm{Id})$ is continuous, if there were $x_1 \in \mathbb{R}$ with $T(-\mathrm{Id})(x_1) \geq 0$, there would be also $x_2 \in \mathbb{R}$ with $F(x_2, -x_2, -1) = T(-\mathrm{Id})(x_2) = 0$. But then, using again (19)

$$\phi_{x_2}(\alpha) = F(x_2, x_2, \alpha) \leq F(-x_2, x_2, -\alpha)F(x_2, -x_2, -1) + S(x_2, x_2, -x_2)$$
$$= S(x_2, x_2, -x_2) ,$$

ϕ_{x_2} would be bounded from above on $\mathbb{R}$, a contradiction. Hence $F(x, -x, -1) < 0$ for all $x \in \mathbb{R}$. Assume now there would be $x_0 \in \mathbb{R}$ with $\phi_{x_0}(-1) = F(x_0, x_0, -1) \geq 0$. Since $F(x_0, -x_0, -1) < 0$, by continuity of $F(x_0, \cdot, -1)$, there would be $z \in \mathbb{R}$ with $F(x_0, z, -1) = 0$. Then

$$\phi_{x_0}(\alpha) = F(x_0, x_0, \alpha) \leq F(z, x_0, -\alpha)F(x_0, z, -1) + S(x_0, x_0, z) = S(x_0, x_0, z) ,$$

i.e. ϕ_{x_0} would be bounded from above, a contradiction. Hence $\phi_x(-1) = F(x, x, -1) < 0$ holds for all $x \in \mathbb{R}$.

We now claim that $\overline{\lim}_{\alpha \to \infty} \phi_x(\alpha) = \infty$ for all $x \in \mathbb{R}$. If not, there would be $x_0 \in \mathbb{R}$ such that $\overline{\lim}_{\alpha \to \infty} \phi_{x_0}(\alpha) < \infty$ and hence $\overline{\lim}_{\alpha \to \infty} \phi_{x_0}(-\alpha) = \infty$ since we know that $\sup_{\alpha \in \mathbb{R}} \phi_{x_0}(\alpha) = \infty$ and that by continuity ϕ_{x_0} is bounded on compact subsets of $\mathbb{R}$. Then by (20)

$$\phi_{x_0}(-\alpha) \leq \phi_{x_0}(\alpha)\phi_{x_0}(-1) + d_{x_0} .$$

Since $\phi_{x_0}(-1) < 0$, this implies that $\underline{\lim}_{\alpha \to \infty} \phi_{x_0}(\alpha) = -\infty$. Thus there would be $\alpha_0 > 0$ with $\phi_{x_0}(\alpha_0) < 0$. But $\phi_{x_0}(1) = T(\mathrm{Id})(x_0) = 1 > 0$ and by continuity of ϕ_{x_0} there would be $\alpha_1 > 0$ with $\phi_{x_0}(\alpha_1) = 0$. Hence by (20)

$$\phi_{x_0}(\beta) \leq \phi_{x_0}(\alpha_1)\phi_{x_0}(\frac{\beta}{\alpha_1}) + d_{x_0} = d_{x_0} \,,$$

i.e. $\sup_{\beta \in \mathbb{R}} \phi_{x_0}(\beta) < \infty$, a contradiction. This shows that $\overline{\lim}_{\alpha \to \infty} \phi_x(\alpha) = \infty$ and that Assumption 1 is satisfied for all ϕ_x.

Hence by Theorem 1 there are $p(x) > 0$ and $A(x) \geq 1$ such that for all $\alpha > 0$

$$\phi_x(\alpha) = \alpha^{p(x)} \quad, \quad -A(x)\alpha^{p(x)} \leq \phi_x(-\alpha) \leq \min(-\frac{1}{A(x)}\alpha^{p(x)}, -A(x)\alpha^{p(x)} + d_x)$$

and $A(x) = \lim_{\alpha \to \infty} \frac{\phi_x(-\alpha)}{-\alpha^{p(x)}} \geq 1$.

(b) Let $e_{x,z} := S(x, z, x)$ and $g_{x,z} := S(x, z, z)$. Putting $y = x$ in (19), we find that

$$\psi_{x,z}(\alpha\beta) \leq \psi_{x,z}(\alpha)\phi_x(\beta) + e_{x,z} \,, \quad \alpha, \beta \in \mathbb{R} \,.$$

For $y = z$ in (19), we get after exchanging α and β,

$$\psi_{x,z}(\alpha\beta) \leq \psi_{x,z}(\alpha)\phi_z(\beta) + g_{x,z} \,, \quad \alpha, \beta \in \mathbb{R} \,.$$

Let $f_{x,z} := F(z, x, 1)$. Replacing z by x and putting $y = z$, $\alpha = 1$ in (19), we get after renaming β by α

$$F(x, x, \alpha) \leq F(z, x, 1)F(x, z, \alpha) + S(x, x, z) \,,$$

$$\phi_x(\alpha) \leq f_{x,z}\psi_{x,z}(\alpha) + h_{x,z} \,, \quad \alpha \in \mathbb{R} \,,$$

with $h_{x,z} := S(x, x, z)$. Replacing x by z and putting $y = x$, $\beta = 1$ in (19), we get with the same value $f_{x,z}$

$$\phi_z(\alpha) \leq f_{x,z}\psi_{x,z}(\alpha) + h_{z,x} \,, \quad \alpha \in \mathbb{R} \,.$$

Note that $f := f_{x,z} = F(z, x, 1) > 0$. Therefore the assumptions of Proposition 8 are satisfied for $\phi = \phi_x$, $\psi = \psi_{x,z}$, $e = e_{x,z} = S(x, z, x)$ and $g = h_{x,z} = S(x, x, z)$, and also for $\phi = \phi_z$, $\psi = \psi_{x,z}$, $e = g_{x,z} = S(x, z, z)$ and $g = h_{z,x} = S(z, z, x)$. Therefore there are $C(x, z) > 0$ and $\tilde{C}(x, z) > 0$ such that for any $\alpha > 0$

$$F(x, z, \alpha) = \psi_{x,z}(\alpha) = C(x, z)\alpha^{p(x)} = \tilde{C}(x, z)\alpha^{p(z)} \,.$$

Obviously this implies that $p := p(x) = p(z)$ is independent of x and z and that $C(x, z) = \tilde{C}(x, z)$,

$$F(x, z, \alpha) = C(x, z)\alpha^p .$$

Also by Proposition 8

$$\lim_{\alpha \to \infty} \frac{\psi_{x,z}(-\alpha)}{-\alpha^p} = A(x)C(x, z) = A(z)C(x, z) ,$$

since the assumptions are satisfied for both pairs of functions ϕ, ψ. Therefore also $A(x) = A(z) =: A \geq 1$ is independent of $x, z \in \mathbb{R}$. Moreover, we have

$$-AC(x, z)\alpha^p \leq F(x, z, -\alpha) \leq \min(-\frac{1}{A}C(x, z)\alpha^p, -AC(x, z)\alpha^p + \min[S(x, z, x), S(x, z, z)])$$

and $\lim_{\alpha \to \infty} \frac{F(x,z,-\alpha)}{-\alpha^p} = \lim_{\alpha \to \infty} \frac{\psi_{x,z}(-\alpha)}{-\alpha^p} = AC(x, z)$.

(c) Inserting $F(x, z, \alpha) = C(x, z)\alpha^p$ into (19), we find for all $\alpha, \beta > 0$

$$C(x, z)(\alpha\beta)^p \leq C(y, z)\alpha^p C(x, y)\beta^p + S(x, z, y) .$$

For $\alpha, \beta \to \infty$ it follows that $C(x, z) \leq C(y, z)C(x, y)$ for all $x, y, z \in \mathbb{R}$. On the other hand, by (19) and part (b)

$$AC(x, z) = \lim_{\alpha,\beta \to \infty} \frac{\psi_{x,z}(-\alpha\beta)}{-(\alpha\beta)^p} \geq \lim_{\alpha \to \infty} \frac{\psi_{y,z}(-\alpha)}{-\alpha^p} \cdot \lim_{\beta \to \infty} \frac{\psi_{x,y}(\beta)}{\beta^p} = AC(y, z)C(x, y) ,$$

i.e. $C(x, z) \geq C(y, z)C(x, y)$. We conclude that for all $x, y, z \in \mathbb{R}$, $C(x, z) = C(y, z)C(x, y)$.

For $x = z$, $\phi_x(\alpha) = \psi_{x,x}(\alpha) = \alpha^p$, $\alpha > 0$. Hence $C(x, x) = 1$ and $1 = C(x, x) = C(0, x)C(x, 0)$ for all $x \in \mathbb{R}$. Let $H(x) := C(0, x)$. Then

$$C(x, z) = C(0, z)C(x, 0) = \frac{H(z)}{H(x)} .$$

The function H is continuous since $H(z) = C(0, z) = F(0, z, 1)$ is continuous in $z \in \mathbb{R}$ as we have seen. We find for all $x, z \in \mathbb{R}, \alpha > 0$

$$F(x, z, \alpha) = \frac{H(z)}{H(x)}\alpha^p ,$$

$$(Tf)(x) = \frac{H(f(x))}{H(x)} f'(x)^p$$

provided that $f'(x) \geq 0$.

(d) For $x, z \in \mathbb{R}$ and $\alpha > 0$, define

$$K(x, z, -\alpha) := \frac{F(x, z, -\alpha)}{C(x, z)} = \frac{H(x)}{H(z)} F(x, z, -\alpha) < 0 .$$

By (a), F and hence also K is continuous in the second and third variable. Then for any $f \in C^1(\mathbb{R})$ with $f(x) = z$ and $f'(x) < 0$

$$Tf(x) = F(x, f(x), f'(x)) = \frac{H(f(x))}{H(x)} K(x, f(x), f'(x)) .$$

By part (b), $\lim_{\alpha \to \infty} \frac{K(x, z, -\alpha)}{-\alpha^p} = A$ exists and is independent of $x, z \in \mathbb{R}$ and for all $\alpha > 0$, $x, z \in \mathbb{R}$

$$-A\alpha^p \leq K(x, z, -\alpha) \leq \min(-\frac{1}{A}\alpha^p, -A\alpha^p + \frac{H(x)}{H(z)} \min[S(x, z, x), S(x, z, z)]) .$$

This proves Theorem 6. □

Proof of Theorem 4 We have by assumption (3)

$$(Tf)(g(x)) Tg(x) - S(x, (f \circ g)(x), g(x)) \leq T(f \circ g)(x)$$
$$\leq (Tf)(g(x)) Tg(x) + S(x, (f \circ g)(x), g(x)) .$$

Proposition 5 and Theorem 6 thus implies the result of Theorem 4 for all $f \in C^1(\mathbb{R})$, $x \in \mathbb{R}$ with $f'(x) \geq 0$. Proposition 5, Theorem 6 and its supermultiplicative analogue yield for $f \in C^1(\mathbb{R})$, $x \in \mathbb{R}$ with $f'(x) < 0$ that for the same functions H and K as in Theorem 6

$$Tf(x) = \frac{H(f(x))}{H(x)} K(x, f(x), f'(x)) ,$$

where

$$1 \leq A = \lim_{\beta \to \infty} \frac{K(x, y, -\beta)}{-\beta^p} = B \leq 1 ,$$

i.e. $A = B = 1$. Therefore $-A\alpha^p \leq K(x, y, -\alpha) \leq \frac{1}{A}\alpha^p$ yields $K(x, y, -\alpha) = -\alpha^p$ for all $x, y \in \mathbb{R}$, $\alpha > 0$. Therefore, if $f'(x) < 0$,

$$Tf(x) = -\frac{H(f(x))}{H(x)} |f'(x)|^p ,$$

or generally for all $f \in C^1(\mathbb{R})$, $x \in \mathbb{R}$

$$Tf(x) = \frac{H(f(x))}{H(x)} \operatorname{sgn} f'(x) \, |f'(x)|^p \, .$$

This operator clearly satisfies

$$T(f \circ g)(x) = (Tf)(g(x))(Tg)(x) \, ,$$

i.e. the function S a posteriori can be chosen to be $S = 0$. $\square$

Acknowledgements Hermann König was supported in part by Minerva. Vitali Milman was supported in part by the Alexander von Humboldt Foundation, by Minerva, by ISF grant 826/13 and by BSF grant 0361-4561.

References

1. J. Aczél, *Lectures on Functional Equations and Their Applications* (Academic Press, New York, 1966)
2. S. Alesker, S. Artstein-Avidan, D. Faifman, V. Milman, A characterization of product preserving maps with applications to a characterization of the Fourier transform. Ill. J. Math. **54**, 1115–1132 (2010)
3. S. Artstein-Avidan, H. König, V. Milman, The chain rule as a functional equation. J. Funct. Anal. **259**, 2999–3024 (2010)
4. H. König, V. Milman, Characterizing the derivative and the entropy function by the Leibniz rule, with an appendix by D. Faifman. J. Funct. Anal. **261**, 1325–1344 (2011)
5. H. König, V. Milman, Rigidity and stability of the Leibniz and the chain rule. Proc. Steklov Inst. **280**, 191–207 (2013)
6. H. König, V. Milman, Submultiplicative functions and operator inequalities. Stud. Math. **223**, 217–231 (2014)

Royen's Proof of the Gaussian Correlation Inequality

Rafał Latała and Dariusz Matlak

Abstract We present in detail Thomas Royen's proof of the Gaussian correlation inequality which states that $\mu(K \cap L) \geq \mu(K)\mu(L)$ for any centered Gaussian measure μ on $\mathbb{R}^d$ and symmetric convex sets K, L in $\mathbb{R}^d$.

1 Introduction

The aim of this note is to present in a self contained way the beautiful proof of the Gaussian correlation inequality, due to Thomas Royen [7]. Although the method is rather simple and elementary, we found the original paper not too easy to follow. One of the reasons behind it is that in [7] the correlation inequality was established for more general class of probability measures. Moreover, the author assumed that the reader is familiar with properties of certain distributions and may justify some calculations by herself/himself. We decided to reorganize a bit Royen's proof, restrict it only to the Gaussian case and add some missing details. We hope that this way a wider readership may appreciate the remarkable result of Royen.

The statement of the Gaussian correlation inequality is as follows.

Theorem 1 *For any closed symmetric sets K, L in $\mathbb{R}^d$ and any centered Gaussian measure μ on $\mathbb{R}^d$ we have*

$$\mu(K \cap L) \geq \mu(K)\mu(L). \tag{1}$$

For $d = 2$ the result was proved by Pitt [5]. In the case when one of the sets K, L is a symmetric strip (which corresponds to $\min\{n_1, n_2\} = 1$ in Theorem 2 below) inequality (1) was established independently by Khatri [3] and Šidák [9]. Hargé [2] generalized the Khatri-Šidák result to the case when one of the sets is a symmetric ellipsoid. Some other partial results may be found in papers of Borell [1] and Schechtman et al. [8].

R. Latała (✉) • D. Matlak
Institute of Mathematics, University of Warsaw, Banacha 2, 02-097 Warszawa, Poland
e-mail: rlatala@mimuw.edu.pl; ddmatlak@gmail.com

© Springer International Publishing AG 2017
B. Klartag, E. Milman (eds.), *Geometric Aspects of Functional Analysis*,
Lecture Notes in Mathematics 2169, DOI 10.1007/978-3-319-45282-1_17

Up to our best knowledge Thomas Royen was the first to present a complete proof of the Gaussian correlation inequality. Some other recent attempts may be found in [4] and [6], however both papers are very long and difficult to check. The first version of [4], placed on the arxiv before Royen's paper, contained a fundamental mistake (Lemma 6.3 there was wrong).

Since any symmetric closed set is a countable intersection of symmetric strips, it is enough to show (1) in the case when

$$K = \{x \in \mathbb{R}^d \colon \forall_{1 \leq i \leq n_1} |\langle x, v_i \rangle| \leq t_i\} \quad \text{and}$$

$$L = \{x \in \mathbb{R}^d \colon \forall_{n_1+1 \leq i \leq n_1+n_2} |\langle x, v_i \rangle| \leq t_i\},$$

where v_i are vectors in $\mathbb{R}^d$ and t_i nonnegative numbers. If we set $n = n_1 + n_2$, $X_i := \langle v_i, G \rangle$, where G is the Gaussian random vector distributed according to μ, we obtain the following equivalent form of Theorem 1.

Theorem 2 *Let $n = n_1 + n_2$ and X be an n-dimensional centered Gaussian vector. Then for any $t_1, \ldots, t_n > 0$,*

$$\mathbb{P}(|X_1| \leq t_1, \ldots, |X_n| \leq t_n)$$

$$\geq \mathbb{P}(|X_1| \leq t_1, \ldots, |X_{n_1}| \leq t_{n_1}) \mathbb{P}(|X_{n_1+1}| \leq t_{n_1+1}, \ldots, |X_n| \leq t_n).$$

Remark 3

(i) The standard approximation argument shows that the Gaussian correlation inequality holds for centered Gaussian measures on separable Banach spaces.
(ii) Theorem 1 has the following functional form:

$$\int_{\mathbb{R}^d} fg d\mu \geq \int_{\mathbb{R}^d} f d\mu \int_{\mathbb{R}^d} g d\mu$$

for any centered Gaussian measure μ on $\mathbb{R}^d$ and even functions $f, g \colon \mathbb{R}^d \to [0, \infty)$ such that sets $\{f \geq t\}$ and $\{g \geq t\}$ are convex for all $t > 0$.
(iii) Thomas Royen established Theorem 2 for a more general class of random vectors X such that $X^2 = (X_1^2, \ldots, X_n^2)$ has an n-variate gamma distribution with appropriately chosen parameters (see [7] for details).

Notation By $\mathcal{N}(0, C)$ we denote the centered Gaussian measure with the covariance matrix C. We write $M_{n \times m}$ for a set of $n \times m$ matrices and $|A|$ for the determinant of a square matrix A. For a matrix $A = (a_{ij})_{i,j \leq n}$ and $J \subset [n] := \{1, \ldots, n\}$ by A_J we denote the square matrix $(a_{ij})_{i,j \in J}$ and by $|J|$ the cardinality of J.

2 Proof of Theorem 2

Without loss of generality we may and will assume that the covariance matrix C of X is nondegenerate (i.e. positive-definite). We may write C as

$$C = \begin{pmatrix} C_{11} & C_{12} \\ C_{21} & C_{22} \end{pmatrix},$$

where C_{ij} is the $n_i \times n_j$ matrix. Let

$$C(\tau) := \begin{pmatrix} C_{11} & \tau C_{12} \\ \tau C_{21} & C_{22} \end{pmatrix}, \quad 0 \le \tau \le 1.$$

Set $Z_i(\tau) := \frac{1}{2} X_i(\tau)^2$, $1 \le i \le n$, where $X(\tau) \sim \mathcal{N}(0, C(\tau))$.

We may restate the assertion as

$$\mathbb{P}(Z_1(1) \le s_1, \ldots, Z_n(1) \le s_n) \ge \mathbb{P}(Z_1(0) \le s_1, \ldots, Z_n(0) \le s_n),$$

where $s_i = \frac{1}{2} t_i^2$. Therefore it is enough to show that the function

$$\tau \mapsto \mathbb{P}(Z_1(\tau) \le s_1, \ldots, Z_n(\tau) \le s_n) \text{ is nondecreasing on } [0, 1].$$

Let $f(x, \tau)$ denote the density of the random vector $Z(\tau)$ and $K = [0, s_1] \times \cdots \times [0, s_n]$. We have

$$\frac{\partial}{\partial \tau} \mathbb{P}(Z_1(\tau) \le s_1, \ldots, Z_n(\tau) \le s_n) = \frac{\partial}{\partial \tau} \int_K f(x, \tau) dx = \int_K \frac{\partial}{\partial \tau} f(x, \tau) dx,$$

where the last equation follows by Lemma 6 applied to $\lambda_1 = \ldots = \lambda_n = 0$. Therefore it is enough to show that $\int_K \frac{\partial}{\partial \tau} f(x, \tau) \ge 0$.

To this end we will compute the Laplace transform of $\frac{\partial}{\partial \tau} f(x, \tau)$. By Lemma 6, applied to $K = [0, \infty)^n$, we have for any $\lambda_1 \ldots, \lambda_n \ge 0$,

$$\int_{[0,\infty)^n} e^{-\sum_{i=1}^n \lambda_i x_i} \frac{\partial}{\partial \tau} f(x, \tau) dx = \frac{\partial}{\partial \tau} \int_{[0,\infty)^n} e^{-\sum_{i=1}^n \lambda_i x_i} f(x, \tau) dx.$$

However by Lemma 4 we have

$$\int_{[0,\infty)^n} e^{-\sum_{i=1}^n \lambda_i x_i} f(x, \tau) dx = \mathbb{E} \exp\left(-\frac{1}{2} \sum_{i=1}^n \lambda_i X_i^2(\tau) \right) = |I + \Lambda C(\tau)|^{-1/2},$$

where $\Lambda = \mathrm{diag}(\lambda_1, \ldots, \lambda_n)$.

Formula (2) below yields

$$|I + \Lambda C(\tau)| = 1 + \sum_{\emptyset \neq J \subset [n]} |(\Lambda C(\tau))_J| = 1 + \sum_{\emptyset \neq J \subset [n]} |C(\tau)_J| \prod_{j \in J} \lambda_j.$$

Fix $\emptyset \neq J \subset [n]$. Then $J = J_1 \cup J_2$, where $J_1 := [n_1] \cap J$, $J_2 := J \setminus [n_1]$ and $C(\tau)_J = \begin{pmatrix} C_{J_1} & \tau C_{J_1 J_2} \\ \tau C_{J_2 J_1} & C_{J_2} \end{pmatrix}$. If $J_1 = \emptyset$ or $J_2 = \emptyset$ then $C(\tau)_J = C_J$, otherwise by (3) we get

$$|C(\tau)_J| = |C_{J_1}||C_{J_2}| \left| I_{|J_1|} - \tau^2 C_{J_1}^{-1/2} C_{J_1 J_2} C_{J_2}^{-1} C_{J_2 J_1} C_{J_1}^{-1/2} \right|$$

$$= |C_{J_1}||C_{J_2}| \prod_{i=1}^{|J_1|} (1 - \tau^2 \mu_{J_1, J_2}(i)),$$

where $\mu_{J_1, J_2}(i)$, $1 \leq i \leq |J_1|$ denote the eigenvalues of $C_{J_1}^{-1/2} C_{J_1 J_2} C_{J_2}^{-1} C_{J_2 J_1} C_{J_1}^{-1/2}$ (by (4) they belong to $[0, 1]$). Thus for any $\emptyset \neq J \subset [n]$ and $\tau \in [0, 1]$ we have

$$a_J(\tau) := -\frac{\partial}{\partial \tau} |C(\tau)_J| \geq 0.$$

Therefore

$$\frac{\partial}{\partial \tau} |I + \Lambda C(\tau)|^{-1/2} = -\frac{1}{2} |I + \Lambda C(\tau)|^{-3/2} \sum_{\emptyset \neq J \subset [n]} \frac{\partial}{\partial \tau} |C(\tau)_J||\Lambda_J|$$

$$= \frac{1}{2} |I + \Lambda C(\tau)|^{-3/2} \sum_{\emptyset \neq J \subset [n]} a_J(\tau) \prod_{j \in J} \lambda_j.$$

We have thus shown that

$$\int_{[0,\infty)^n} e^{-\sum_{i=1}^n \lambda_i x_i} \frac{\partial}{\partial \tau} f(x, \tau) dx = \sum_{\emptyset \neq J \subset [n]} \frac{1}{2} a_J(\tau) |I + \Lambda C(\tau)|^{-3/2} \prod_{j \in J} \lambda_j.$$

Let $h_\tau := h_{3, C(\tau)}$ be the density function on $(0, \infty)^n$ defined by (5). By Lemmas 8 and 7 (iii) we know that

$$|I + \Lambda C(\tau)|^{-3/2} \prod_{j \in J} \lambda_j = \int_{(0,\infty)^n} e^{-\sum_{i=1}^n \lambda_i x_i} \frac{\partial^{|J|}}{\partial x_J} h_\tau.$$

This shows that

$$\frac{\partial}{\partial \tau} f(x, \tau) = \sum_{\emptyset \neq J \subset [n]} \frac{1}{2} a_J(\tau) \frac{\partial^{|J|}}{\partial x_J} h_\tau(x).$$

Finally recall that $a_J(\tau) \geq 0$ and observe that by Lemma 7 (ii),

$$\lim_{x_i \to 0+} \frac{\partial^{|I|}}{\partial x_I} h_\tau(x) = 0 \quad \text{for } i \notin I \subset [n],$$

thus

$$\int_K \frac{\partial^{|J|}}{\partial x_J} h_\tau(x) dx = \int_{\prod_{j \in J^c} [0, s_j]} h_\tau(s_J, x_{J^c}) dx_{J^c} \geq 0,$$

where $J^c = [n] \setminus J$ and $y = (s_J, x_{J^c})$ if $y_i = s_i$ for $i \in J$ and $y_i = x_i$ for $i \in J^c$. $\square$

3   Auxiliary Lemmas

Lemma 4 *Let X be an n dimensional centered Gaussian vector with the covariance matrix C. Then for any* $\lambda_1, \ldots, \lambda_n \geq 0$ *we have*

$$\mathbb{E} \exp \left(-\sum_{i=1}^n \lambda_i X_i^2 \right) = |I_n + 2\Lambda C|^{-1/2},$$

where $\Lambda := \text{diag}(\lambda_1, \ldots, \lambda_n)$.

Proof Let A be a symmetric positive-definite matrix. Then $A = UDU^T$ for some $U \in O(n)$ and $D = \text{diag}(d_1, d_2, \ldots, d_n)$. Hence

$$\int_{\mathbb{R}^n} \exp(-\langle Ax, x \rangle) dx = \int_{\mathbb{R}^n} \exp(-\langle Dx, x \rangle) dx = \prod_{k=1}^n \sqrt{\frac{\pi}{d_k}} = \pi^{n/2} |D|^{-1/2} = \pi^{n/2} |A|^{-1/2}.$$

Therefore for a canonical Gaussian vector $Y \sim \mathcal{N}(0, I_n)$ and a symmetric matrix B such that $2B < I_n$ we have

$$\mathbb{E} \exp(\langle BY, Y \rangle) = (2\pi)^{-n/2} \int_{\mathbb{R}^n} \exp \left(-\left\langle \left(\frac{1}{2} I_n - B \right) x, x \right\rangle \right) dx = 2^{-n/2} \left| \frac{1}{2} I_n - B \right|^{-1/2}$$

$$= |I_n - 2B|^{-1/2}.$$

We may represent $X \sim \mathcal{N}(0, C)$ as $X \sim AY$ with $Y \sim \mathcal{N}(0, I_n)$ and $C = AA^T$. Thus

$$\mathbb{E} \exp\left(-\sum_{i=1}^{n} \lambda_i X_i^2\right) = \mathbb{E} \exp(-\langle \Lambda X, X \rangle)$$

$$= \mathbb{E} \exp(-\langle \Lambda AY, AY \rangle) = \mathbb{E} \exp(-\langle A^T \Lambda AY, Y \rangle)$$

$$= |I_n + 2A^T \Lambda A|^{-1/2} = |I_n + 2\Lambda C|^{-1/2},$$

where to get the last equality we used the fact that $|I_n + A_1 A_2| = |I_n + A_2 A_1|$ for $A_1, A_2 \in M_{n \times n}$. □

Lemma 5

(i) *For any matrix $A \in M_{n \times n}$,*

$$|I_n + A| = 1 + \sum_{\emptyset \neq J \subset [n]} |A_J|. \tag{2}$$

(ii) *Suppose that $n = n_1 + n_2$ and $A \in M_{n \times n}$ is symmetric and positive-definite with a block representation $A = \begin{pmatrix} A_{11} & A_{12} \\ A_{21} & A_{22} \end{pmatrix}$, where $A_{ij} \in M_{n_i \times n_j}$. Then*

$$|A| = |A_{11}||A_{22}| \left| I_{n_1} - A_{11}^{-1/2} A_{12} A_{22}^{-1} A_{21} A_{11}^{-1/2} \right|. \tag{3}$$

Moreover,

$$0 \leq A_{11}^{-1/2} A_{12} A_{22}^{-1} A_{21} A_{11}^{-1/2} \leq I_{n_1}. \tag{4}$$

Proof

(i) This formula may be verified in several ways—e.g. by induction on n or by using the Leibniz formula for the determinant.

(ii) We have

$$\begin{pmatrix} A_{11} & A_{12} \\ A_{21} & A_{22} \end{pmatrix} = \begin{pmatrix} A_{11}^{1/2} & 0 \\ 0 & A_{22}^{1/2} \end{pmatrix} \begin{pmatrix} I_{n_1} & A_{11}^{-1/2} A_{12} A_{22}^{-1/2} \\ A_{22}^{-1/2} A_{21} A_{11}^{-1/2} & I_{n_2} \end{pmatrix} \begin{pmatrix} A_{11}^{1/2} & 0 \\ 0 & A_{22}^{1/2} \end{pmatrix}$$

and

$$\left| \begin{pmatrix} I_{n_1} & A_{11}^{-1/2} A_{12} A_{22}^{-1/2} \\ A_{22}^{-1/2} A_{21} A_{11}^{-1/2} & I_{n_2} \end{pmatrix} \right| = \left| \begin{pmatrix} I_{n_1} - A_{11}^{-1/2} A_{12} A_{22}^{-1} A_{21} A_{11}^{-1/2} & 0 \\ A_{22}^{-1/2} A_{21} A_{11}^{-1/2} & I_{n_2} \end{pmatrix} \right|$$

$$= \left| I_{n_1} - A_{11}^{-1/2} A_{12} A_{22}^{-1} A_{21} A_{11}^{-1/2} \right|.$$

To show the last part of the statement notice that $A_{11}^{-1/2} A_{12} A_{22}^{-1} A_{21} A_{11}^{-1/2} = B^T B \geq 0$, where $B := A_{22}^{-1/2} A_{21} A_{11}^{-1/2}$. Since A is positive-definite, for any $t \in \mathbb{R}$, $x \in \mathbb{R}^{n_1}$ and $y \in \mathbb{R}^{n_2}$ we have $t^2 \langle A_{11}x, x \rangle + 2t \langle A_{21}x, y \rangle + \langle A_{22}y, y \rangle \geq 0$. This implies $\langle A_{21}x, y \rangle^2 \leq \langle A_{11}x, x \rangle \langle A_{22}y, y \rangle$. Replacing x by $A_{11}^{-1/2}x$ and y by $A_{22}^{-1/2}y$ we get $\langle Bx, y \rangle^2 \leq |x|^2 |y|^2$. Choosing $y = Bx$ we get $\langle B^T Bx, x \rangle \leq |x|^2$, i.e. $B^T B \leq I_{n_1}$.

$\square$

Lemma 6 *Let $f(x, \tau)$ be the density of the random vector $Z(\tau)$ defined above. Then for any Borel set K in $[0, \infty)^n$ and any $\lambda_1, \ldots, \lambda_n \geq 0$,*

$$\int_K e^{-\sum_{i=1}^n \lambda_i x_i} \frac{\partial}{\partial \tau} f(x, \tau) dx = \frac{\partial}{\partial \tau} \int_K e^{-\sum_{i=1}^n \lambda_i x_i} f(x, \tau) dx.$$

Proof The matrix C is nondegenerate, therefore matrices C_{11} and C_{22} are nondegenerate and $C(\tau)$ is nondegenerate for any $\tau \in [0, 1]$. Random vector $X(\tau) \sim \mathcal{N}(0, C(\tau))$ has the density $|C(\tau)|^{-1/2} (2\pi)^{-n/2} \exp(-\frac{1}{2} \langle C(\tau)^{-1}x, x \rangle)$. Standard calculation shows that $Z(\tau)$ has the density

$$f(x, \tau) = |C(\tau)|^{-1/2} (4\pi)^{-n/2} \frac{1}{\sqrt{x_1 \cdots x_n}} \sum_{\varepsilon \in \{-1, 1\}^n} e^{-\langle C(\tau)^{-1} \varepsilon \sqrt{x}, \varepsilon \sqrt{x} \rangle} \mathbf{1}_{(0,\infty)^n}(x),$$

where for $\varepsilon \in \{-1, 1\}^n$ and $x \in (0, \infty)^n$ we set $\varepsilon \sqrt{x} := (\varepsilon_i \sqrt{x_i})_i$.
The function $\tau \mapsto |C(\tau)|^{-1/2}$ is smooth on $[0, 1]$, in particular

$$\sup_{\tau \in [0,1]} |C(\tau)|^{-1/2} + \sup_{\tau \in [0,1]} \frac{\partial}{\partial \tau} |C(\tau)|^{-1/2} =: M < \infty.$$

Since $C(\tau) = \tau C(1) + (1 - \tau) C(0)$ we have $\frac{\partial}{\partial \tau} C(\tau) = C(1) - C(0)$ and

$$\frac{\partial}{\partial \tau} e^{-\langle C(\tau)^{-1} \varepsilon \sqrt{x}, \varepsilon \sqrt{x} \rangle} = -\langle C(\tau)^{-1} (C(1) - C(0)) C(\tau)^{-1} \varepsilon \sqrt{x}, \varepsilon \sqrt{x} \rangle e^{-\langle C(\tau)^{-1} \varepsilon \sqrt{x}, \varepsilon \sqrt{x} \rangle}.$$

The continuity of the function $\tau \mapsto C(\tau)$ gives

$$\langle C(\tau)^{-1} \varepsilon \sqrt{x}, \varepsilon \sqrt{x} \rangle \geq a \langle \varepsilon \sqrt{x}, \varepsilon \sqrt{x} \rangle = a \sum_{i=1}^n |x_i|$$

and

$$|\langle C(\tau)^{-1} (C(1) - C(0)) C(\tau)^{-1} \varepsilon \sqrt{x}, \varepsilon \sqrt{x} \rangle| \leq b \langle \varepsilon \sqrt{x}, \varepsilon \sqrt{x} \rangle = b \sum_{i=1}^n |x_i|$$

for some $a > 0, b < \infty$. Hence for $x \in (0, \infty)^n$

$$\sup_{\tau \in [0,1]} \left| \frac{\partial}{\partial \tau} f(x, \tau) \right| \le g(x) := M \pi^{-n/2} \frac{1}{\sqrt{x_1 \cdots x_n}} \left(1 + b \sum_{i=1}^n |x_i| \right) e^{-a \sum_{i=1}^n |x_i|}.$$

Since $g(x) \in L_1((0, \infty)^n)$ and $e^{-\sum_{i=1}^n \lambda_i x_i} \le 1$ the statement easily follows by the Lebesgue dominated convergence theorem. □

Let for $\alpha > 0$,

$$g_\alpha(x, y) := e^{-x-y} \sum_{k=0}^{\infty} \frac{x^{k+\alpha-1}}{\Gamma(k+\alpha)} \frac{y^k}{k!} \quad x > 0, y \ge 0.$$

For $\mu, \alpha_1, \ldots, \alpha_n > 0$ and a random vector $Y = (Y_1, \ldots, Y_n)$ such that $\mathbb{P}(Y_i \ge 0) = 1$ we set

$$h_{\alpha_1, \ldots, \alpha_n, \mu, Y}(x_1, \ldots, x_n) := \mathbb{E} \left[\prod_{i=1}^n \frac{1}{\mu} g_{\alpha_i} \left(\frac{x_i}{\mu}, Y_i \right) \right], \quad x_1, \ldots, x_n > 0.$$

Lemma 7 *Let $\mu > 0$ and Y be a random n-dimensional vector with nonnegative coordinates. For $\alpha = (\alpha_1, \ldots, \alpha_n) \in (0, \infty)^n$ set $h_\alpha := h_{\alpha_1, \ldots, \alpha_n, \mu, Y}$.*

(i) *For any $\alpha \in (0, \infty)^n$, $h_\alpha \ge 0$ and $\int_{(0,\infty)^n} h_\alpha(x) dx = 1$.*

(ii) *If $\alpha \in (0, \infty)^n$ and $\alpha_i > 1$ then $\lim_{x_i \to 0+} h_\alpha(x) = 0$, $\frac{\partial}{\partial x_i} h_\alpha(x)$ exists and*

$$\frac{\partial}{\partial x_i} h_\alpha(x) = h_{\alpha - e_i} - h_\alpha.$$

(iii) *If $\alpha \in (1, \infty)^n$ then for any $J \subset [n]$, $\frac{\partial^{|J|}}{\partial x_J} h_\alpha(x)$ exists and belongs to $L_1((0, \infty)^n)$. Moreover for $\lambda_1, \ldots, \lambda_n \ge 0$,*

$$\int_{(0,\infty)^n} e^{-\sum_{i=1}^n \lambda_i x_i} \frac{\partial^{|J|}}{\partial x_J} h_\alpha(x) dx = \prod_{i \in J} \lambda_i \int_{(0,\infty)^n} e^{-\sum_{i=1}^n \lambda_i x_i} h_\alpha(x) dx.$$

Proof

(i) Obviously $h_\alpha \in [0, \infty]$. We have for any $y \ge 0$ and $\alpha > 0$,

$$\int_0^\infty \frac{1}{\mu} g_\alpha \left(\frac{x}{\mu}, y \right) dx = \int_0^\infty g_\alpha(x, y) dx = 1.$$

Hence by the Fubini theorem,

$$\int_{(0,\infty)^n} h_\alpha(x)dx = \mathbb{E} \prod_{i=1}^{n} \int_0^\infty \frac{1}{\mu} g_{\alpha_i}\left(\frac{x_i}{\mu}, Y_i\right) dx_i = 1.$$

(ii) It is well known that $\Gamma(x)$ is decreasing on $(0, x_0]$ and increasing on $[x_0, \infty)$, where $1 < x_0 < 2$ and $\Gamma(x_0) > 1/2$. Therefore for $k = 1, \ldots$ and $\alpha > 0$, $\Gamma(k + \alpha) \geq \frac{1}{2}\Gamma(k) = \frac{1}{2}(k - 1)!$ and

$$g_\alpha(x, y) \leq e^{-x} \sum_{k=0}^{\infty} \frac{x^{k+\alpha-1}}{\Gamma(k+\alpha)} \leq 2\left(x^{\alpha-1}e^{-x} + x^\alpha \sum_{k=1}^{\infty} \frac{x^{k-1}}{(k-1)!}e^{-x}\right)$$

$$= 2x^{\alpha-1}(e^{-x} + x).$$

This implies that for $\alpha > 0$ and $0 < a < b < \infty$, $g_\alpha(x, y) \leq C(\alpha, a, b) < \infty$ for $x \in (a, b)$ and $y \geq 0$. Moreover,

$$h_\alpha(x) \leq \left(\frac{2}{\mu}\right)^n \prod_{i=1}^{n} \left(\frac{x_i}{\mu}\right)^{\alpha_i-1} \left(1 + \frac{x_i}{\mu}\right).$$

In particular $\lim_{x_i \to 0+} h_\alpha(x) = 0$ if $\alpha_i > 1$. Observe that for $\alpha \geq 1$, $\frac{\partial}{\partial x} g_\alpha = g_{\alpha-1} - g_\alpha$. Standard application of the Lebesgue dominated convergence theorem concludes the proof of part (ii).

(iii) By (ii) we get

$$\frac{\partial^{|J|}}{\partial x_J} h_\alpha = \sum_{K \subset J} (-1)^{|J|-|K|} h_{\alpha-\sum_{i \in K} e_i} \in L_1((0, \infty)^n).$$

Moreover $\lim_{x_j \to 0+} \frac{\partial^{|J|}}{\partial x_J} h_\alpha(x) = 0$ for $j \notin J$. We finish the proof by induction on $|J|$ using integration by parts.

□

Let C be a positive-definite symmetric $n \times n$ matrix. Then there exists $\mu > 0$ such that $B := C - \mu I_n$ is positive-definite. Let $X^{(l)} := (X_i^{(l)})_{i \leq n}$ be independent Gaussian vectors $\mathcal{N}(0, \frac{1}{2\mu}B)$,

$$Y_i = \sum_{l=1}^{k} (X_i^{(l)})^2 \quad 1 \leq i \leq n$$

and

$$h_{k,C} := h_{\frac{k}{2}, \ldots, \frac{k}{2}, \mu, Y}. \tag{5}$$

Lemma 8 *For any $\lambda_1, \ldots, \lambda_n \geq 0$ we have*

$$\int_{(0,\infty)^n} e^{-\sum_{i=1}^n \lambda_i x_i} h_{k,C}(x) = |I_n + \Lambda C|^{-\frac{k}{2}},$$

where $\Lambda = \mathrm{diag}(\lambda_1, \ldots, \lambda_n)$.

Proof We have for any $\alpha, \mu > 0$ and $\lambda, y \geq 0$

$$\int_0^\infty \frac{1}{\mu} e^{-\lambda x} g_\alpha \left(\frac{x}{\mu}, y \right) dx = e^{-y} \sum_{k=0}^\infty \frac{y^k}{k! \Gamma(k+\alpha)} \int_0^\infty e^{-(\lambda + \frac{1}{\mu})x} \frac{x^{k+\alpha-1}}{\mu^{k+\alpha}} dx$$

$$= e^{-y} \sum_{k=0}^\infty \frac{y^k}{k!(1+\mu\lambda)^{k+\alpha}} = (1+\mu\lambda)^{-\alpha} e^{-\frac{\mu\lambda}{1+\mu\lambda} y}.$$

By the Fubini theorem we have

$$\int_{(0,\infty)^n} e^{-\sum_{i=1}^n \lambda_i x_i} h_{k,C}(x) dx = \mathbb{E} \prod_{i=1}^n \int_0^\infty e^{-\lambda_i x_i} \frac{1}{\mu} g_{k/2} \left(\frac{x_i}{\mu}, Y_i \right) dx_i$$

$$= |I_n + \mu\Lambda|^{-\frac{k}{2}} \mathbb{E} e^{-\sum_{i=1}^n \frac{\mu\lambda_i}{1+\mu\lambda_i} Y_i}.$$

Therefore by Lemma 4 we have

$$\int_{(0,\infty)^n} e^{-\sum_{i=1}^n \lambda_i x_i} h_{k,C}(x) dx = |I_n + \mu\Lambda|^{-\frac{k}{2}} \left| I + 2\mu\Lambda(I+\mu\Lambda)^{-1} \frac{1}{2\mu} B \right|^{-\frac{k}{2}}$$

$$= |I_n + \Lambda C|^{-\frac{k}{2}}.$$

$\square$

References

1. C. Borell, A Gaussian correlation inequality for certain bodies in R^n. Math. Ann. **256**, 569–573 (1981)
2. G. Hargé, A particular case of correlation inequality for the Gaussian measure. Ann. Probab. **27**, 1939–1951 (1999)
3. C.G. Khatri, On certain inequalities for normal distributions and their applications to simultaneous confidence bounds. Ann. Math. Statist. **38**, 1853–1867 (1967)
4. Y. Memarian, The Gaussian correlation conjecture proof. arXiv:1310.8099 (2013)
5. L.D. Pitt, A Gaussian correlation inequality for symmetric convex sets. Ann. Probab. **5**, 470–474 (1977)
6. G. Qingyang, The Gaussian correlation inequality for symmetric convex sets. arXiv:1012.0676 (2010)

7. T. Royen, A simple proof of the Gaussian correlation conjecture extended to multivariate gamma distributions. Far East J. Theor. Stat. **48**, 139–145 (2014)
8. G. Schechtman, T. Schlumprecht, J. Zinn, On the Gaussian measure of the intersection. Ann. Probab. **26**, 346–357 (1998)
9. Z. Šidák, Rectangular confidence regions for the means of multivariate normal distributions. J. Amer. Statist. Assoc. **62**, 626–633 (1967)

A Simple Tool for Bounding the Deviation of Random Matrices on Geometric Sets

Christopher Liaw, Abbas Mehrabian, Yaniv Plan, and Roman Vershynin

Abstract Let A be an isotropic, sub-gaussian $m \times n$ matrix. We prove that the process $Z_x := \|Ax\|_2 - \sqrt{m}\,\|x\|_2$ has sub-gaussian increments, that is, $\|Z_x - Z_y\|_{\psi_2} \le C\|x - y\|_2$ for any $x, y \in \mathbb{R}^n$. Using this, we show that for any bounded set $T \subseteq \mathbb{R}^n$, the deviation of $\|Ax\|_2$ around its mean is uniformly bounded by the Gaussian complexity of T. We also prove a local version of this theorem, which allows for unbounded sets. These theorems have various applications, some of which are reviewed in this paper. In particular, we give a new result regarding model selection in the constrained linear model.

1 Introduction

Recall that a random variable Z is *sub-gaussian* if its distribution is dominated by a normal distribution. One of several equivalent ways to define this rigorously is to require the Orlicz norm

$$\|Z\|_{\psi_2} := \inf\{K > 0 : \mathbb{E}\psi_2(|Z|/K) \le 1\}$$

C. Liaw
Department of Computer Science, University of British Columbia, 2366 Main Mall, Vancouver, BC, Canada V6T 1Z4
e-mail: cvliaw@cs.ubc.ca

A. Mehrabian
Department of Computer Science, University of British Columbia, 2366 Main Mall, Vancouver, BC, Canada V6T 1Z4

School of Computing Science, Simon Fraser University, 8888 University Drive, Burnaby, BC, Canada V5A 1S6
e-mail: abbasmehrabian@gmail.com

Y. Plan
Department of Mathematics, University of British Columbia, 1984 Mathematics Rd, Vancouver, BC, Canada V6T 1Z4
e-mail: yaniv@math.ubc.ca

R. Vershynin (✉)
Department of Mathematics, University of Michigan, 530 Church St., Ann Arbor, MI 48109, USA
e-mail: romanv@umich.edu

© Springer International Publishing AG 2017
B. Klartag, E. Milman (eds.), *Geometric Aspects of Functional Analysis*,
Lecture Notes in Mathematics 2169, DOI 10.1007/978-3-319-45282-1_18

to be finite, for the Orlicz function $\psi_2(x) = \exp(x^2) - 1$. Also recall that a random vector X in $\mathbb{R}^n$ is sub-gaussian if all of its one-dimensional marginals are sub-gaussian random variables; this is quantified by the norm

$$\|X\|_{\psi_2} := \sup_{\theta \in S^{n-1}} \left\| \langle X, \theta \rangle \right\|_{\psi_2}.$$

For basic properties and examples of sub-gaussian random variables and vectors, see e.g. [27].

In this paper we study *isotropic, sub-gaussian random matrices A*. This means that we require the rows A_i of A to be independent, isotropic, and sub-gaussian random vectors:

$$\mathbb{E} A_i A_i^\top = I, \quad \|A_i\|_{\psi_2} \le K. \tag{1}$$

In Remark 1 below we show how to remove the isotropic assumption.

Suppose A is an $m \times n$ isotropic, sub-gaussian random matrix, and $T \subset \mathbb{R}^n$ is a given set. We are wondering when A acts as an approximate isometry on T, that is, when $\|Ax\|_2$ concentrates near the value $(\mathbb{E}\|Ax\|_2^2)^{1/2} = \sqrt{m}\|x\|_2$ uniformly over vectors $x \in T$.

Such a uniform deviation result must somehow depend on the "size" of the set T. A simple way to quantify the size of T is through the *Gaussian complexity*

$$\gamma(T) := \mathbb{E} \sup_{x \in T} |\langle g, x \rangle| \quad \text{where } g \sim N(0, I_n). \tag{2}$$

One can often find in the literature the following translation-invariant cousin of Gaussian complexity, called the *Gaussian width* of T:

$$w(T) := \mathbb{E} \sup_{x \in T} \langle g, x \rangle = \frac{1}{2} \mathbb{E} \sup_{x \in T-T} \langle g, x \rangle.$$

These two quantities are closely related. Indeed, a standard calculation shows that

$$\frac{1}{3}\left[w(T) + \|y\|_2\right] \le \gamma(T) \le 2\left[w(T) + \|y\|_2\right] \quad \text{for every } y \in T. \tag{3}$$

The reader is referred to [19, Sect. 2], [28, Sect. 3.5] for other basic properties of Gaussian width. Our main result is that the deviation of $\|Ax\|_2$ over T is uniformly bounded by the Gaussian complexity of T.

Theorem 1 (Deviation of Random Matrices on Sets) *Let A be an isotropic, sub-gaussian random matrix as in (1), and T be a bounded subset of $\mathbb{R}^n$. Then*

$$\mathbb{E} \sup_{x \in T} \left| \|Ax\|_2 - \sqrt{m}\|x\|_2 \right| \le CK^2 \cdot \gamma(T).$$

(Throughout, c and C denote absolute constants that may change from line to line). For Gaussian random matrices A, this theorem follows from a result of Schechtman [23]. For sub-gaussian random matrices A, one can find related results in [4, 10, 13]. Comparisons with these results can be found in Sect. 3.

The dependence of the right-hand-side of this theorem on T is essentially optimal. This is not hard to see for $m = 1$ by a direct calculation. For general m, optimality follows from several consequences of Theorem 1 that are known to be sharp; see Sect. 2.5.

We do not know if the dependence on K in the theorem is optimal or if the dependence can be improved to linear. However, none of the previous results have shown a linear dependence on K even in partial cases.

Remark 1 (Removing Isotropic Condition) Theorem 1 and the results below may also be restated without the assumption that A is isotropic using a simple linear transformation. Indeed, suppose that instead of being isotropic, each row of A satisfies $\mathbb{E}A_i A_i^T = \Sigma$ for some invertible covariance matrix Σ. Consider the whitened version $B_i := \sqrt{\Sigma^{-1}}A_i$. Note that $\|B_i\|_{\psi_2} \leq \|\sqrt{\Sigma^{-1}}\| \cdot \|A_i\|_{\psi_2} \leq \|\sqrt{\Sigma^{-1}}\| \cdot K$. Let B be the random matrix whose ith row is B_i. Then

$$\mathbb{E}\sup_{x \in T}\left| \|Ax\|_2 - \sqrt{m}\|\sqrt{\Sigma}x\|_2 \right| = \mathbb{E}\sup_{x \in T}\left| \|B\sqrt{\Sigma}x\|_2 - \sqrt{m}\|\sqrt{\Sigma}x\|_2 \right|$$

$$= \mathbb{E}\sup_{x \in \sqrt{\Sigma}T}\left| \|Bx\|_2 - \sqrt{m}\|x\|_2 \right|$$

$$\leq C\|\Sigma^{-1}\|K^2\gamma(\sqrt{\Sigma}T).$$

The last line follows from Theorem 1. Note also that $\gamma(\sqrt{\Sigma}T) \leq \|\sqrt{\Sigma}\|\gamma(T) = \sqrt{\|\Sigma\|}\gamma(T)$, which follows from Sudakov-Fernique's inequality. Summarizing, our bounds can be extended to anisotropic distributions by including in them the smallest and largest eigenvalues of the covariance matrix Σ.

Our proof of Theorem 1 given in Sect. 4.1 is particularly simple, and is inspired by the approach of Schechtman [23]. He showed that for Gaussian matrices A, the random process $Z_x := \|Ax\|_2 - (\mathbb{E}\|Ax\|_2^2)^{1/2}$ indexed by points $x \in \mathbb{R}^n$, has sub-gaussian increments, that is

$$\|Z_x - Z_y\|_{\psi_2} \leq C\|x - y\|_2 \quad \text{for every } x, y \in \mathbb{R}^n. \tag{4}$$

Then Talagrand's Majorizing Measure Theorem implies the desired conclusion that[1] $\mathbb{E}\sup_{x \in T}|Z_x| \lesssim \gamma(T)$.

However, it should be noted that G. Schechtman's proof of (4) makes heavy use of the rotation invariance property of the Gaussian distribution of A. When A is only sub-gaussian, there is no rotation invariance to rely on, and it was unknown if one

---

[1] In this paper, we sometimes hide absolute constants in the inequalities marked $\lesssim$.

can transfer G. Schechtman's argument to this setting. This is precisely what we do here: we show that, perhaps surprisingly, the sub-gaussian increment property (4) holds for general sub-gaussian matrices A.

Theorem 2 (Sub-gaussian Process) *Let A be an isotropic, sub-gaussian random matrix as in* (1). *Then the random process*

$$Z_x := \|Ax\|_2 - (\mathbb{E}\|Ax\|_2^2)^{1/2} = \|Ax\|_2 - \sqrt{m}\|x\|_2$$

has sub-gaussian increments:

$$\|Z_x - Z_y\|_{\psi_2} \le CK^2\|x - y\|_2 \quad \text{for every } x, y \in \mathbb{R}^n. \tag{5}$$

The proof of this theorem, given in Sect. 5, essentially consists of a couple of non-trivial applications of Bernstein's inequality; parts of the proof are inspired by G. Schechtman's argument. Applying Talagrand's Majorizing Measure Theorem (see Theorem 8 below), we immediately obtain Theorem 1.

We also prove a high-probability version of Theorem 1.

Theorem 3 (Deviation of Random Matrices on Sets: Tail Bounds) *Under the assumptions of Theorem 1, for any $u \ge 0$ the event*

$$\sup_{x \in T} \big| \|Ax\|_2 - \sqrt{m}\|x\|_2 \big| \le CK^2\big[w(T) + u \cdot \mathrm{rad}(T)\big]$$

holds with probability at least $1 - \exp(-u^2)$. Here $\mathrm{rad}(T) := \sup_{x \in T}\|x\|_2$ denotes the radius of T.

This result will be deduced in Sect. 4.1 from a high-probability version of Talagrand's theorem.

In the light of the equivalence (3), notice that Theorem 3 implies the following simpler but weaker bound

$$\sup_{x \in T} \big| \|Ax\|_2 - \sqrt{m}\|x\|_2 \big| \le CK^2 u \cdot \gamma(T) \tag{6}$$

if $u \ge 1$. Note that even in this simple bound, $\gamma(T)$ cannot be replaced with the Gaussian width $w(T)$, e.g. the result would fail for a singleton T. This explains why the radius of T appears in Theorem 3.

Restricting the set T to the unit sphere, we obtain the following corollary.

Corollary 1 *Under the assumptions of Theorem 1, for any $u \ge 0$ the event*

$$\sup_{x \in T \cap S^{n-1}} \big| \|Ax\|_2 - \sqrt{m} \big| \le CK^2\big[w(T \cap S^{n-1}) + u\big]$$

holds with probability at least $1 - \exp(-u^2)$.

In Theorems 1 and 3, we assumed that the set T is bounded. For unbounded sets, we can still prove a 'local version' of Theorem 3. Let us state a simpler form of this result here. In Sect. 6, we will prove a version of the following theorem with a better probability bound.

Theorem 4 (Local Version) *Let $(Z_x)_{x \in \mathbb{R}^n}$ be a random process with sub-gaussian increments as in (5). Assume that the process is homogeneous, that is, $Z_{\alpha x} = \alpha Z_x$ for any $\alpha \geq 0$. Let T be a star-shaped[2] subset of $\mathbb{R}^n$, and let $t \geq 1$. With probability at least $1 - \exp(-t^2)$, we have*

$$|Z_x| \leq t \cdot CK^2 \gamma \left(T \cap \|x\|_2 B_2^n \right) \quad \text{for all } x \in T. \tag{7}$$

Combining with Theorem 2, we immediately obtain the following result.

Theorem 5 (Local Version of Theorem 3) *Let A be an isotropic, sub-gaussian random matrix as in (1), and let T be a star-shaped subset of $\mathbb{R}^n$, and let $t \geq 1$. With probability at least $1 - \exp(-t^2)$, we have*

$$\left| \|Ax\|_2 - \sqrt{m}\|x\|_2 \right| \leq t \cdot CK^2 \gamma \left(T \cap \|x\|_2 B_2^n \right) \quad \text{for all } x \in T. \tag{8}$$

Remark 2 We note that Theorems 4 and 5 can also apply when T is not a star-shaped set, simply by considering the smallest star-shaped set that contains T:

$$\text{star}(T) := \bigcup_{\lambda \in [0,1]} \lambda T.$$

Then one only needs to replace T by $\text{star}(T)$ in the right-hand side of Eqs. (7) and (8).

Results of the type of Theorems 1, 3 and 5 have been useful in a variety of applications. For completeness, we will review some of these applications in the next section.

2 Applications

Random matrices have proven to be useful both for modeling data and transforming data in a variety of fields. Thus, the theory of this paper has implications for several applications. A number of classical theoretical discoveries as well as some new results follow directly from our main theorems. In particular, the local version of our theorem (Theorem 5), allows a new result in model selection under the constrained linear model, with applications in *compressed sensing*. We give details below.

---

[2]Recall that a set T is called star-shaped if $t \in T$ implies $\lambda t \in T$ for all $\lambda \in [0, 1]$.

2.1 Singular Values of Random Matrices

The singular values of a random matrix are an important topic of study in random matrix theory. A small sample includes covariance estimation [26], stability in numerical analysis [29], and quantum state tomography [8].

Corollary 1 may be specialized to bound the singular values of a sub-gaussian matrix. Indeed, take $T = S^{n-1}$ and note that $w(T) \le \sqrt{n}$. Then the corollary states that, with high probability,

$$\big| \|Ax\|_2 - \sqrt{m} \big| \le CK^2 \sqrt{n} \qquad \text{for all } x \in S^{n-1}.$$

This recovers the well-known result that, with high probability, all of the singular values of A reside in the interval $[\sqrt{m} - CK^2\sqrt{n}, \sqrt{m} + CK^2\sqrt{n}]$ (see [27]). When $nK^4 \ll m$, all of the singular values concentrate around $\sqrt{m}$. In other words, a tall random matrix is well conditioned with high probability.

2.2 Johnson-Lindenstrauss Lemma

The Johnson-Lindenstrauss lemma [9] describes a simple and effective method of dimension reduction. It shows that a (finite) set of data vectors $\mathcal{X}$ belonging to a very high-dimensional space, $\mathbb{R}^n$, can be mapped to a much lower dimensional space while roughly preserving pairwise distances. This is useful from a computational perspective since the storage space and the speed of computational tasks both improve in the lower dimensional space. Further, the mapping can be done simply by multiplying each vector by the random matrix $A/\sqrt{m}$.

The classic Johnson-Lindenstrauss lemma follows immediately from our results. Indeed, take $T' = \mathcal{X} - \mathcal{X}$. To construct T, remove the 0 vector from T' and project all of the remaining vectors onto S^{n-1} (by normalizing). Since T belongs to the sphere and has fewer than $|\mathcal{X}|^2$ elements, it is not hard to show that $\gamma(T) \le C\sqrt{\log |\mathcal{X}|}$. Then by Corollary 1, with high probability,

$$\sup_{x \in T} \left| \frac{1}{\sqrt{m}} \|Ax\|_2 - 1 \right| \le \frac{CK^2\sqrt{\log |\mathcal{X}|}}{\sqrt{m}}.$$

Equivalently, for all $x, y \in \mathcal{X}$

$$(1 - \delta)\|x - y\|_2 \le \frac{1}{\sqrt{m}} \|A(x - y)\|_2 \le (1 + \delta)\|x - y\|_2, \qquad \delta = \frac{CK^2\sqrt{\log |\mathcal{X}|}}{\sqrt{m}}.$$

This is the classic Johnson-Lindenstrauss lemma. It shows that as long as $m \gg K^4 \log |\mathcal{X}|$, the mapping $x \to Ax/\sqrt{m}$ nearly preserves pair-wise distances. In other words, $\mathcal{X}$ may be embedded into a space of dimension slightly larger than $\log |\mathcal{X}|$ while preserving distances.

In contrast to the classic Johnson-Lindenstrauss lemma that applies only to finite sets $\mathcal{X}$, the argument above based on Corollary 1 allows $\mathcal{X}$ to be infinite. In this case, the size of $\mathcal{X}$ is quantified using the notion of Gaussian width instead of cardinality.

To get even more precise control of the geometry of $\mathcal{X}$ in Johnson-Lindenstrauss lemma, we may use the local version of our results. To this end, apply Theorem 5 combined with Remark 2 to the set $T = \mathcal{X} - \mathcal{X}$. This shows that with high probability, for all $x, y \in \mathcal{X}$,

$$\left| \frac{1}{\sqrt{m}} \|A(x-y)\|_2 - \|x-y\|_2 \right| \leq \frac{CK^2 \gamma \left(\text{star}(\mathcal{X}-\mathcal{X}) \cap \|x-y\|_2 B_2^n \right)}{\sqrt{m}}. \tag{9}$$

One may recover the classic Johnson-Lindenstrauss lemma from the above bound using the containment $\text{star}(\mathcal{X}-\mathcal{X}) \subset \text{cone}(\mathcal{X}-\mathcal{X})$. However, the above result also applies to infinite sets, and further can benefit when $\mathcal{X} - \mathcal{X}$ has different structure at different scales, e.g., when $\mathcal{X}$ has clusters.

2.3 Gordon's Escape Theorem

In [7], Gordon answered the following question: *Let T be an arbitrary subset of S^{n-1}. What is the probability that a random subspace has nonempty intersection with T?* Gordon showed that this probability is small provided that the codimension of the subspace exceeds $w(T)$. This result also follows from Corollary 1 for a general model of random subspaces.

Indeed, let A be an isotropic, sub-gaussian $m \times n$ random matrix as in (1). Then its kernel $\ker A$ is a *random subspace* in $\mathbb{R}^n$ of dimension at least $n - m$. Corollary 1 implies that, with high probability,

$$\ker A \cap T = \emptyset \tag{10}$$

provided that $m \geq CK^4 w(T)^2$. To see this, note that in this case Corollary 1 yields that $\left| \|Ax\|_2 - \sqrt{m} \right| < \sqrt{m}$ for all $x \in T$, so $\|Ax\|_2 > 0$ for all $x \in T$, which in turn is equivalent to (10).

We also note that there is an equivalent version of the above result when T is a cone. Then, with high probability,

$$\ker A \cap T = \{0\} \quad \text{provided that} \quad m \geq CK^4 \gamma(T \cap S^{n-1})^2. \tag{11}$$

The conical version follows from the spherical version by expanding the sphere into a cone.

2.4   Sections of Sets by Random Subspaces: The M* Theorem

The M^* theorem [14, 15, 18] answers the following question: *Let T be an arbitrary subset of $\mathbb{R}^n$. What is the diameter of the intersection of a random subspace with T?* We may bound the radius of this intersection (which of course bounds the diameter) using our main results, and again for a general model of random subspaces.

Indeed, let us consider the kernel of an $m \times n$ random matrix A as in the previous section. By Theorem 3 (see (6)), we have

$$\sup_{x \in T} \left| \|Ax\|_2 - \sqrt{m}\|x\|_2 \right| \leq CK^2 \gamma(T) \tag{12}$$

with high probability. On the event that the above inequality holds, we may further restrict the supremum to $\ker A \cap T$, giving

$$\sup_{x \in \ker A \cap T} \sqrt{m}\|x\|_2 \leq CK^2 \gamma(T).$$

The left-hand side is $\sqrt{m}$ times the radius of $T \cap \ker A$. Thus, with high probability,

$$\mathrm{rad}(\ker A \cap T) \leq \frac{CK^2 \gamma(T)}{\sqrt{m}}. \tag{13}$$

This is a classical form of the so-called M^* estimate. It is typically used for sets T that contain the origin. In these cases, the Gaussian complexity $\gamma(T)$ can be replaced by Gaussian width $w(T)$. Indeed, (3) with $y = 0$ implies that these two quantities are equivalent.

2.5   The Size of Random Linear Images of Sets

Another question that can be addressed using our main results is how the size of a set T in $\mathbb{R}^n$ changes under the action of a random linear transformation $A : \mathbb{R}^n \to \mathbb{R}^m$. Applying (6) and the triangle inequality, we obtain

$$\mathrm{rad}(AT) \leq \sqrt{m} \cdot \mathrm{rad}(T) + CK^2 \gamma(T) \tag{14}$$

with high probability. This result has been known for random projections, where $A = \sqrt{n}P$ and P is the orthogonal projection onto a random m-dimensional subspace in $\mathbb{R}^n$ drawn according to the Haar measure on the Grassmanian, see [2, Proposition 5.7.1].

It is also known that the bound (14) is sharp (up to absolute constant factor) even for random projections, see [2, Sect. 5.7.1]. This in particular implies optimality of the bound in our main result, Theorem 1.

2.6 Signal Recovery from the Constrained Linear Model

The constrained linear model is the backbone of many statistical and signal processing problems. It takes the form

$$y = Ax + z, \qquad x \in T, \tag{15}$$

where $x \in T \subset \mathbb{R}^n$ is unknown, $y \in \mathbb{R}^m$ is a vector of known observations, the measurement matrix $A \in \mathbb{R}^{m \times n}$ is known, and $z \in \mathbb{R}^m$ is unknown noise which can be either fixed or random and independent of A.

For example, in the statistical linear model, A is a matrix of explanatory variables, and x is a coefficient vector. It is common to assume, or enforce, that only a small percentage of the explanatory variables are significant. This is encoded by taking T to be the set of vectors with less than s non-zero entries, for some $s \le n$. In other words, T encodes *sparsity*. In another example, y is a vector of MRI measurements [12], in which case x is the image to be constructed. Natural images have quite a bit of structure, which may be enforced by bounding the total variation, or requiring sparsity in a certain dictionary, each of which gives a different constraint set T. There are a plethora of other applications, with various constraint sets T, including low-rank matrices, low-rank tensors, non-negative matrices, and structured sparsity. In general, a goal of interest is to estimate x.

When T is a linear subspace, it is standard to estimate x via least squares regression, and the performance of such an estimator is well known. However, when T is non-linear, the problem can become quite complicated, both in designing a tractable method to estimate x and also analyzing the performance. The field of compressed sensing [5, 6] gives a comprehensive treatment of the case when T encodes sparsity, showing that convex programming can be used to estimate x, and that enforcing the sparse structure this way gives a substantial improvement over least squares regression. A main idea espoused in compressed sensing is that random matrices A give near optimal recovery guarantees.

Predating, but especially following, the works in compressed sensing, there have also been several works which tackle the general case, giving results for arbitrary T [1, 3, 11, 16, 17, 20, 21, 25]. The deviation inequalities of this paper allow for a general treatment as well. We will first show how to recover several known signal recovery results, and then give a new result in Sect. 2.7.

Consider the constrained linear model (15). A simple and natural way to estimate the unknown signal x is to solve the optimization problem

$$\hat{x} := \arg\min_{x' \in T} \|Ax' - y\|_2^2 \tag{16}$$

We note that depending on T, the constrained least squares problem (16) may be computationally tractable or intractable. We do not focus on algorithmic issues here, but just note that T may be replaced by a larger tractable set (e.g., convexified) to aid computation.

Our goal is to bound the Euclidean norm of the error

$$h := \hat{x} - x.$$

Since $\hat{x}$ minimizes the squared error, we have $\|A\hat{x} - y\|_2^2 \le \|Ax - y\|_2^2$. Simplifying this, we obtain

$$\|Ah\|_2^2 \le 2\langle h, A^T z \rangle. \tag{17}$$

We now proceed to control $\|h\|_2$ depending on the structure of T.

2.6.1 Exact Recovery

In the noiseless case where $z = 0$, inequality (17) simplifies and we have

$$h \in \ker A \cap (T - x). \tag{18}$$

(The second constraint here follows since $h = \hat{x} - x$ and $\hat{x} \in T$.)

In many cases of interest, $T - x$ is a cone, or is contained in a cone, which is called the *tangent cone* or *descent cone*. Gordon-type inequality (11) then implies that $h = 0$, and thus we have *exact recovery* $\hat{x} = x$, provided that the number of observations m significantly exceeds the Gaussian complexity of this cone: $m \ge CK^4 \gamma((T - x) \cap S^{n-1})^2$.

For example, if x is a sparse vector with s non-zero entries, and T is an appropriately scaled ℓ_1 ball, then $T - x$ is contained in a tangent cone, D, satisfying $\gamma(D)^2 \le Cs \log(n/s)$. This implies that $\hat{x} = x$ with high probability, provided $m \ge CK^4 s \log(n/s)$.

2.6.2 Approximate Recovery

In the cases where $T - x$ is not a cone or cannot be extended to a narrow cone (for example, when x lies in the interior of T), we can use the M^* Theorem for the analysis of the error. Indeed, combining (18) with (13), we obtain

$$\|h\|_2 \le \frac{CK^2 w(T)}{\sqrt{m}}.$$

Here we also used that since $T - T$ contains the origin, we have $\gamma(T - T) \sim w(T)$ according to (3). In particular, this means that x can be estimated up to an additive error of ε in the Euclidean norm provided that the number of observations satisfies $m \ge CK^4 w(T)^2/\varepsilon^2$.

For a more detailed description of the M^* Theorem, Gordon's Escape Theorem, and their implications for the constrained linear model, see [28].

2.7 Model Selection for Constrained Linear Models

It is often unknown precisely what constraint set to use for the constrained linear model, and practitioners often experiment with different constraint sets to see which gives the best performance. This is a form of model selection. We focus on the case when the form of the set is known, but the scaling is unknown. For example, in compressed sensing, it is common to assume that x is *compressible*, i.e., that it can be well approximated by setting most of its entries to 0. This can be enforced by assuming that x belongs to a scaled ℓ_p ball for some $p \in (0, 1]$. However, generally it is not known what scaling to use for this ℓ_p ball.

Despite this need, previous theory concentrates on controlling the error for one fixed choice of the scaling. Thus, a practitioner who tries many different scalings cannot be sure that the error bounds will hold uniformly over all such scalings. In this subsection, we remove this uncertainty by showing that the error in constrained least squares can be controlled simultaneously for an infinite number of scalings of the constraint set.

Assume $x \in T$, but the precise scaling of T is unknown. Thus, x is estimated using a scaled version of T:

$$\hat{x}_\lambda := \arg\min_{x' \in \lambda T} \|Ax' - y\|_2^2, \qquad \lambda \geq 1. \tag{19}$$

The following corollary controls the estimation error.

Corollary 2 *Let T be a convex, symmetric set. Given $\lambda \geq 1$, let $\hat{x}_\lambda$ be the solution to* (19). *Let $h_\lambda := \hat{x}_\lambda - x$, let $v_\lambda = h_\lambda/(1 + \lambda)$, and let $\delta = \|v_\lambda\|_2$. Then with probability at least* 0.99, *the following occurs. For every $\lambda \geq 1$,*

$$\delta \leq \frac{CK^2\gamma(T \cap \delta B_2^n)}{\sqrt{m}} + CK\sqrt{\frac{\gamma(T \cap \delta B_2^n) \cdot \|z\|_2}{m(1 + \lambda)}}. \tag{20}$$

The corollary is proven using Theorem 5. To our knowledge, this corollary is new. It recovers previous results that only apply to a single, fixed λ, as in [11, 20]. It is known to be nearly minimax optimal for many constraint sets of interest and for stochastic noise term z, in which case $\|z\|_2$ would be replaced by its expected value [21].

The rather complex bound of Eq. (20) seems necessary in order to allow generality. To aid understanding, we specialize the result to a very simple set— a linear subspace—for which the behaviour of constrained least squares is well known, the scaling becomes irrelevant, and the result simplifies significantly. When T is a d-dimensional subspace, we may bound the Gaussian complexity as $\gamma(T \cap \delta B_2) \leq \delta\sqrt{d}$. Plugging in the bound on $\gamma(T \cap \delta B_2^n)$ into (20), substituting h_λ back in, and massaging the equation gives

$$\|h_\lambda\|_2^2 \leq CK^4 \cdot \frac{d\|z\|_2^2}{m^2} \qquad \text{as long as} \qquad m \geq CK^4 d.$$

If z is Gaussian noise with standard deviation σ, then it's norm concentrates around $\sqrt{m}\sigma$, giving (with high probability)

$$\|h_\lambda\|_2^2 \leq CK^4 \cdot \frac{d\sigma^2}{m} \qquad \text{as long as} \quad m \geq CK^4 d.$$

In other words, the performance of least squares is proportional to the noise level multiplied by the dimension of the subspace, and divided by the number of observations, m. This is well known.

In this corollary, for simplicity we assumed that T is convex and symmetric. Note that this already allows constraint sets of interest, such as the ℓ_1 ball. However, this assumption can be weakened. All that is needed is for $T - \lambda T$ to be contained in a scaled version of T, and to be star shaped. This also holds, albeit for more complex scalings, for arbitrary ℓ_p balls with $p > 0$.

Proof (of Corollary 2) For simplicity of notation, we assume $K \leq 10$ (say), and absorb K into other constants. The general case follows the same proof. First note that $h_\lambda \in \lambda T - T$. Since T is convex and symmetric, we have $\lambda T - T \subset (1 + \lambda)T$ and as $v_\lambda = h_\lambda/(1 + \lambda)$, we get

$$v_\lambda \in T. \tag{21}$$

Moreover, (17) gives

$$\|Av_\lambda\|_2^2 \leq \frac{\langle v_\lambda, A^T z \rangle}{1 + \lambda}, \qquad v_\lambda \in T. \tag{22}$$

We will show that, with high probability, any vector v_λ satisfying (21) and (22) has a small norm, thus completing the proof. We will do this by upper bounding $\langle v_\lambda, A^T z \rangle$ and lower bounding $\|Av_\lambda\|_2$ by $\|v_\lambda\|_2$ minus a deviation term.

For the former goal, let $w := A^T z/\|z\|_2$. Recall that the noise vector z is fixed (and in case z random and independent of A, condition on z to make it fixed). Then w is a sub-gaussian vector with independent entries whose sub-gaussian norm is upper-bounded by a constant; see [27]. Thus, the random process $Z_x := \langle x, w \rangle$ has the sub-gaussian increments required in Theorem 4 (again, see [27]). By this theorem, with probability ≥ 0.995,

$$|Z_x| \leq C\gamma(T \cap \|x\|_2 B_2^n) \qquad \text{for all } x \in T.$$

Let F be the 'good' event that the above equation holds.

To control $\|Av_\lambda\|_2$, consider the 'good' event G that

$$\|Ax\|_2 \geq \sqrt{m}\|x\|_2 - C\gamma(T \cap \|x\|_2 B_2^n) \qquad \text{for all } x \in T.$$

By Theorem 5, G holds with probability at least 0.995.

Now, suppose that both G and F hold (which occurs with probability at least 0.99 by the union bound). We will show that for every $\lambda > 1$, v_λ is controlled. The event G gives

$$\langle v_\lambda, A^T z \rangle \leq C\gamma(T \cap \|v_\lambda\|_2 B_2^n) \cdot \|z\|_2.$$

The event F gives

$$\|Av_\lambda\|_2 \geq \sqrt{m}\|v_\lambda\|_2 - C\gamma(T \cap \|v_\lambda\|_2 B_2^n).$$

Taking square roots of both sides of (22) and plugging in these two inequalities gives (20). ☐

3 Comparison with Known Results

Several partial cases of our main results have been known. As we already mentioned, the special case of Theorem 1 where the entries of A have standard normal distribution follows from the main result of the paper by Schechtman [23].

Generalizing the result of [23], Klartag and Mendelson proved the following theorem.

Theorem 6 (Theorem 4.1 in [10]) *Let A be an isotropic, sub-gaussian random matrix as in (1), and let $T \subseteq S^{n-1}$. Assume that $w(T) \geq C'(K)$.[3] Then with probability larger than $1/2$,*

$$\sup_{x \in T} \left| \|Ax\|_2 - \sqrt{m} \right| \leq C(K)w(T). \tag{23}$$

Here $C'(K)$ and $C(K)$ may depend on K only.

A similar but slightly more informative statement follows from our main results. Indeed, Corollary 1 gives the same conclusion, but with explicit dependence on K (the sub-gaussian norms of the rows of A) as well as probability of success. Moreover, our general results, Theorems 1 and 3, do not require the set T to lie on the unit sphere.

Another related result was proved by S. Mendelson, A. Pajor, and N. Tomczak-Jaegermann.

---

[3]This restriction is not explicitly mentioned in the statement of Theorem 4.1 in [10], but it is used in the proof. Indeed, this result is derived from their Theorem 1.3, which explicitly requires that $\gamma(T)$ be large enough. Without such requirement, Theorem 4.1 in [10] fails e.g. when T is a singleton, since in that case we have $w(T) = 0$.

Theorem 7 (Theorem 2.3 in [13]) *Let A be an isotropic, sub-gaussian random matrix as in* (1), *and T be a star-shaped subset of* $\mathbb{R}^n$. *Let* $0 < \theta < 1$. *Then with probability at least* $1 - \exp(-c\theta^2 m/K^4)$ *we have that all vectors* $x \in T$ *with*

$$\|x\|_2 \geq r^* := \inf\left\{\rho > 0 : \rho \geq CK^2\gamma\left(T \cap \rho \cdot S^{n-1}\right)/(\theta\sqrt{m})\right\}$$

satisfy

$$(1 - \theta)\|x\|_2^2 \leq \frac{\|Ax\|_2^2}{m} \leq (1 + \theta)\|x\|_2^2.$$

Applying our Theorem 3 to the bounded set $T \cap r^* \cdot S^{n-1}$ precisely implies Theorem 7 with the same failure probability (up to the values of the absolute constants c, C). Moreover, our Theorem 3 treats all $x \in T$ uniformly, whereas Theorem 7 works only for x with large norm.

Yet another relevant result was proved by Dirksen [4, Theorem 5.5]. He showed that the inequality

$$\left|\|Ax\|_2^2 - m\|x\|_2^2\right| \lesssim K^2w(T)^2 + \sqrt{m}K^2\operatorname{rad}(T)w(T)$$
$$+ u\sqrt{m}K^2\operatorname{rad}(T)^2 + u^2K^2\operatorname{rad}(T)^2 \qquad (24)$$

holds uniformly over $x \in T$ with probability at least $1 - \exp(-u^2)$. To compare with our results, one can see that Theorem 3 implies that, with the same probability,

$$\left|\|Ax\|_2^2 - m\|x\|_2^2\right| \lesssim K^4w(T)^2 + \sqrt{m}K^2\|x\|_2 w(T)$$
$$+ u\sqrt{m}K^2\operatorname{rad}(T)\|x\|_2 + uK^4\operatorname{rad}(T)w(T) + u^2K^4\operatorname{rad}(T)^2,$$

which is stronger than (24) when $K = O(1)$ and $m \gtrsim n$, since then $\|x\|_2 \leq \operatorname{rad}(T)$ and $w(T) \lesssim \sqrt{m}\operatorname{rad}(T)$.

4 Preliminaries

4.1 Majorizing Measure Theorem, and Deduction of Theorems 1 and 3

As we mentioned in the Introduction, Theorems 1 and 3 follow from Theorem 2 via Talagrand's Majorizing Measure Theorem (and its high-probability counterpart). Let us state this theorem specializing to processes that are indexed by points in $\mathbb{R}^n$. For $T \subset \mathbb{R}^n$, let $\operatorname{diam}(T) := \sup_{x,y \in T} \|x - y\|_2$.

Theorem 8 (Majorizing Measure Theorem) *Consider a random process $(Z_x)_{x \in T}$ indexed by points x in a bounded set $T \subset \mathbb{R}^n$. Assume that the process has sub-gaussian increments, that is there exists $M \geq 0$ such that*

$$\|Z_x - Z_y\|_{\psi_2} \leq M \|x - y\|_2 \quad \text{for every } x, y \in T. \tag{25}$$

Then

$$\mathbb{E} \sup_{x,y \in T} |Z_x - Z_y| \leq CM \mathbb{E} \sup_{x \in T} \langle g, x \rangle,$$

where $g \sim N(0, I_n)$. Moreover, for any $u \geq 0$, the event

$$\sup_{x,y \in T} |Z_x - Z_y| \leq CM \big[\mathbb{E} \sup_{x \in T} \langle g, x \rangle + u \operatorname{diam}(T) \big]$$

holds with probability at least $1 - \exp(-u^2)$.

The first part of this theorem can be found e.g. in [24, Theorems 2.1.1, 2.1.5]. The second part, a high-probability bound, is borrowed from [4, Theorem 3.2].

Let us show how to deduce Theorems 1 and 3. According to Theorem 2, the random process $Z_x := \|Ax\|_2 - \sqrt{m}\|x\|_2$ satisfies the hypothesis (25) of the Majorizing Measure Theorem 8 with $M = CK^2$. Fix an arbitrary $y \in T$ and use the triangle inequality to obtain

$$\mathbb{E} \sup_{x \in T} |Z_x| \leq \mathbb{E} \sup_{x \in T} |Z_x - Z_y| + \mathbb{E}|Z_y|. \tag{26}$$

Majorizing Measure Theorem bounds the first term: $\mathbb{E} \sup_{x \in T} |Z_x - Z_y| \lesssim K^2 w(T)$. (We suppress absolute constant factors in this inequality and below.) The second term can be bounded more easily as follows: $\mathbb{E}|Z_y| \lesssim \|Z_y\|_{\psi_2} \lesssim K^2 \|y\|_2$, where we again used Theorem 2 with $x = 0$. Using (3), we conclude that

$$\mathbb{E} \sup_{x \in T} |Z_x| \lesssim K^2 (w(T) + \|y\|_2) \lesssim K^2 \gamma(T),$$

as claimed in Theorem 1.

We now prove Theorem 3. Since adding 0 to a set does not change its radius, we may assume that $0 \in T$. Let $Z_x := \|Ax\|_2 - \sqrt{m}\|x\|_2$. Since $Z_0 = 0$, and since Z_x has sub-gaussian increments by Theorems 2, 8 gives that with probability at least $1 - \exp(-u^2)$,

$$\sup_{x \in T} |Z_x| = \sup_{x \in T} |Z_x - Z_0| \lesssim K^2 \big[\mathbb{E} \sup_{x \in T} \langle g, x \rangle + u \cdot \operatorname{diam}(T) \big]$$

$$\lesssim K^2 \big[\mathbb{E} \sup_{x \in T} \langle g, x \rangle + u \cdot \operatorname{rad}(T) \big]. \qquad \square$$

4.2 Sub-exponential Random Variables, and Bernstein's Inequality

Our argument will make an essential use of Bernstein's inequality for sub-exponential random variables. Let us briefly recall the relevant notions, which can be found, e.g., in [27]. A random variable Z is *sub-exponential* if its distribution is dominated by an exponential distribution. More formally, Z is sub-exponential if the Orlicz norm

$$\|Z\|_{\psi_1} := \inf\{K > 0 : \mathbb{E}\psi_1(|Z|/K) \le 1\}$$

is finite, for the Orlicz function $\psi_1(x) = \exp(x) - 1$. Every sub-gaussian random variable is sub-exponential. Moreover, an application of Young's inequality implies the following relation for any two sub-gaussian random variables X and Y:

$$\|XY\|_{\psi_1} \le \|X\|_{\psi_2}\|Y\|_{\psi_2}. \tag{27}$$

The classical Bernstein's inequality states that a sum of independent sub-exponential random variables is dominated by a mixture of sub-gaussian and sub-exponential distributions.

Theorem 9 (Bernstein-Type Deviation Inequality, See e.g. [27]) *Let $X_1,\ldots,X_m$ be independent random variables, which satisfy $\mathbb{E}X_i = 0$ and $\|X_i\|_{\psi_1} \le L$. Then*

$$\mathbb{P}\left\{\left|\frac{1}{m}\sum_{i=1}^m X_i\right| > t\right\} \le 2\exp\left[-cm\min\left(\frac{t^2}{L^2},\frac{t}{L}\right)\right], \quad t \ge 0.$$

5 Proof of Theorem 2

Proposition 1 (Concentration of the Norm) *Let $X \in \mathbb{R}^m$ be a random vector with independent coordinates X_i that satisfy $\mathbb{E}X_i^2 = 1$ and $\|X_i\|_{\psi_2} \le K$. Then*

$$\left\|\|X\|_2 - \sqrt{m}\right\|_{\psi_2} \le CK^2.$$

Remark 3 If $\mathbb{E}X_i = 0$, this proposition follows from [22, Theorem 2.1], whose proof uses the Hanson-Wright inequality.

Proof Let us apply Bernstein's deviation inequality (Theorem 9) for the sum of independent random variables $\|X\|_2^2 - m = \sum_{i=1}^m(X_i^2 - 1)$. These random variables have zero means and sub-exponential norms

$$\|X_i^2 - 1\|_{\psi_1} \le 2\|X_i^2\|_{\psi_1} \le 2\|X_i\|_{\psi_2}^2 \le 2K^2.$$

(Here we used a simple centering inequality which can be found e.g. in [27, Remark 5.18] and the inequality (27).) Bernstein's inequality implies that

$$\mathbb{P}\left\{\left|\|X\|_2^2 - m\right| > tm\right\} \leq 2 \exp\left[-cm \min\left(\frac{t^2}{K^4}, \frac{t}{K^2}\right)\right], \quad t \geq 0. \tag{28}$$

To deduce a concentration inequality for $\|X\|_2 - \sqrt{m}$ from this, let us employ the numeric bound $|x^2 - m| \geq \sqrt{m}\,|x - \sqrt{m}|$ valid for all $x \geq 0$. Using this together with (28) for $t = s/\sqrt{m}$, we obtain

$$\mathbb{P}\left\{\left|\|X\|_2 - \sqrt{m}\right| > s\right\} \leq \mathbb{P}\left\{\left|\|X\|_2^2 - m\right| > s\sqrt{m}\right\}$$
$$\leq 2\exp(-cs^2/K^4) \quad \text{for } s \leq K^2\sqrt{m}.$$

To handle large s, we proceed similarly but with a different numeric bound, namely $|x^2 - m| \geq (x - \sqrt{m})^2$ which is valid for all $x \geq 0$. Using this together with (28) for $t = s^2/m$, we obtain

$$\mathbb{P}\left\{\left|\|X\|_2 - \sqrt{m}\right| > s\right\} \leq \mathbb{P}\left\{\left|\|X\|_2^2 - m\right| > s^2\right\}$$
$$\leq 2\exp(-cs^2/K^2) \quad \text{for } s \geq K\sqrt{m}.$$

Since $K \geq 1$, in both cases we bounded the probability in question by $2\exp(-cs^2/K^4)$. This completes the proof. □

Lemma 1 (Concentration of a Random Matrix on a Single Vector) *Let A be an isotropic, sub-gaussian random matrix as in* (1). *Then*

$$\left\|\,\|Ax\|_2 - \sqrt{m}\,\right\|_{\psi_2} \leq CK^2 \quad \text{for every } x \in S^{n-1}.$$

Proof The coordinates of the vector $Ax \in \mathbb{R}^m$ are independent random variables $X_i := \langle A_i, x \rangle$. The assumption that $\mathbb{E}A_i A_i^\top = I$ implies that $\mathbb{E}X_i^2 = 1$, and the assumption that $\|A_i\|_{\psi_2} \leq K$ implies that $\|X_i\|_{\psi_2} \leq K$. The conclusion of the lemma then follows from Proposition 1. □

Lemma 1 can be viewed as a partial case of the increment inequality of Theorem 2 for $x \in S^{n-1}$ and $y = 0$, namely

$$\|Z_x\|_{\psi_2} \leq CK^2 \quad \text{for every } x \in S^{n-1}. \tag{29}$$

Our next intermediate step is to extend this by allowing y to be an arbitrary unit vector.

Lemma 2 (Sub-Gaussian Increments for Unit Vectors) *Let A be an isotropic, sub-gaussian random matrix as in* (1). *Then*

$$\left\|\,\|Ax\|_2 - \|Ay\|_2\,\right\|_{\psi_2} \leq CK^2 \|x - y\|_2 \quad \text{for every } x, y \in S^{n-1}.$$

Proof Given $s \geq 0$, we will bound the tail probability

$$p := \mathbb{P}\left\{ \frac{|\|Ax\|_2 - \|Ay\|_2|}{\|x - y\|_2} > s \right\}. \tag{30}$$

Case 1: $s \geq 2\sqrt{m}$. Using the triangle inequality we have $|\|Ax\|_2 - \|Ay\|_2| \leq \|A(x - y)\|_2$. Denoting $u := (x - y)/\|x - y\|_2$, we find that

$$p \leq \mathbb{P}\left\{ \|Au\|_2 > s \right\} \leq \mathbb{P}\left\{ \|Au\|_2 - \sqrt{m} > s/2 \right\} \leq \exp(-Cs^2/K^4).$$

Here the second bound holds since $s \geq 2\sqrt{m}$, and the last bound follows by Lemma 1.

Case 2: $s \leq 2\sqrt{m}$. Multiplying both sides of the inequality defining p in (30) by $\|Ax\|_2 + \|Ay\|_2$, we can write p as

$$p = \mathbb{P}\left\{ |Z| > s(\|Ax\|_2 + \|Ay\|_2) \right\} \quad \text{where} \quad Z := \frac{\|Ax\|_2^2 - \|Ay\|_2^2}{\|x - y\|_2}.$$

In particular,

$$p \leq \mathbb{P}\left\{ |Z| > s\|Ax\|_2 \right\} \leq \mathbb{P}\left\{ |Z| > \frac{s\sqrt{m}}{2} \right\}$$

$$+ \mathbb{P}\left\{ \|Ax\|_2 \leq \frac{\sqrt{m}}{2} \right\} =: p_1 + p_2.$$

We may bound p_2 using Lemma 1:

$$p_2 \leq 2\exp\left(-\frac{(\sqrt{m}/2)^2}{C^2 K^4} \right) = 2\exp\left(-\frac{m}{4C^2 K^4} \right) \leq 2\exp\left(-\frac{s^2}{16C^2 K^4} \right). \tag{31}$$

Next, to bound p_1, it will be useful to write Z as

$$Z = \frac{\langle A(x - y), A(x + y)\rangle}{\|x - y\|_2} = \langle Au, Av\rangle, \quad \text{where} \quad u := \frac{x - y}{\|x - y\|_2}, \quad v := x + y.$$

Since the coordinates of Au and Av are $\langle A_i, u\rangle$ and $\langle A_i, v\rangle$ respectively, Z can be represented as a sum of independent random variables:

$$Z = \sum_{i=1}^{m} \langle A_i, u\rangle \langle A_i, v\rangle. \tag{32}$$

Note that each of these random variables $\langle A_i, u\rangle \langle A_i, v\rangle$ has zero mean, since

$$\mathbb{E}\langle A_i, x - y\rangle \langle A_i, x + y\rangle = \mathbb{E}\left[\langle A_i, x\rangle^2 - \langle A_i, y\rangle^2 \right] = 1 - 1 = 0.$$

(Here we used the assumptions that $\mathbb{E}A_i A_i^\top = I$ and $\|x\|_2 = \|y\|_2 = 1$.) Moreover, the assumption that $\|A_i\|_{\psi_2} \leq K$ implies that $\|\langle A_i, u \rangle\|_{\psi_2} \leq K\|u\|_2 = K$ and $\|\langle A_i, v \rangle\|_{\psi_2} \leq K\|v\|_2 \leq 2K$. Recalling inequality (27), we see that $\langle A_i, u \rangle \langle A_i, v \rangle$ are sub-exponential random variables with $\|\langle A_i, u \rangle \langle A_i, v \rangle\|_{\psi_1} \leq CK^2$. Thus we can apply Bernstein's inequality (Theorem 9) to the sum of mean zero, sub-exponential random variables in (32), and obtain

$$p_1 = \mathbb{P}\left\{|Z| > \frac{s\sqrt{m}}{2}\right\} \leq 2\exp(-cs^2/K^4), \quad \text{since } s \leq 2K^2\sqrt{m}.$$

Combining this with the bound on p_2 obtained in (31), we conclude that

$$p = p_1 + p_2 \leq 2\exp(-cs^2/K^4).$$

This completes the proof. $\qquad\square$

Finally, we are ready to prove the increment inequality in full generality, for all $x, y \in \mathbb{R}^n$.

Proof (of Theorem 2) Without loss of generality we may assume that $\|x\|_2 = 1$ and $\|y\|_2 \geq 1$. Consider the unit vector $\bar{y} := y/\|y\|_2$ and apply the triangle inequality to get

$$\|Z_x - Z_y\|_{\psi_2} \leq \|Z_x - Z_{\bar{y}}\|_{\psi_2} + \|Z_{\bar{y}} - Z_y\|_{\psi_2} =: R_1 + R_2.$$

By Lemma 2, $R_1 \leq CK^2\|x - \bar{y}\|_2$. Next, since $\bar{y}$ and y are collinear, we have $R_2 = \|\bar{y} - y\|_2 \cdot \|Z_{\bar{y}}\|_{\psi_2}$. Since $\bar{y} \in S^{n-1}$, inequality (29) states that $\|Z_{\bar{y}}\|_{\psi_2} \leq CK^2$, and we conclude that $R_2 \leq CK^2\|\bar{y} - y\|_2$. Combining the bounds on R_1 and R_2, we obtain

$$\|Z_x - Z_y\|_{\psi_2} \leq CK^2(\|x - \bar{y}\|_2 + \|\bar{y} - y\|_2).$$

It is not difficult to check that since $\|y\|_2 \geq 1$, we have $\|x - \bar{y}\|_2 \leq \|x - y\|_2$ and $\|\bar{y} - y\|_2 \leq \|x - y\|_2$. This completes the proof. $\qquad\square$

6 Proof of Theorem 4

We will prove a slightly stronger statement. For $r > 0$, define

$$E_r := \sup_{x \in \frac{1}{r}T \cap B_2^n} |Z_x|.$$

Set $W := \lim_{r \to \mathrm{rad}(T)_-} \gamma\left(\frac{1}{r}T \cap B_2^n\right)$. Since $\frac{1}{r}T \cap B_2^n$ contains at least one point on the boundary for every $r < \mathrm{rad}(T)$, it follows that $W \geq \sqrt{2/\pi}$. We will show that, with probability at least $1 - \exp\left(-c't^2W^2\right)$, one has

$$E_r \leq t \cdot CK^2\gamma\left(\frac{1}{r}T \cap B_2^n\right) \text{ for all } r \in (0,\infty),$$

which, when combined with the assumption of homogeneity, will clearly imply the theorem with a stronger probability.

Fix $\varepsilon > 0$. Let $\varepsilon = r_0 < r_1 < \ldots < r_N$ be a sequence of real numbers satisfying the following conditions:

- $\gamma\left(\frac{1}{r_i}T \cap B_2^n\right) = 2 \cdot \gamma\left(\frac{1}{r_{i+1}}T \cap B_2^n\right)$ for $i = 0,1,\ldots,N-1$, and
- $\gamma\left(\frac{1}{r_N}T \cap B_2^n\right) \leq 2 \cdot W$.

The quantities $r_1,\ldots,r_N$ exist since the map $r \mapsto \gamma\left(\frac{1}{r}T \cap B_2^n\right)$ is decreasing and continuous when T is star-shaped.

Applying the Majorizing Measure Theorem 8 to the set $\frac{1}{r}T \cap B_2^n$ and noting that $Z_0 = 0$, we obtain that

$$E_r \lesssim K^2\left[\gamma\left(\frac{1}{r}T \cap B_2^n\right) + u\right]$$

with probability at least $1 - \exp(-u^2)$. Set $c := 10 \cdot \sqrt{\frac{\pi}{2}} \geq 10/W$ and use the above inequality for $u = ct\gamma\left(\frac{1}{r}T \cap B_2^n\right)$. We get

$$E_r \lesssim t \cdot K^2\gamma\left(\frac{1}{r}T \cap B_2^n\right) \tag{33}$$

holds with probability at least $1 - \exp\left(-c^2t^2\gamma\left(\frac{1}{r}T \cap B_2^n\right)^2\right)$. Thus for each $i \in \{0,1,\ldots,N\}$, we have

$$E_{r_i} \lesssim t \cdot K^2\gamma\left(\frac{1}{r_i}T \cap B_2^n\right) \tag{34}$$

with probability at least

$$1 - \exp\left(-c^2t^24^{N-i}\gamma\left(\frac{1}{r_N}T \cap B_2^n\right)^2\right) \geq 1 - \exp\left(-c^2t^24^{N-i}W^2\right).$$

By our choice of c and the union bound, (34) holds for all i simultaneously with probability at least

$$1 - \sum_{i=0}^{N}\exp\left(-c^2t^24^{N-i}W^2\right) \geq 1 - 2 \cdot \exp(-100t^2W^2) =: 1 - \exp(-c't^2W^2).$$

We now show that if (34) holds for all i, then (33) holds for all $r \in (\varepsilon, \infty)$. This is done via an approximation argument. To this end, assume that (34) holds and let $r \in (r_{i-1}, r_i)$ for some $i \in [N]$. Since T is star-shaped, we have $\frac{1}{r}T \cap B_2^n \subseteq \frac{1}{r_{i-1}}T \cap B_2^n$, so

$$E_r \leq E_{r_{i-1}} \lesssim t \cdot K^2 \gamma \left(\frac{1}{r_{i-1}} T \cap B_2^n \right) = 2t \cdot K^2 \gamma \left(\frac{1}{r_i} T \cap B_2^n \right)$$

$$\leq 2t \cdot K^2 \gamma \left(\frac{1}{r} T \cap B_2^n \right).$$

Also, for $\text{rad}(T) \geq r > r_N$ we have

$$E_r \lesssim t \cdot K^2 \gamma \left(\frac{1}{r_N} T \cap B_2^n \right) \leq 2t \cdot K^2 W \leq 2t \cdot K^2 \gamma \left(\frac{1}{r} T \cap B_2^n \right).$$

Let F_k be the event that (33) holds for all $r \in (1/k, \infty)$. We have just shown that $\mathbb{P}\{F_k\} \geq 1 - \exp\left(-c't^2 W^2\right)$ for all $k \in \mathbb{N}$. As $F_1 \supseteq F_2 \supseteq \ldots$ and $\cap_k F_k =: F_\infty$ is the event that (33) holds for all $r \in (0, \infty)$, it follows by continuity of measure that $\mathbb{P}\{F_\infty\} \geq 1 - \exp\left(-c't^2 W^2\right)$, thus completing the proof.

7   Further Thoughts

In the definition of Gaussian complexity $\gamma(T) = \mathbb{E}\sup_{x \in T} |\langle g, x \rangle|$, the absolute value is essential to make Theorem 1 hold. In other words, the bound would fail if we replace $\gamma(T)$ by the Gaussian width $w(T) = \mathbb{E}\sup_{x \in T} \langle g, x \rangle$. This can be seen by considering a set T that consists of a single point.

However, *one-sided* deviation inequalities do hold for Gaussian width. Thus a one-sided version of Theorem 1 states that

$$\mathbb{E}\sup_{x \in T} \left(\|Ax\|_2 - \sqrt{m}\|x\|_2 \right) \leq CK^2 \cdot w(T), \tag{35}$$

and the same bound holds for $\mathbb{E}\sup_{x \in T} \left(-\|Ax\|_2 + \sqrt{m}\|x\|_2 \right)$. To prove (35), one modifies the argument in Sect. 4.1 as follows. Fix a $y \in T$. Since $\mathbb{E}\|Ay\|_2 \leq \left(\mathbb{E}\|Ay\|_2^2\right)^{1/2} = \sqrt{m}\|y\|_2$, we have $\mathbb{E}Z_y \leq 0$, thus

$$\mathbb{E}\sup_{x \in T} Z_x \leq \mathbb{E}\sup_{x \in T}(Z_x - Z_y) \leq \mathbb{E}\sup_{x \in T} |Z_x - Z_y| \lesssim K^2 w(T)$$

where the last bound follows by Majorizing Measure Theorem 8. Thus in this argument there is no need to separate the term $\mathbb{E}|Z_y|$ as was done before in Eq. (26).

Acknowledgements Christopher Liaw is partially supported by an NSERC graduate scholarship. Abbas Mehrabian is supported by an NSERC Postdoctoral Fellowship. Yaniv Plan is partially supported by NSERC grant 22R23068. Roman Vershynin is partially supported by NSF grant DMS 1265782 and USAF Grant FA9550-14-1-0009.

References

1. D. Amelunxen, M. Lotz, M.B. McCoy, J.A. Tropp, Living on the edge: phase transitions in convex programs with random data. Inf. Inference **3**(3), 224–294 (2014). doi:10.1093/imaiai/iau005. http://dx.doi.org/10.1093/imaiai/iau005
2. S. Artstein-Avidan, A. Giannopoulos, V.D. Milman, Asymptotic geometric analysis. Part I, in *Mathematical Surveys and Monographs*, vol. 202 (American Mathematical Society, Providence, RI, 2015)
3. V. Chandrasekaran, B. Recht, P.A. Parrilo, A.S. Willsky, The convex geometry of linear inverse problems. Found. Comput. Math. **12**(6), 805–849 (2012)
4. S. Dirksen, Tail bounds via generic chaining. Electron. J. Probab. **20**(53), 1–29 (2015). doi:10.1214/EJP.v20-3760. http://dx.doi.org/10.1214/EJP.v20-3760
5. Y.C. Eldar, G. Kutyniok, *Compressed Sensing: Theory and Applications* (Cambridge University Press, Cambridge, 2012)
6. S. Foucart, H. Rauhut, A mathematical introduction to compressive sensing, in *Applied and Numerical Harmonic Analysis* (Birkhäuser/Springer, New York, 2013) doi:10.1007/978-0-8176-4948-7. http://dx.doi.org/10.1007/978-0-8176-4948-7
7. Y. Gordon, On Milman's inequality and random subspaces which escape through a mesh in **R**^{n}, in *Geometric Aspects of Functional Analysis (1986/87)*. Lecture Notes in Mathematics, vol. 1317 (Springer, Berlin, 1988), pp. 84–106. doi:10.1007/BFb0081737. http://dx.doi.org/10.1007/BFb0081737
8. D. Gross, Y.K. Liu, S.T. Flammia, S. Becker, J. Eisert, Quantum state tomography via compressed sensing. Phys. Rev. Lett. **105**(15), 150,401 (2010)
9. W.B. Johnson, J. Lindenstrauss, Extensions of Lipschitz mappings into a Hilbert space. Contemp. Math. **26**(1), 189–206 (1984)
10. B. Klartag, S. Mendelson, Empirical processes and random projections. J. Funct. Anal. **225**(1), 229–245 (2005). doi:10.1016/j.jfa.2004.10.009. http://dx.doi.org/10.1016/j.jfa.2004.10.009
11. G. Lecué, S. Mendelson, Learning subgaussian classes: upper and minimax bounds (2013). Available at http://arxiv.org/abs/1305.4825
12. M. Lustig, D. Donoho, J.M. Pauly, Sparse MRI: The application of compressed sensing for rapid MR imaging. Magn. Reson. Med. **58**(6), 1182–1195 (2007)
13. S. Mendelson, A. Pajor, N. Tomczak-Jaegermann, Reconstruction and subgaussian operators in asymptotic geometric analysis. Geom. Funct. Anal. **17**(4), 1248–1282 (2007). doi:10.1007/s00039-007-0618-7. http://dx.doi.org/10.1007/s00039-007-0618-7
14. V.D. Milman, Geometrical inequalities and mixed volumes in the local theory of Banach spaces. Astérisque **131**, 373–400 (1985)
15. V.D. Milman, Random subspaces of proportional dimension of finite dimensional normed spaces: approach through the isoperimetric inequality, in *Banach Spaces* (Springer, Berlin, 1985), pp. 106–115
16. S. Oymak, C. Thrampoulidis, B. Hassibi, Simple bounds for noisy linear inverse problems with exact side information (2013). Available at http://arxiv.org/abs/1312.0641
17. S. Oymak, C. Thrampoulidis, B. Hassibi, The squared-error of generalized lasso: a precise analysis, in *51st Annual Allerton Conference on Communication, Control, and Computing*, IEEE (2013), pp. 1002–1009
18. A. Pajor, N. Tomczak-Jaegermann, Subspaces of small codimension of finite-dimensional banach spaces. Proc. Am. Math. Soc. **97**(4), 637–642 (1986)

19. Y. Plan, R. Vershynin, Robust 1-bit compressed sensing and sparse logistic regression: a convex programming approach. IEEE Trans. Inform. Theory **59**(1), 482–494 (2013). doi:10.1109/TIT.2012.2207945. http://dx.doi.org/10.1109/TIT.2012.2207945
20. Y. Plan, R. Vershynin, The generalized lasso with non-linear observations. IEEE Trans. Inform. Theory **62**(3), 1528–1537 (2016). doi:10.1109/TIT.2016.2517008
21. Y. Plan, R. Vershynin, E. Yudovina, High-dimensional estimation with geometric constraints (2014). Available at http://arxiv.org/abs/1404.3749
22. M. Rudelson, R. Vershynin, Hanson-Wright inequality and sub-Gaussian concentration. Electron. Commun. Probab. **18**(82), 1–9 (2013). doi:10.1214/ECP.v18-2865. http://dx.doi.org/10.1214/ECP.v18-2865
23. G. Schechtman, Two observations regarding embedding subsets of Euclidean spaces in normed spaces. Adv. Math. **200**(1), 125–135 (2006). doi:10.1016/j.aim.2004.11.003. http://dx.doi.org/10.1016/j.aim.2004.11.003
24. M. Talagrand, The generic chaining: upper and lower bounds of stochastic processes, in *Springer Monographs in Mathematics* (Springer, Berlin, 2005)
25. C. Thrampoulidis, S. Oymak, B. Hassibi, Simple error bounds for regularized noisy linear inverse problems, in *IEEE International Symposium on Information Theory (ISIT)*, IEEE (2014), pp. 3007–3011
26. R. Vershynin, How close is the sample covariance matrix to the actual covariance matrix? J. Theor. Probab. **25**(3), 655–686 (2012)
27. R. Vershynin, Introduction to the non-asymptotic analysis of random matrices, in *Compressed Sensing* (Cambridge University Press, Cambridge, 2012), pp. 210–268
28. R. Vershynin, Estimation in high dimensions: a geometric perspective, in *Sampling Theory, A Renaissance* (Birkhauser, Basel, 2015), pp. 3–66
29. J. Von Neumann, *Collected Works*, ed. by A.H. Taub (Pergamon, Oxford, 1961)

On Multiplier Processes Under Weak Moment Assumptions

Shahar Mendelson

Abstract We show that if $V \subset \mathbb{R}^n$ satisfies a certain symmetry condition that is closely related to unconditionality, and if X is an isotropic random vector for which $\|\langle X, t\rangle\|_{L_p} \leq L\sqrt{p}$ for every $t \in S^{n-1}$ and every $1 \leq p \lesssim \log n$, then the suprema of the corresponding empirical and multiplier processes indexed by V behave as if X were L-subgaussian.

1 Introduction

The motivation for this work comes from various problems in Learning Theory, in which one encounters the following family of random processes.

Let $X = (x_1, \ldots, x_n)$ be a random vector in $\mathbb{R}^n$ (whose coordinates $(x_i)_{i=1}^n$ *need not* be independent) and let ξ be a random variable that need not be independent of X. Set $(X_i, \xi_i)_{i=1}^N$ to be N independent copies of (X, ξ), and for $V \subset \mathbb{R}^n$ the supremum of the centred multiplier process is

$$\sup_{v \in V} \left| \frac{1}{\sqrt{N}} \sum_{i=1}^{N} \left(\xi_i \langle X_i, v \rangle - \mathbb{E}\xi \langle X, v \rangle \right) \right|. \tag{1}$$

Multiplier processes are often studied in a more general context, in which the indexing set need not be a class of linear functionals on $\mathbb{R}^n$. Instead, one may consider an arbitrary probability space (Ω, μ) and a class of real-valued functions F defined on Ω. If $X_1, \ldots, X_N$ are independent, distributed according to μ, then the supremum of the multiplier process indexed by F is

$$\sup_{f \in F} \left| \frac{1}{\sqrt{N}} \sum_{i=1}^{N} \left(\xi_i f(X_i) - \mathbb{E}\xi f(X_i) \right) \right|. \tag{2}$$

S. Mendelson (✉)
Department of Mathematics, Technion - I.I.T., Haifa, Israel

Mathematical Sciences Institute, The Australian National University, Canberra, ACT, Australia
e-mail: shahar@tx.technion.ac.il

© Springer International Publishing AG 2017   301
B. Klartag, E. Milman (eds.), *Geometric Aspects of Functional Analysis*,
Lecture Notes in Mathematics 2169, DOI 10.1007/978-3-319-45282-1_19

Naturally, the simplest multiplier process is when $\xi \equiv 1$ and (2) is just the supremum of the standard empirical process.

Controlling a multiplier process is relatively straightforward when $\xi \in L_q$ for some $q > 2$ and is independent of X. For example, one may show (see, e.g., [20], Chap. 2.9) that if $(\xi_i)_{i=1}^N$ are independent copies of a mean-zero random variable $\xi \in L_{2,1}$, and are independent of $(X_i)_{i=1}^N$, then

$$\mathbb{E} \sup_{f \in F} \left| \frac{1}{\sqrt{N}} \sum_{i=1}^N (\xi_i f(X_i) - \mathbb{E}\xi f(X_i)) \right| \leq C\|\xi\|_{L_{2,1}} \mathbb{E} \sup_{f \in F} \left| \frac{1}{\sqrt{N}} \sum_{i=1}^N \varepsilon_i f(X_i) \right|;$$

here and throughout the article $(\varepsilon_i)_{i=1}^N$ denote independent, symmetric $\{-1, 1\}$-valued random variables that are independent of $(X_i, \xi_i)_{i=1}^N$, and C is an absolute constant.

This estimate and others of its kind show that multiplier processes are as 'complex' as their seemingly simpler empirical counterparts. However, the results we are looking for are of a different nature: estimates on multiplier processes that are based on some natural complexity parameter of the underlying class F which exhibits the class' geometry.

It turns out that chaining methods lead to such estimates, and the structure of F may be captured by a parameter that is a close relative of Talagrand's γ-functionals (see [19] for a detailed study on generic chaining and the γ functionals).

Definition 1.1 For a random variable Z and $p \geq 1$, set

$$\|Z\|_{(p)} = \sup_{1 \leq q \leq p} \frac{\|Z\|_{L_q}}{\sqrt{q}}.$$

Given a class of functions F, $u \geq 1$ and $s_0 \geq 0$, put

$$\Lambda_{s_0,u}(F) = \inf \sup_{f \in F} \sum_{s \geq s_0} 2^{s/2} \|f - \pi_s f\|_{(u^2 2^s)}, \tag{3}$$

where the infimum is taken with respect to all sequences $(F_s)_{s \geq 0}$ of subsets of F, and of cardinality $|F_s| \leq 2^{2^s}$. $\pi_s f$ is the nearest point in F_s to f with respect to the $(u^2 2^s)$ norm.

Let

$$\tilde{\Lambda}_{s_0,u}(F) = \Lambda_{s_0,u}(F) + 2^{s_0/2} \sup_{f \in F} \|\pi_{s_0} f\|_{(u^2 2^{s_0})}.$$

To put these definitions in some perspective, $\|Z\|_{(p)}$ measures the local-subgaussian behaviour of Z, and the meaning of 'local' is that $\| \ \|_{(p)}$ takes into account the growth of Z's moments up to a fixed level p. In comparison,

$$\|Z\|_{\psi_2} \sim \sup_{q \geq 2} \frac{\|Z\|_{L_q}}{\sqrt{q}},$$

implying that for $2 \leq p < \infty$, $\|Z\|_{(p)} \lesssim \|Z\|_{\psi_2}$; hence, for every $u \geq 1$ and $s \geq s_0$,

$$\Lambda_{s_0,u}(F) \lesssim \inf_{f \in F} \sup_{s \geq s_0} \sum 2^{s/2} \|f - \pi_s f\|_{\psi_2},$$

and $\tilde{\Lambda}_{0,u}(F) \leq c\gamma_2(F, \psi_2)$.

Recall that the canonical gaussian process indexed by F is defined by assigning to each $f \in F$ a centred gaussian random variables G_f, and the covariance structure of the process is endowed by the inner product in $L_2(\mu)$. Let

$$\mathbb{E} \sup_{f \in F} G_f = \sup\{\mathbb{E} \sup_{f \in F'} G_f : F' \subset F, F' \text{ is finite}\}$$

and note that if the class $F \subset L_2(\mu)$ is L-subgaussian, that is, if for every $f, h \in F \cup \{0\}$,

$$\|f - h\|_{\psi_2(\mu)} \leq L\|f - h\|_{L_2(\mu)},$$

then $\tilde{\Lambda}_{s_0,u}(F)$ may be bounded in terms of the process $\{G_f : f \in F\}$. Indeed, by Talagrand's Majorizing Measures Theorem [18, 19], for every $s_0 \geq 0$,

$$\tilde{\Lambda}_{s_0,u}(F) \lesssim L\left(\mathbb{E} \sup_{f \in F} G_f + 2^{s_0/2} \sup_{f \in F} \|f\|_{L_2(\mu)}\right).$$

As an example, let $V \subset \mathbb{R}^n$ and set $F = \{\langle v, \cdot \rangle : v \in V\}$ to be the class of linear functionals endowed by V. If X is an isotropic, L-subgaussian vector, it follows that for every $t \in \mathbb{R}^n$,

$$\|\langle X, t \rangle\|_{\psi_2} \leq L\|\langle X, t \rangle\|_{L_2} = L\|t\|_{\ell_2^n}.$$

Therefore, if $G = (g_1, \ldots, g_n)$ is the standard gaussian vector in $\mathbb{R}^n$, $\ell_*(V) = \mathbb{E} \sup_{v \in V} |\langle G, v \rangle|$ and $d_2(V) = \sup_{v \in V} \|v\|_{\ell_2^n}$, one has

$$\tilde{\Lambda}_{s_0,u}(F) \lesssim L\left(\mathbb{E} \sup_{v \in V} \langle G, v \rangle + 2^{s_0/2} \sup_{v \in V} \|\langle X, v \rangle\|_{L_2}\right)$$

$$\lesssim L\left(\ell_*(V) + 2^{s_0/2} d_2(V)\right).$$

As the following estimate from [11] shows, $\tilde{\Lambda}$ can be used to control a multiplier process in a relatively general situation.

Theorem 1.2 *For $q > 2$, there are constants c_0, c_1, c_2, c_3 and c_4 that depend only on q for which the following holds. Let $\xi \in L_q$ (that need not be independent of X) and set $(X_i, \xi_i)_{i=1}^N$ to be independent copies of (X, ξ). Fix an integer $s_0 \geq 0$ and $w, u > c_0$. Then, with probability at least*

$$1 - c_1 w^{-q} N^{-((q/2)-1)} \log^q N - 2\exp(-c_2 u^2 2^{s_0}),$$

$$\sup_{f \in F} \left| \frac{1}{\sqrt{N}} \sum_{i=1}^N (\xi_i f(X_i) - \mathbb{E}\xi f) \right| \leq c_3 w u \|\xi\|_{L_q} \tilde{\Lambda}_{s_0, c_4 u}(F).$$

It follows from Theorem 1.2 that if X is an isotropic, L-subgaussian random vector, $V \subset \mathbb{R}^n$ and

$$D(V) = \left(\frac{\ell_*(V)}{d_2(V)} \right)^2$$

then with probability at least

$$1 - c_2 w^{-q} N^{-((q/2)-1)} \log^q N - 2\exp(-c_3 u^2 D(V)),$$

$$\sup_{v \in V} \left| \frac{1}{\sqrt{N}} \sum_{i=1}^N (\xi_i \langle v, X_i \rangle - \mathbb{E}\xi \langle v, X \rangle) \right| \lesssim L w u \|\xi\|_{L_q} \ell_*(V). \tag{4}$$

There are other generic situations in which $\tilde{\Lambda}_{s_0, u}(F)$ may be controlled using the geometry of F; for example, when F is a class of linear functionals on $\mathbb{R}^n$ and X is an isotropic, unconditional, log-concave random vector [11, 13]. However, these are rather special cases, and there is no satisfactory theory that describes $\tilde{\Lambda}_{s_0, u}(F)$ for arbitrary F and μ. Moreover, because the definition of $\Lambda_{s_0, u}(F)$ involves $\| \ \|_{(p)}$ for every p, class members must have arbitrarily high moments for $\Lambda_{s_0, u}(F)$ to even be well defined.

In the context of classes of linear functionals on $\mathbb{R}^n$, one expects an analogous result to Theorem 1.2 to be true even if the functionals $\langle X, t \rangle$ do not have arbitrarily high moments. A realistic conjecture is that if for each $t \in S^{n-1}$

$$\|\langle X, t \rangle\|_{L_q} \leq L\sqrt{q} \|\langle X, t \rangle\|_{L_2} \text{ for every } 2 \leq q \lesssim n$$

then a subgaussian-type estimate like (4) should still be true, because Euclidean entropy numbers of bounded sets in $\mathbb{R}^n$ decay very quickly once one uses more than $\exp(cn)$ points in a cover.

In what follows we will not focus on such a general result that is likely to hold for *every* $V \subset \mathbb{R}^n$. Rather, we will focus our attention on situations in which linear functionals only satisfy that

$$\|\langle X, t \rangle\|_{L_q} \leq L\sqrt{q} \|\langle X, t \rangle\|_{L_2} \text{ for every } 2 \leq q \lesssim \log n.$$

The obvious example in which only $\sim \log n$ moments should suffice is for $V = B_1^n$ (or similar sets that have $\sim n$ extreme points). Having said that, the applications that motivated this work require that a broad spectrum of sets exhibit a subgaussian behaviour as in (4).

Question 1.3 Let $X = (x_1, \ldots, x_n)$ be an isotropic random vector and assume that $\|x_i\|_{L_q} \leq L\sqrt{q}$ for every $2 \leq q \leq p$ and $1 \leq i \leq n$. If $\xi \in L_{q_0}$ for some $q_0 > 2$, how small can p be and still ensure that

$$\mathbb{E} \sup_{v \in V} \left| \frac{1}{\sqrt{N}} \sum_{i=1}^{N} \left(\xi_i \langle X_i, v \rangle - \mathbb{E} \xi \langle X, v \rangle \right) \right| \leq C(L, q_0) \|\xi\|_{L_{q_0}} \ell_*(V)?$$

We will show $p \sim \log n$ suffices for a positive answer to Question 1.3 if the norm $\|z\|_{V^\circ} = \sup_{v \in V} |\langle v, z \rangle|$ satisfies the following unconditionality property:

Definition 1.4 Given a vector $x = (x_i)_{i=1}^n$, let $(x_i^*)_{i=1}^n$ be the non-increasing rearrangement of $(|x_i|)_{i=1}^n$.

The normed space $(\mathbb{R}^n, \| \ \|)$ is K-unconditional with respect to the basis $\{e_1, \ldots, e_n\}$ if for every $x \in \mathbb{R}^n$ and every permutation of $\{1, \ldots, n\}$

$$\left\| \sum_{i=1}^n x_i e_i \right\| \leq K \left\| \sum_{i=1}^n x_{\pi(i)} e_i \right\|,$$

and if $y \in \mathbb{R}^n$ and $x_i^* \leq y_i^*$ for $1 \leq i \leq n$ then

$$\left\| \sum_{i=1}^n x_i e_i \right\| \leq K \left\| \sum_{i=1}^n y_i e_i \right\|.$$

Remark 1.5 This is not the standard definition of an unconditional basis, though every unconditional basis (in the classical sense) of an infinite dimensional space satisfies Definition 1.4 for some constant K (see, e.g., [1]).

There are many natural examples of K-unconditional spaces, most notably, all the ℓ_p spaces. Moreover, the norm $\|z\| = \sup_{v \in V} \sum_{i=1}^n v_i^* z_i^*$ is 1-unconditional. In fact, if $V \subset \mathbb{R}^n$ is closed under permutations and reflections (sign-changes), then $\| \cdot \|_{V^\circ}$ is 1-unconditional.

We will show the following:

Theorem 1.6 *There exists an absolute constant c_1 and for $K \geq 1, L \geq 1$ and $q_0 > 2$ there exists a constant c_2 that depends only on K, L and q_0 for which the following holds. Consider*

- $V \subset \mathbb{R}^n$ *for which the norm* $\| \cdot \|_{V^\circ} = \sup_{v \in V} |\langle v, \cdot \rangle|$ *is K-unconditional with respect to the basis* $\{e_1, \ldots, e_n\}$.
- $\xi \in L_{q_0}$ *for some $q_0 > 2$.*

- An isotropic random vector $X \in \mathbb{R}^n$ that satisfies

$$\max_{1 \le j \le n} \|\langle X, e_j \rangle\|_{(p)} \le L \text{ for } p = c_1 \log n.$$

If $(X_i, \xi_i)_{i=1}^N$ are independent copies of (X, ξ) then

$$\mathbb{E} \sup_{v \in V} \left| \frac{1}{\sqrt{N}} \sum_{i=1}^N \left(\xi_i \langle X_i, v \rangle - \mathbb{E}\xi \langle X, v \rangle \right) \right| \le c_2 \|\xi\|_{L_{q_0}} \ell_*(V).$$

The proof of Theorem 1.6 is based on properties of a conditioned Bernoulli process. Indeed, a standard symmetrization argument (see, e.g., [8, 20]) shows that if $(\varepsilon_i)_{i=1}^N$ are independent, symmetric, $\{-1, 1\}$-valued random variables that are independent of $(X_i, \xi_i)_{i=1}^N$ then

$$\mathbb{E} \sup_{v \in V} \left| \frac{1}{\sqrt{N}} \sum_{i=1}^N \left(\xi_i \langle X_i, v \rangle - \mathbb{E}\xi \langle X, v \rangle \right) \right| \le C\mathbb{E} \sup_{v \in V} \left| \frac{1}{\sqrt{N}} \sum_{i=1}^N \varepsilon_i \xi_i \langle X_i, v \rangle \right|$$

for an absolute constant C; a similar bound hold with high probability, showing that it suffices to study the supremum of the Bernoulli process

$$\sup_{v \in V} \left| \frac{1}{\sqrt{N}} \sum_{i=1}^N \varepsilon_i \xi_i \langle X_i, v \rangle \right| = (*)$$

conditioned on $(X_i, \xi_i)_{i=1}^N$.

Put $x_i(j) = \langle X_i, e_j \rangle$ and set $Z_j = N^{-1/2} \sum_{i=1}^N \varepsilon_i \xi_i x_i(j)$, which is a sum of iid random variables. Therefore, if $Z = (Z_1, \ldots, Z_n)$ then

$$(*) = \sup_{v \in V} |\langle Z, v \rangle|.$$

The proof of Theorem 1.6 follows by showing that for a well-chosen constant $C = C(L, q)$ the event

$$\{Z_j^* \le CEg_j^* \text{ for every } 1 \le j \le n\}$$

is of high probability, and if the norm $\| \cdot \|_{V^\circ} = \sup_{v \in V} |\langle \cdot, v \rangle|$ is K-unconditional then

$$\sup_{v \in V} |\langle Z, v \rangle| \le C_1(K, L, q) \mathbb{E} \sup_{v \in V} |\langle G, v \rangle|.$$

Before presenting the proof of Theorem 1.6, let us turn to one of its outcomes—estimates on the random Gelfand widths of a convex body. We will present another application, motivated by a question in the rapidly developing area of *Spare Recovery* in Sect. 3.

Let $V \subset \mathbb{R}^n$ be a convex, centrally symmetric set. A well known question in Asymptotic Geometric Analysis has to do with the diameter of a random m-codimensional section of V (see, e.g., [2, 14–16]). In the past, the focus was on obtaining such estimates for subspaces selected uniformly according to the Haar measure, or alternatively, according to the measure endowed by the kernel of an $m \times n$ gaussian matrix (see, e.g. [17]). More recently, there has been a growing interest in other notions of randomness, in particular randomness generated by kernels of other random matrix ensembles. For example, the following was established in [12]:

Theorem 1.7 *Let $X_1, \dots, X_m$ be distributed according to an isotropic, L-subgaussian random vector on $\mathbb{R}^n$, set $\Gamma = \sum_{i=1}^{m} \langle X_i, \cdot \rangle e_i$ and put*

$$r_G(V, \gamma) = \inf\{r > 0 : \ell_*(V \cap rB_2^n) \le \gamma r \sqrt{m}\}.$$

Then, with probability at least $1 - 2\exp(-c_1(L)m)$

$$diam(ker(\Gamma) \cap V) \le r_G(V, c_2(L)),$$

for constants c_1 and c_2 that depends only on L.

A version of Theorem 1.7 was obtained under a much weaker assumption: the random vector need not be L-subgaussian; rather, it suffices that it satisfies a weak small-ball condition.

Definition 1.8 The isotropic random vector X satisfies a small-ball condition with constants $\kappa > 0$ and $0 < \varepsilon \le 1$ if for every $t \in S^{n-1}$,

$$Pr(|\langle X, t \rangle| \ge \kappa) \ge \varepsilon.$$

The analog of gaussian parameter r_G for a general random vector X turns out to be

$$r_X(V, \gamma) = \inf\left\{r > 0 : \mathbb{E} \sup_{v \in V \cap rB_2^n} \left|\frac{1}{\sqrt{m}} \sum_{i=1}^{m} \langle X_i, v \rangle\right| \le \gamma r \sqrt{m}\right\}.$$

Clearly, if X is L-subgaussian then $r_X(V, \gamma) \le r_G(V, cL\gamma)$ for a suitable absolute constant c.

Theorem 1.9 ([9, 10]) *Let X be an isotropic random vector that satisfies the small-ball condition with constants κ and ε. If $X_1, \dots X_m$ are independent copies of X and $\Gamma = \sum_{i=1}^{m} \langle X_i, \cdot \rangle e_i$, then with probability at least $1 - 2\exp(-c_0(\varepsilon)m)$*

$$diam(ker(\Gamma) \cap V) \le r_X(V, c_1(\kappa, \varepsilon)).$$

Therefore, if the norms $\sup_{v \in V \cap rB_2^n} \langle v, \cdot \rangle$ are K-unconditional for every $r > 0$, and if the growth of the moments of the coordinate linear functionals $\langle X, e_i \rangle$ is 'L-subgaussian' up to the level $p \sim \log n$, then the small-ball condition depends only on L and $r_X(V, c_1(L)) \leq r_G(V, c_2(L, K))$. Hence, with probability at least $1 - 2 \exp(-c_0(L)m)$,

$$\operatorname{diam}(\ker(\Gamma) \cap V) \leq r_G\big(V, c_2(L, K)\big),$$

even when the choice of a subspace is made according to an ensemble that could be very far from a subgaussian one.

We end this introduction with a word about notation. Throughout, absolute constants are denoted by $c, c_1 \ldots$, etc. Their value may change from line to line or even within the same line. When a constant depends on a parameter α it will be denoted by $c(\alpha)$. $A \lesssim B$ means that $A \leq cB$ for an absolute constant c, and the analogous two-sided inequality is denoted by $A \sim B$. In a similar fashion, $A \lesssim_\alpha B$ implies that $A \leq c(\alpha)B$, etc.

2 Proof of Theorem 1.6

There are two substantial difficulties in the proof of Theorem 1.6. First, $Z_1, \ldots, Z_n$ are not independent random variables, not only because of the Bernoulli random variables $(\varepsilon_i)_{i=1}^N$ that appear in all the Z_i's, but also because the coordinates of $X = (x_1, \ldots, x_n)$ need not be independent. Second, while there is some flexibility in the moment assumptions on the coordinates of X, there is no flexibility in the moment assumption on ξ, which is only 'slightly better' than square-integrable.

As a starting point, let us address the fact that the coordinates of Z need not be independent.

Lemma 2.1 *There exist absolute constants c_1 and c_2 for which the following holds. Let $\beta \geq 1$ and set $p = 2\beta \log(en)$. If $(W_j)_{j=1}^n$ are random variables and satisfy that $\|W_j\|_{(p)} \leq L$, then for every $t \geq e$, with probability at least $1 - c_1 t^{-2\beta}$,*

$$W_j^* \leq c_2 tL \sqrt{\beta \log(en/j)} \quad \text{for every} \ 1 \leq j \leq n.$$

Proof Let $a_1, \ldots, a_k \in \mathbb{R}$ and by the convexity of $t \to t^q$,

$$\Big(\frac{1}{k} \sum_{j=1}^k a_j^2\Big)^q \leq \frac{1}{k} \sum_{j=1}^k a_j^{2q}.$$

Thus, given $(a_i)_{i=1}^n$, and taking the maximum over subsets of $\{1,\ldots,n\}$ of cardinality k,

$$\max_{|J_1|=k}\Big(\frac{1}{k}\sum_{j\in J_1}a_j^2\Big)^q \leq \max_{|J_1|=k}\frac{1}{k}\sum_{j\in J_1}a_j^{2q} \leq \frac{1}{k}\sum_{j=1}^n a_j^{2q}.$$

When applied to $a_j = W_j$, it follows that point-wise,

$$\Big(\frac{1}{k}\sum_{j=1}^k (W_j^*)^2\Big)^q \leq \frac{1}{k}\sum_{i=1}^n W_j^{2q}. \tag{5}$$

Since $\|W_j\|_{(p)} \leq L$ it is evident that $\mathbb{E}W_j^{2q} \leq L^{2q}(2q)^q$ for $2q \leq p$. Hence, taking the expectation in (5),

$$\Big(\mathbb{E}\big(\frac{1}{k}\sum_{j=1}^k (W_j^*)^2\big)^q\Big)^{1/q} \leq qL^2 \cdot \Big(\frac{n}{k}\Big)^{1/q} \leq c_1 qL^2$$

for $q = \beta\log(en/k)$ (which does satisfy $2q \leq p$). Therefore, by Chebyshev's inequality, for $t \geq 1$,

$$Pr\Big(\frac{1}{k}\sum_{j\leq k}(W_j^*)^2 \geq (et)^2 c_1^2 L^2 q\Big) \leq \frac{1}{t^{2q}}\cdot e^{-2q} \leq \Big(\frac{k}{en}\Big)^{2\beta}\cdot\frac{1}{t^{2\beta}}. \tag{6}$$

Using (6) for $k = 2^j$ and applying the union bound, it is evident that for $t \geq e$, with probability at least $1 - 2t^{-2\beta}\sum(2^j/n)^{2\beta} \geq 1 - ct^{-2\beta}$, for every $1 \leq k \leq n$,

$$(W_k^*)^2 \leq \frac{1}{k}\sum_{j\leq k}(W_j^*)^2 \lesssim t^2 L^2 \beta\log(en/k).$$

■

Recall that $q_0 > 2$ and set $\eta = (q_0 - 2)/4$. Let $u \geq 2$ and consider the event

$$\mathcal{A}_u = \{\xi_i^* \leq u\|\xi\|_{L_{q_0}}(eN/i)^{1/q_0} \text{ for every } 1 \leq i \leq N\}.$$

A standard binomial estimate combined with Chebyshev's inequality for $|\xi|^{q_0}$ shows that $\mathcal{A}_u$ is a nontrivial event. Indeed,

$$Pr\big(\xi_i^* \geq u\|\xi\|_{L_{q_0}}(eN/i)^{1/q_0}\big) \leq \binom{N}{i}Pr^i\big(|\xi| \geq u\|\xi\|_{L_{q_0}}(eN/i)^{1/q_0}\big) \leq \frac{1}{u^{iq_0}},$$

and by the union bound for $1 \leq i \leq n$, $Pr(\mathcal{A}_u) \leq 2/u^{q_0}$.

The random variables we shall use in Lemma 2.1 are

$$W_j = Z_j \mathbb{1}_{A_u},$$

for $u \geq 2$ and $1 \leq j \leq n$.

The following lemma is the crucial step in the proof of Theorem 1.6.

Lemma 2.2 *There exists an absolute constant c for which the following holds. Let X be a random variable that satisfies $\|X\|_{(p)} \leq L$ for some $p > 2$ and set $X_1, \ldots, X_N$ to be independent copies if X. If*

$$W = \left| \frac{1}{\sqrt{N}} \sum_{i=1}^{N} \varepsilon_i \xi_i X_i \right| \mathbb{1}_{A_u},$$

then $\|W\|_{(p)} \leq cuL\|\xi\|_{L_{q_0}}$.

The proof of Lemma 2.2 requires two preliminary estimates on the 'gaussian' behaviour of the monotone rearrangement of N independent copies of a random variable.

Lemma 2.3 *There exists an absolute constant c for which the following holds. Assume that $\|X\|_{(2p)} \leq L$. If $X_1, \ldots, X_N$ are independent copies of X, then for every $1 \leq k \leq N$ and $2 \leq q \leq p$,*

$$\Big\| \Big(\sum_{i \leq k} (X_i^*)^2 \Big)^{1/2} \Big\|_{L_q} \leq cL \big(\sqrt{k \log(eN/k)} + \sqrt{q} \big).$$

Proof The proof follows from a comparison argument, showing that up to the p-th moment, the 'worst case' is when X is a gaussian variable.

Let $V_1, \ldots, V_k$ be independent, nonnegative random variables and set $V_1', \ldots, V_k'$ to be independent and nonnegative as well. Observe that if $\|V_i\|_{L_q} \leq L\|V_i'\|_{L_q}$ for every $1 \leq q \leq p$ and $1 \leq i \leq N$, then

$$\Big\| \sum_{i=1}^{k} V_i \Big\|_{L_p} \leq L \Big\| \sum_{i=1}^{k} V_i' \Big\|_{L_p}. \tag{7}$$

Indeed, consider all the integer-valued vectors $\vec{\alpha} = (\alpha_1, \ldots, \alpha_k)$, where $\alpha_i \geq 0$ and $\sum_{i=1}^{k} \alpha_i = p$. There are constants $c_{\vec{\alpha}}$ for which

$$\Big\| \sum_{i=1}^{k} V_i \Big\|_{L_p}^{p} = \mathbb{E} \Big(\sum_{i=1}^{k} V_i \Big)^p = \mathbb{E} \sum_{\vec{\alpha}} c_{\vec{\alpha}} \prod_{i=1}^{k} V_i^{\alpha_i} = \sum_{\vec{\alpha}} c_{\vec{\alpha}} \prod_{i=1}^{k} \mathbb{E} V_i^{\alpha_i},$$

and an identical type of estimate holds for (V_i'). Equation (7) follows if

$$\prod_{i=1}^{k} \mathbb{E} V_i^{\alpha_i} \leq L^p \prod_{i=1}^{k} \mathbb{E}(V_i')^{\alpha_i},$$

and the latter may be verified because $\|V_i\|_{L_q} \leq L\|V_i'\|_{L_q}$ for every $1 \leq q \leq p$.

Let $G = (g_i)_{i=1}^{k}$ be a vector whose coordinates are independent standard gaussian random variables. If $V_i = X_i^2$ and $V_i' = c^2 L^2 g_i^2$, then by (7), for every $1 \leq q \leq p$,

$$\|\sum_{i=1}^{k} X_i^2\|_{L_q} \leq c^2 L^2 \|\sum_{i=1}^{k} g_i^2\|_{L_q} = c^2 L^2 \left(\mathbb{E}\|G\|_{\ell_2^k}^{2q}\right)^{1/q}.$$

It is standard to verify that

$$\mathbb{E}\|G\|_{\ell_2^k}^{2q} \leq c^{2q}(\sqrt{k} + \sqrt{q})^{2q},$$

and therefore,

$$\|\sum_{i=1}^{k} X_i^2\|_{L_q} \lesssim L^2 \max\{k, q\}.$$

By a binomial estimate,

$$Pr\left(\sum_{i \leq k}(X_i^*)^2 \geq t^2\right) \leq \binom{N}{k} Pr\left(\sum_{i \leq k} X_i^2 \geq t^2\right)$$

$$\leq \binom{N}{k} t^{-2q} \|\sum_{i \leq k} X_i^2\|_{L_q}^{q} \lesssim \left(\frac{eN}{k}\right)^{k} t^{-2q} \cdot L^{2q}(\max\{k, q\})^q,$$

and if $q \geq k \log(eN/k)$ and $t = euL\sqrt{q}$ for $u \geq 1$ then

$$Pr\left(\left(\sum_{i \leq k}(X_i^*)^2\right)^{1/2} \geq euL\sqrt{q}\right) \leq u^{-2q}. \tag{8}$$

Hence, setting $q = k \log(eN/k)$, tail integration implies that

$$\|(\sum_{i \leq k}(X_i^*)^2)^{1/2}\|_{L_q} \lesssim L\sqrt{k \log(eN/k)},$$

and if $q \geq k \log(eN/k)$, one has

$$\|(\sum_{i \leq k}(X_i^*)^2)^{1/2}\|_{L_q} \lesssim L\sqrt{q},$$

as claimed.   ∎

The second preliminary result we require also follows from a straightforward binomial estimate:

Lemma 2.4 *Assume that* $\|X\|_{(p)} \leq L$ *and let* $X_1, \ldots, X_N$ *be independent copies of X. Consider* $s \geq 1$, $1 \leq q \leq p$ *and* $1 \leq k \leq N$ *that satisfies that* $k \log(eN/k) \geq q$. *Then*

$$\|(\sum_{i>k}(X_i^*)^s)^{1/s}\|_{L_q} \leq c(s)LN^{1/s},$$

for a constant $c(s)$ *that depends only on s.*

Proof Clearly, for every $1 \leq i \leq N$ and $2 \leq r \leq p$,

$$Pr\left(X_i^* \geq t\right) \leq \binom{N}{i} Pr^i\left(X \geq t\right) \leq \binom{N}{i}\left(\frac{\|X\|_{L_r}^r}{t^r}\right)^i \leq \left(\frac{eN}{i} \cdot \frac{L^r r^{r/2}}{t^r}\right)^i.$$

Hence, if $t = L\sqrt{r} \cdot eu$ for $u \geq 4$ and $r = 3\log(eN/i)$, then

$$Pr\left(X_i^* \geq u \cdot eL\sqrt{3\log(eN/i)}\right) \leq u^{-3i\log(eN/i)}. \tag{9}$$

Applying the union bound for every $i \geq k$, it follows that with probability at least $1 - (u/2)^{-3k\log(eN/k)}$,

$$X_i^* \leq u \cdot eL\sqrt{3\log(eN/i)}, \quad \text{for every } k \leq i \leq N. \tag{10}$$

On that event

$$\left(\sum_{i \geq k}(X_i^*)^s\right)^{1/s} \leq c(s)uLN^{1/s},$$

and since $k\log(eN/k) \geq q$, tail integration shows that

$$\|(\sum_{i \geq k}(X_i^*)^s)^{1/s}\|_{L_q} \leq c_1(s)LN^{1/s}.$$

                                                                                            ∎

Proof of Lemma 2.2 Recall that $q_0 = 2 + 4\eta$, that $\xi \in L_{q_0}$ and that

$$W = \left| \frac{1}{\sqrt{N}} \sum_{i=1}^{N} \varepsilon_i \xi_i X_i \right| \mathbb{1}_{\mathcal{A}_u}.$$

Note that for every $(a_i)_{i=1}^{N} \in \mathbb{R}^N$ and any integer $0 \le k \le N$,

$$\left\| \sum_{i=1}^{N} \varepsilon_i a_i \right\|_{L_q} \lesssim \sum_{i \le k} a_i^* + \sqrt{q} \Big(\sum_{i>k} (a_i^*)^2 \Big)^{1/2} \tag{11}$$

where the two extreme cases of $k = 0$ and $k = N$ mean that one of the terms in (11) is 0.

Set $r = 1 + \eta$ and put $\theta = 1/q_0$. Since $(\varepsilon_i)_{i=1}^{N}$ are independent of $(X_i, \xi)_{i=1}^{N}$ and using the definition of the event $\mathcal{A}_u$,

$$N^{q/2} \mathbb{E} W^q = N^{q/2} \mathbb{E}(\mathbb{1}_{\mathcal{A}_u} \mathbb{E}_\varepsilon W^q) \le c^q \mathbb{E} \mathbb{1}_{\mathcal{A}_u} \Big(\Big(\sum_{i \le k} \xi_i^* X_i^* \Big)^q + q^{q/2} \Big(\sum_{i>k} (\xi_i^*)^2 (X_i^*)^2 \Big)^{q/2} \Big)$$

$$\le c^q u^q \| \xi \|_{L_{q_0}} \cdot \mathbb{E}_X \Big(\Big(\sum_{i \le k} (N/i)^\theta X_i^* \Big)^q + q^{q/2} \Big(\sum_{i>k} (N/i)^{2\theta} (X_i^*)^2 \Big)^{q/2} \Big).$$

By the Cauchy-Schwarz inequality,

$$\Big(\sum_{i \le k} (N/i)^\theta X_i^* \Big)^q \le \Big(\sum_{i \le k} (N/i)^{2\theta} \Big)^{q/2} \cdot \Big(\sum_{i \le k} (X_i^*)^2 \Big)^{q/2},$$

and

$$\sum_{i \le k} (N/i)^{2\theta} = \sum_{i \le k} (N/i)^{1/1+2\eta} \le \frac{c_1}{\eta} N^{1/(1+2\eta)} k^{2\eta/(1+2\eta)} \le \frac{c_1}{\eta} N.$$

Therefore,

$$\mathbb{E} \Big(\sum_{i \le k} (N/i)^\theta X_i^* \Big)^q \lesssim \eta^{-q/2} N^{q/2} \mathbb{E} \Big(\sum_{i \le k} (X_i^*)^2 \Big)^{q/2} = (*).$$

Also, by Hölder's inequality for $r = 1 + \eta$ and its conjugate index r',

$$\Big(\sum_{i>k} (N/i)^{2\theta} (X_i^*)^2 \Big)^{q/2} \le \Big(\sum_{i \ge k} (N/i)^{2\theta r} \Big)^{q/2r} \cdot \Big(\sum_{i \ge k} (X_i^*)^{2r'} \Big)^{q/2r'}$$

and

$$\sum_{i \ge k} (N/i)^{2\theta r} = \sum_{i \ge k} (N/i)^{(1+\eta)/(1+2\eta)} \le \frac{c_1}{\eta} N.$$

Hence,

$$\mathbb{E}\Big(\sum_{i>k}(N/i)^{2\theta}(X_i^*)^2\Big)^{q/2} \lesssim \eta^{-q/2r}N^{q/2r}\mathbb{E}\Big(\sum_{i>k}(X_i^*)^{2r'}\Big)^{q/2r'} = (**).$$

Let $k \in \{1, \ldots, N\}$ be the smallest that satisfies $k\log(eN/k) \geq q$ (and without loss of generality we will assume that such a k exists; if it does not, the modifications to the proof are straightforward and are omitted).

Applying Lemma 2.3 for that choice of k,

$$(*) \leq c^q \eta^{-q/2}N^{q/2} \cdot L^q(\sqrt{k\log(eN/k)} + \sqrt{q})^q \leq c_1^q \eta^{-q/2}L^q N^{q/2}q^{q/2}.$$

Turning to $(**)$, set $s = 2r' \sim \max\{\eta^{-1}, 2\}$ and one has to control

$$\mathbb{E}\Big(\sum_{i>k}(X_i^*)^s\Big)^{q/s}$$

for the choice of k as above. By Lemma 2.4,

$$\mathbb{E}\Big(\sum_{i>k}(X_i^*)^s\Big)^{q/s} \leq c^q(s)L^q N^{q/s} = c_1^q(\eta)L^q N^{q/2r'}.$$

Therefore,

$$(**) \leq c^q(\eta)L^q N^{q/2r} \cdot N^{q/2r'} = c^q(\eta)L^q N^{q/2}.$$

Combining the two estimates,

$$N^{q/2}\mathbb{E}W^q \leq N^{q/2}u\|\xi\|_{L_{q_0}} \cdot c^q(\eta)L^q q^{q/2},$$

implying that $\|W\|_{L_q}/\sqrt{q} \leq c(\eta)uL\|\xi\|_{L_{q_0}}$.   ∎

Proof of Theorem 1.6 By Lemma 2.2, for every $1 \leq j \leq n$, $\|W_j\|_{(p)} \leq c(\eta)L\|\xi\|_{L_{q_0}}$, and thus, by Lemma 2.1, with probability at least $1 - c_1 t^{-2\beta}$,

$$W_j^* \leq c(\eta)tL\|\xi\|_{L_{q_0}}\sqrt{\beta\log(en/j)} \quad \text{for every} \quad 1 \leq j \leq n.$$

Moreover, $Pr(\mathcal{A}_u) \geq 1 - 2/u^{q_0}$; therefore, with probability at least $1 - c_1 t^{-2\beta} - 2u^{-q_0}$, for every $1 \leq j \leq n$,

$$Z_j^* \leq c(\eta)tuL\|\xi\|_{L_{q_0}}\sqrt{\beta\log(eN/j)}.$$

Hence, on that event and because the norm $\sup_{v \in V}|\langle v, \cdot\rangle|$ is K unconditional,

$$\sup_{v \in V}|\langle Z, v\rangle| \leq Kc(\eta)\sqrt{\beta}tuL\|\xi\|_{L_{q_0}}\sup_{v \in V}|\langle Z_0, v\rangle|,$$

for a fixed vector Z_0 whose coordinates are $(\sqrt{\log(en/j)})_{j=1}^n$. Observe that $|\langle Z_0, e_j \rangle| \lesssim \mathbb{E} g_j^*$, and thus

$$\sup_{v \in V} |\langle Z_0, v \rangle| \leq K \sup_{v \in V} |\sum_{i=1}^n v_i \mathbb{E} g_i^*|.$$

Therefore, by Jensen's inequality, with probability at least $1 - t^{-2\beta} - 2u^{-q_0}$,

$$\sup_{v \in V} |\langle Z, v \rangle| \leq c(\eta, K) \sqrt{\beta} tuL \|\xi\|_{L_{q_0}} \mathbb{E} \sup_{v \in V} |\langle G, v \rangle|.$$

And, fixing β and integrating the tails,

$$\mathbb{E} \sup_{v \in V} |\langle Z, v \rangle| \leq c(K, \eta, L) \|\xi\|_{L_{q_0}} \ell_*(V),$$

as claimed.   ∎

3 Applications in Sparse Recovery

Spare recovery is a central topic in modern Statistics and Signal Processing, though our outline of the sparse recovery problem is far from its most general form. Because a detailed description of the subtleties of sparse recovery would be unreasonably lengthy, some statements may appear a little vague; for more information on the topic we refer the reader to the books [3–5].

The question in sparse recovery is to identify, or at least approximate, an unknown vector $v_0 \in \mathbb{R}^n$, and to do so using relatively few linear measurements. The measurements one is given are 'noisy', of the form

$$Y_i = \langle v_0, X_i \rangle - \xi_i \text{ for } 1 \leq i \leq N;$$

$X_1, \ldots, X_N$ are independent copies of a random, isotropic vector $X \in \mathbb{R}^n$ and $\xi_1, \ldots, \xi_N$ are independent copies of a random variable ξ that belongs to L_q for some $q > 2$.

The reason for the name "sparse recovery" is the underlying assumption that v_0 is sparse: it is supported on at most s coordinates, though the identity of the support itself is not known. Thus, one would like to use the given random data $(X_i, Y_i)_{i=1}^N$ and select $\hat{v}$ in a wise way, leading to a high probability estimate on the *error rate* $\|\hat{v} - v_0\|_{\ell_2^n}$ as a function of the number of measurements N and of the 'degree of sparsity' s.

In the simplest recovery problem, $\xi = 0$ and the data is noise-free. Alternatively, one may assume that the ξ_i's are independent of $X_1, \ldots, X_N$, or, in a more general formulation, very little is assumed on the ξ_i's.

The standard method of producing $\hat{v}$ in a noise-free problem and when v_0 is assumed to be sparse is the *basis pursuit algorithm*. The algorithm produces $\hat{v}$ which is the point in $\mathbb{R}^n$ with the smallest ℓ_1^n norm that satisfies $\langle X_i, v_0 \rangle = \langle X_i, v \rangle$ for every $1 \leq i \leq N$.

It is well known [12] that if X is isotropic and L-subgaussian, v_0 is supported on at most s coordinates, and one is given

$$N = c(L) s \log \left(\frac{en}{s} \right) \tag{12}$$

random measurements $(\langle X_i, v_0 \rangle)_{i=1}^N$, then with high probability, the basis pursuit algorithm has a unique solution, and that solution is v_0.

Recently, it has been observed in [6] that the subgaussian assumption can be relaxed: the same number of measurements as in (12) suffices for v_0 to be the unique solution of Basis Pursuit if

$$\max_{1 \leq j \leq n} \|\langle X, e_j \rangle\|_{(p)} \leq L \text{ for } p \sim \log n.$$

Moreover, the estimate of $p \sim \log n$ happens to be almost optimal: there is an example of an isotropic vector X with iid coordinates for which

$$\max_{1 \leq j \leq n} \|\langle X, e_j \rangle\|_{(p)} \leq L \text{ for } p \sim (\log n)/(\log \log n) \tag{13}$$

but still, with probability $1/2$, Basis Pursuit does not recover even a 1-sparse vector when given the same number of random measurements as in (12).

Since 'real world' data is not noise-free, some effort has been invested in producing analogs of the basis pursuit algorithm in a 'noisy' setup. The most well known among these procedures is the LASSO (see, e.g. the books [3, 5] for more details) in which $\hat{v}$ is selected to be the minimizer in $\mathbb{R}^n$ of the functional

$$v \to \frac{1}{N} \sum_{i=1}^N (\langle v, X_i \rangle - Y_i)^2 + \lambda \|v\|_{\ell_1^n}, \tag{14}$$

for a well-chosen of λ.

Following the introduction of the LASSO, there have been many variations on the same theme—by changing the penalty $\| \ \|_{\ell_1^n}$ and replacing it with other norms. Until very recently, the behaviour of most of these procedures has been studied under rather strong assumptions on X and ξ—usually, that X and ξ are independent and gaussian, or at best, subgaussian.

One may show that Theorem 1.6 can be used to extend the estimates on $\|\hat{v} - v_0\|_{\ell_2^n}$ beyond the gaussian case thanks to two significant facts:

- The norms used in the LASSO and in many of its modifications happen to be well behaved under permutations and sign changes: for example, among these norms are weighted ℓ_1^n norms and mixtures of the ℓ_1^n and the ℓ_2^n norms.
- As noted in [7], if Ψ is a norm, B_Ψ is its unit ball and $\hat{v}$ is the minimizer in $\mathbb{R}^n$ of the functional

$$v \to \frac{1}{N} \sum_{i=1}^{N} (\langle v, X_i \rangle - Y_i)^2 + \lambda \Psi(v), \tag{15}$$

then the key to controlling $\|\hat{v} - v\|_{\ell_2^n}$ is the behaviour of

$$\sup_{v \in B_\Psi \cap rB_2^n} \left| \frac{1}{\sqrt{N}} \sum_{i=1}^{N} \xi_i \langle X_i, v \rangle - \mathbb{E}\xi \langle X, v \rangle \right|, \tag{16}$$

which is precisely the type of question that Theorem 1.6 deals with.

It follows from Theorem 1.6 that if $\xi \in L_q$ for some $q > 2$, and linear forms have $\sim \log n$ subgaussian moments, then the expectation of (16) is, up to a multiplicative constant, the same as if ξ and X were independent and gaussian. Thus, under those conditions, one can expect the 'gaussian' error estimate in procedures like (15). Moreover, because of (13), the condition that linear forms exhibit a subgaussian growth of moments up to $p \sim \log n$ is necessary, making the outcome of Theorem 1.6 optimal in this context.

The following is a simplified version of an application of Theorem 1.6. We refer the reader to [7] for its general formulation, as well as for other examples of a similar nature.

Let X be an isotropic measure on $\mathbb{R}^n$ that satisfies $\max_{1 \le j \le n} \|\langle X, e_j \rangle\|_{(p)} \le L$ for $p \le c_0 \log n$. Set $\xi \in L_q$ for $q > 2$ that is mean-zero and independent of X and put $Y = \langle X, v_0 \rangle - \xi$.

Given an independent sample $(X_i, Y_i)_{i=1}^{N}$ selected according to (X, Y), let $\hat{v}$ be the minimizer of the functional (14).

Theorem 3.1 *Assume that v_0 is supported on at most s coordinates and let $0 < \delta < 1$. If $\lambda = c_1(L, \delta)\|\xi\|_{L_q}\sqrt{\log(en)/N}$, then with probability at least $1 - \delta$, for every $1 \le p \le 2$*

$$\|\hat{v} - v_0\|_p \le c_2(L, \delta)\|\xi\|_{L_q} s^{1/p} \sqrt{\frac{\log(ed)}{N}}.$$

The proof of Theorem 3.1 follows by combining Theorem 3.2 from [7] with Theorem 1.6.

Acknowledgements S. Mendelson is supported in part by the Israel Science Foundation.

References

1. F. Albiac, N.J. Kalton, *Topics in Banach Space Theory*. Graduate Texts in Mathematics, vol. 233 (Springer, New York, 2006)
2. S. Artstein-Avidan, A. Giannopoulos, V.D. Milman, *Asymptotic Geometric Analysis. Part I*. Mathematical Surveys and Monographs, vol. 202 (American Mathematical Society, Providence, RI, 2015)
3. P. Bühlmann, S. van de Geer, *Statistics for High-Dimensional Data. Methods, Theory and Applications*. Springer Series in Statistics (Springer, Heidelberg, 2011)
4. S. Foucart, H. Rauhut, *A Mathematical Introduction to Compressive Sensing*. Applied and Numerical Harmonic Analysis (Birkhäuser/Springer, New York, 2013)
5. V. Koltchinskii, *Oracle Inequalities in Empirical Risk Minimization and Sparse Recovery Problems*. Lecture Notes in Mathematics, vol. 2033 (Springer, Heidelberg, 2011). Lectures from the 38th Probability Summer School held in Saint-Flour, 2008, École d'Été de Probabilités de Saint-Flour. [Saint-Flour Probability Summer School]
6. G. Lecué, S. Mendelson, Sparse recovery under weak moment assumptions. Technical report, CNRS, Ecole Polytechnique and Technion (2014). J. Eur. Math. Soc. **19**(3), 881–904 (2017)
7. G. Lecué, S. Mendelson, Regularization and the small-ball method I: sparse recovery. Technical report, CNRS, ENSAE and Technion, I.I.T. (2015). Ann. Stati. (to appear)
8. M. Ledoux, M. Talagrand, *Probability in Banach Spaces*. Isoperimetry and processes. Ergebnisse der Mathematik und ihrer Grenzgebiete (3) [Results in Mathematics and Related Areas (3)], vol. 23 (Springer, Berlin, 1991)
9. S. Mendelson, Learning without concentration for general loss function. Technical report, Technion, I.I.T. (2013). arXiv:1410.3192
10. S. Mendelson, A remark on the diameter of random sections of convex bodies, in *Geometric Aspects of Functional Analysis*, Lecture Notes in Mathematics, vol. 2116, pp. 395–404 (Springer, Cham, 2014)
11. S. Mendelson, Upper bounds on product and multiplier empirical processes. Stoch. Process. Appl. **126**(12), 3652–3680 (2016)
12. S. Mendelson, A. Pajor, N. Tomczak-Jaegermann, Reconstruction and subgaussian operators in asymptotic geometric analysis. Geom. Funct. Anal. **17**(4), 1248–1282 (2007)
13. S. Mendelson, G. Paouris, On generic chaining and the smallest singular value of random matrices with heavy tails. J. Funct. Anal. **262**(9), 3775–3811 (2012)
14. V.D. Milman, Random subspaces of proportional dimension of finite-dimensional normed spaces: approach through the isoperimetric inequality, in *Banach Spaces (Columbia, MO, 1984)*. Lecture Notes in Mathematics, vol. 1166, pp. 106–115 (Springer, Berlin, 1985)
15. A. Pajor, N. Tomczak-Jaegermann, Nombres de Gel′fand et sections euclidiennes de grande dimension, in *Séminaire d'Analyse Fonctionelle 1984/1985*. Publ. Math. Univ. Paris VII, vol. 26, pp. 37–47 (University of Paris VII, Paris, 1986)
16. A. Pajor, N. Tomczak-Jaegermann, Subspaces of small codimension of finite-dimensional Banach spaces. Proc. Am. Math. Soc. **97**(4), 637–642 (1986)
17. G. Pisier, *The Volume of Convex Bodies and Banach Space Geometry*. Cambridge Tracts in Mathematics, vol. 94 (Cambridge University Press, Cambridge, 1989)
18. M. Talagrand, Regularity of Gaussian processes. Acta Math. **159**(1–2), 99–149 (1987)
19. M. Talagrand, *Upper and Lower Bounds for Stochastic Processes*. Modern Methods and Classical Problems. Ergebnisse der Mathematik und ihrer Grenzgebiete. 3. Folge. A Series of Modern Surveys in Mathematics [Results in Mathematics and Related Areas. 3rd Series. A Series of Modern Surveys in Mathematics], vol. 60. (Springer, Heidelberg, 2014)
20. A.W. van der Vaart, J.A. Wellner, *Weak Convergence and Empirical Processes. With Applications to Statistics*. Springer Series in Statistics (Springer, New York, 1996)

Characterizing the Radial Sum for Star Bodies

Vitali Milman and Liran Rotem

Abstract In this paper we prove two theorems characterizing the radial sum of star bodies. By doing so we demonstrate an interesting phenomenon: essentially the same conditions, on two different spaces, can uniquely characterize very different operations. In our first theorem we characterize the radial sum by its induced homothety, and our list of assumptions is identical to the assumptions of the corresponding theorem which characterizes the Minkowski sum for convex bodies. In our second theorem give a different characterization from a short list of natural properties, without assuming the homothety has any specific form. For this theorem one has to add an assumption to the corresponding theorem for convex bodies, as we demonstrate by a simple example.

1 Introduction

The main goal of this paper is to characterize addition operations on star-shaped sets. Before doing so, however, we will quickly discuss addition of convex sets. To fix some notation, let $\mathcal{K}_0^n$ denote the class of closed convex sets containing the origin.

Definition 1 An *addition operation* on convex sets is a map $\oplus : \mathcal{K}_0^n \times \mathcal{K}_0^n \to \mathcal{K}_0^n$ such that:

1. $\oplus$ is associative: For every $A, B, C \in \mathcal{K}_0^n$ one has $(A \oplus B) \oplus C = A \oplus (B \oplus C)$.
2. $\oplus$ has an identity element: There exists $K \in \mathcal{K}_0^n$ such that $A \oplus K = K \oplus A = A$ for all $A \in \mathcal{K}_0^n$.

We will now describe two natural families of addition operations. Remember that for $A \in \mathcal{K}_0^n$ the support function of A is the convex function $h_A : \mathbb{R}^n \to [0, \infty]$

V. Milman (✉)
School of Mathematical Sciences, Tel Aviv University, Tel Aviv 69978, Israel
e-mail: milman@post.tau.ac.il

L. Rotem
School of Mathematics, University of Minnesota, 206 Church St. SE, Minneapolis, MN 55455, USA
e-mail: lrotem@umn.edu

© Springer International Publishing AG 2017
B. Klartag, E. Milman (eds.), *Geometric Aspects of Functional Analysis*,
Lecture Notes in Mathematics 2169, DOI 10.1007/978-3-319-45282-1_20

defined by $h_A(y) = \sup_{x \in A} \langle x, y \rangle$. Fixing a parameter $p \in [1, \infty)$, the p-addition $A +_p B$ of A and B is implicitly defined by the relation

$$h^p_{A+_pB}(y) = h^p_A(y) + h^p_B(y)$$

for all $y \in \mathbb{R}^n$. For $p = 1$, the 1-addition $A +_1 B$ is just the closure of the classical Minkowski addition,

$$A + B = \{a + b : a \in A, \, b \in B\}$$

(the closure is superfluous if A or B are compact, but may be necessary otherwise). For $p > 1$, p-additions were originally defined by Firey [2], and were first systematically studied by Lutwak [5, 6]. For $p = \infty$ we set $A +_\infty B = \mathrm{conv}\,(A \cup B)$, where conv denotes the convex hull. Notice that $h_{A+_\infty B}(y) = \max\{h_A(y), h_B(y)\}$. It is easy to check that all p-additions are addition operations in the sense of Definition 1, with $\{0\}$ as an identity element.

Using p-additions, we may construct a second family of addition operations. For $A \in \mathcal{K}^n_0$, the polar body A° is defined by

$$A^\circ = \{y \in \mathbb{R}^n : h_A(y) \le 1\}.$$

For $p \in [1, \infty]$ we may now define the p-polar addition by $A +_{-p} B = \left(A^\circ +_p B^\circ\right)^\circ$. All p-polar additions are addition operations in the sense of Definition 1, with $\mathbb{R}^n$ as an identity element. Notice that $A +_{-\infty} B = A \cap B$.

Given an addition operation $\oplus$, we define the induced homothety $\odot : \mathbb{N} \times \mathcal{K}^n_0 \to \mathcal{K}^n_0$ by

$$m \odot A = \underbrace{A \oplus A \oplus \cdots \oplus A}_{m \text{ times}}.$$

For $p \in [-\infty, -1] \cup [1, \infty]$, The induced homothety of $+_p$ is easily seen to be

$$m \cdot_p A = m^{1/p} A = \left\{m^{1/p} a : a \in A\right\}.$$

This formula is one of the reasons for the notation $+_{-p}$ for the p-polar addition.

Let us list a few properties we expect an addition to have. The p-additions and p-polar additions all satisfy these properties:

Definition 2 We say that an addition $\oplus : \mathcal{K}^n_0 \times \mathcal{K}^n_0 \to \mathcal{K}^n_0$ is:

1. *Monotone* if $A_1 \subseteq B_1$ and $A_2 \subseteq B_2$ implies $A_1 \oplus A_2 \subseteq B_1 \oplus B_2$.
2. *Strongly monotone* if it is monotone, and in addition $m \odot A \subseteq m \odot B$ implies $A \subseteq B$.
3. *Divisible* if for every $A \in \mathcal{K}^n_0$ and $m \in \mathbb{N}$ there exists $B \in \mathcal{K}^n_0$ such that $m \odot B = A$.
4. *Subspace preserving* if for every linear subspace $V \subseteq \mathbb{R}^n$ we have $A, B \subseteq V$ implies $A \oplus B \subseteq V$.

In the paper [7], we proved several characterization theorems for the p-addition. The first theorem shows that under mild hypotheses, $\oplus$ is uniquely determined by its homothety operation $\odot$:

Theorem 1 *Let* $\oplus : \mathcal{K}_0^n \times \mathcal{K}_0^n \to \mathcal{K}_0^n$ *be a monotone addition operation. Assume that there exists a function* $f : \mathbb{N} \to (0, \infty)$ *such that* $m \odot A = f(m)A$ *for all* $A \in \mathcal{K}_0^n$ *and* $m \in \mathbb{N}$.

1. *If* f *is not the constant function* 1, *then there exists* $p \neq 0$ *such that* $A \oplus B = A +_p B$ *for every* $A, B \in \mathcal{K}_0^n$. *If* $n \geq 2$ *then* $1 \leq |p| < \infty$.
2. *If* $f \equiv 1$ *and the identity element of* $\oplus$ *is* $\{0\}$, *then* $A \oplus B = A +_\infty B$ *for every* $A, B \in \mathcal{K}_0^n$. *Similarly, if the identity element is* $\mathbb{R}^n$ *then* $A \oplus B = A +_{-\infty} B$.

We also proved the following theorem, characterizing the p-addition without assuming the homothety has any specific form:

Theorem 2 *Assume* $n \geq 2$. *Let* $\oplus : \mathcal{K}_0^n \times \mathcal{K}_0^n \to \mathcal{K}_0^n$ *be an addition operation with* $\{0\}$ *as identity element. Assume that* $\oplus$ *is strongly monotone, divisible and subspace preserving. Then there exists a* $p \geq 1$ *such that* $A \oplus B = A +_p B$ *for all* $A, B \in \mathcal{K}_0^n$.

The main observation of this note is that essentially the same conditions as in Theorems 1 and 2, but on a different domain, can be used to characterize an entirely different operation. Let us denote by $\mathcal{S}_0^n$ the class of closed star bodies in $\mathbb{R}^n$. By a star-shaped set, or a star body, we mean any nonempty set A such that $x \in A$ implies that $\lambda x \in A$ for all $0 \leq \lambda \leq 1$. Every star body A is uniquely characterized by its radial function

$$r_A(x) = \sup\{\lambda \geq 0 : \lambda x \in A\}.$$

The definition of an addition operation $\oplus$ on $\mathcal{S}_0^n$ is the obvious analogue of Definition 1. We also define the induced homothety, and the various properties $\oplus$ may satisfy (monotonicity, divisibility, etc.) in the obvious way.

Given any $p \neq 0$, the p-radial sum $A \tilde{+}_p B$ is defined by the relation

$$r_{A \tilde{+}_p B}^p(x) = r_A^p(x) + r_B^p(x)$$

for all $x \in \mathbb{R}^n$. For $p = \infty$ we set $A \tilde{+}_\infty B = A \cup B$, and for $p = -\infty$ we set $A \tilde{+}_{-\infty} B = A \cap B$.

In [3], Gardner, Hug and Weil proved a characterization theorem for p-radial sums of star-shaped bodies, as well as characterization theorems for p-sums of convex bodies. However, in their work a different set of properties was needed in each case. For convex bodies, the main property assumed was projection covariance, i.e.

$$\text{Proj}_V(A \oplus B) = \text{Proj}_V(A) \oplus \text{Proj}_V(B)$$

for all subspaces V. This property holds for p-additions of convex sets, but not for p-polar additions or for p-radial additions of star shaped sets. Hence for p-radial additions a different property was needed, which is section covariance:

$$(A \oplus V) \cap E = (A \cap V) \oplus (B \cap V)$$

for all subspaces V.

In our case, we have the following perfect analogue of Theorem 1:

Theorem 3 *Let* $\oplus : S_0^n \times S_0^n \to S_0^n$ *be a monotone addition operation. Assume that there exists a function* $f : \mathbb{N} \to (0, \infty)$ *such that* $m \odot A = f(m)A$ *for all* $A \in S_0^n$ *and* $m \in \mathbb{N}$.

1. *If* f *is not the constant function* 1, *then there exists* $p \neq 0$ *such that* $A \oplus B = A \tilde{+}_p B$ *for every* $A, B \in S_0^n$.
2. *If* $f \equiv 1$ *and the identity element of* $\oplus$ *is* $\{0\}$, *then* $A \oplus B = A \tilde{+}_\infty B$ *for every* $A, B \in S_0^n$. *Similarly, if the identity element is* $\mathbb{R}^n$ *then* $A \oplus B = A \tilde{+}_{-\infty} B$.

For Theorem 2, the situation is slightly more complicated, as the conditions in this theorem do not suffice to characterize the p-radial sum. Intuitively, the reason for this is that there is no condition "relating the different directions". Hence we may fix our favorite function $p : S^{n-1} \to (0, \infty)$, where S^{n-1} denotes the unit sphere in $\mathbb{R}^n$, and define an addition $\oplus : S_0^n \times S_0^n \to S_0^n$ by the relation

$$r_{A \oplus B}(\theta) = \left(r_A(\theta)^{p(\theta)} + r_B(\theta)^{p(\theta)} \right)^{1/p(\theta)}$$

for all $\theta \in S^{n-1}$ (this example appears already in [3]). It is easy to check that $\oplus$ satisfies all properties of Theorem 2, without being a p-radial sum.

Therefore, in order to relate the different directions, we add the following assumption: For every convex set $A \in K_0^n \subseteq S_0^n$, and for every $m \in \mathbb{N}$, the set $m \odot A$ is also convex. We therefore have

Theorem 4 *Assume* $n \geq 2$. *Let* $\oplus : S_0^n \times S_0^n \to S_0^n$ *be an addition operation with* $\{0\}$ *as identity element. Assume that* $\oplus$ *is strongly monotone, divisible, subspace preserving, and that* $m \odot A$ *is convex for every convex* A *and every* $m \in \mathbb{N}$. *Then there exists a* $p > 0$ *such that* $A \oplus B = A \tilde{+}_p B$ *for all* $A, B \in S_0^n$.

In fact, we will see in the proof that it is enough to assume that $m \odot H$ is convex for any half-spaces H, and not for arbitrary convex bodies.

In the next section we will prove a main lemma, crucial for the proof of both theorems. In Sect. 3 we will prove both Theorems 3 and 4 using the lemma. Finally, in Sect. 4 we will briefly discuss polynomiality of volume with respect our additions. The proofs of the main theorems are similar, and in some sense "dual", to the proofs of [7]. For the reader's convenience we give self-contained proofs for all new theorems.

2 The Main Lemma

In this section we prove a main lemma that will be used in the proof of both Theorems 3 and 4. For $\theta \in S^{n-1}$ and $c \in [0, \infty]$, let us denote by $R_{\theta,c} \in S_0^n$ the set with radial function

$$r_{R_{\theta,c}}(\eta) = \begin{cases} c & \eta = \theta \\ \infty & \text{otherwise.} \end{cases}$$

Notice that $R_{\theta,c}$ is just a complement of a ray, and that $R_{\theta,\infty} = \mathbb{R}^n$. We will also write $R_\theta = R_{\theta,0}$.

Our main lemma then reads:

Lemma 1 *Let $\oplus : S_0^n \times S_0^n \to S_0^n$ be a monotone addition operation with identity element $\{0\}$. Assume that there exists a function $f : \mathbb{N} \to (0, \infty)$ such that $m \odot R_{\theta,c} = R_{\theta, f(m)c}$ for all $\theta \in S^{n-1}$, $c \in (0, \infty)$ and $m \in \mathbb{N}$. Then $\oplus = \tilde{+}_p$ for some $0 < p < \infty$.*

We will now prove the main lemma, by a sequence of claims.

Claim There exists $0 < q < \infty$ such that $f(m) = m^q$.

Proof First, we prove that f is monotone increasing. Write $S = R_{\theta,1}$ for some fixed $\theta \in S^{n-1}$. Notice that for any m we have

$$f(m+1)S = (m+1) \odot S = (m \odot S) \oplus S \supseteq (m \odot S) \oplus \{0\} = m \odot S = f(m)S.$$

By comparing radial functions in the direction θ it follows that indeed $f(m+1) \geq f(m)$.

Next, we prove that f is multiplicative: For all integers m and k we have

$$f(mk)S = (mk) \odot S = m \odot (k \odot S) = m \odot (f(k)S) = f(m)f(k)S,$$

so again by comparing radial functions $f(mk) = f(m)f(k)$.

However, it is known that every increasing and multiplicative function must be of the form $f(m) = m^q$, so we are done. See [4] for a simple proof of this fact (in fact, a much more general theorem is true and is due to Erdős—see [1]).

From now on we will write $p = \frac{1}{q}$, and prove that $\oplus = \tilde{+}_p$. For brevity we write

$$M_p(a, b) = (a^p + b^p)^{\frac{1}{p}}$$

for every $0 < p < \infty$ and $0 \leq a, b \leq \infty$.

Claim For every $\theta \in S^{n-1}$ and every $0 \leq c, d \leq \infty$ we have $R_{\theta,c} \oplus R_{\theta,d} = R_{\theta, M_p(c,d)}$

Proof First assume that $c^p = \frac{m}{k}$ and $d^p = \frac{s}{t}$ are positive rationals, then

$$R_{\theta,c} \oplus R_{\theta,d} = \left(\frac{m^q}{k^q}R_{\theta,1}\right) \oplus \left(\frac{s^q}{t^q}R_{\theta,1}\right)$$

$$= \left[(mt) \odot \left(\frac{1}{k^q t^q}R_{\theta,1}\right)\right] \oplus \left[(sk) \odot \left(\frac{1}{k^q t^q}R_{\theta,1}\right)\right]$$

$$= (mt + sk) \odot \left(\frac{1}{k^q t^q}R_{\theta,1}\right) = \left(\frac{mt + sk}{kt}\right)^q R_{\theta,1}$$

$$= \left(\frac{m}{k} + \frac{s}{t}\right)^q R_{\theta,1} = (c^p + d^p)^{\frac{1}{p}} R_{\theta,1} = R_{\theta,M_p(c,d)}.$$

Since the rationals are dense in $[0, \infty]$, all the remaining cases can be proven by approximation, using the monotonicity of $\oplus$.

Claim For every $A \in \mathcal{S}_0^n$ we have $A \oplus R_\theta = R_\theta \oplus A = R_{\theta,r_A(\theta)}$.

Proof We will only prove that $A \oplus R_\theta = R_{\theta,r_A(\theta)}$, as the second equality is completely analogous.

For one inclusion, notice that $A \oplus R_\theta \supseteq A \oplus \{0\} = A$, and similarly $A \oplus R_\theta \supseteq R_\theta$. Hence

$$A \oplus R_\theta \supseteq A \cup R_\theta = R_{\theta,r_A(\theta)}.$$

For the opposite inclusion we obviously have $A \subseteq R_{\theta,r_A(\theta)}$, so by monotonicity

$$A \oplus R_\theta \subseteq R_{\theta,h_A(\theta)} \oplus R_{\theta,0} = R_{\theta,M_p(r_A(\theta),0)} = R_{\theta,r_A(\theta)}$$

Claim We have $\oplus = \tilde{+}_p$.

Proof Fix $A, B \in \mathcal{S}_0^n$ and $\theta \in S^{n-1}$. Our goal is to prove that

$$r_{A\oplus B}(\theta) = M_p(r_A(\theta), r_B(\theta)). \tag{1}$$

On the one hand, using the previous claim, we know that

$$(A \oplus B) \oplus R_\theta = R_{\theta,r_{A\oplus B}(\theta)}.$$

On the other hand, we have

$$(A \oplus B) \oplus R_\theta = (A \oplus B) \oplus (R_\theta \oplus R_\theta) = A \oplus (B \oplus R_\theta) \oplus R_\theta$$

$$= A \oplus (R_\theta \oplus B) \oplus R_\theta = (A \oplus R_\theta) \oplus (B \oplus R_\theta) = R_{\theta,r_A(\theta)} \oplus R_{\theta,r_B(\theta)}$$

$$= R_{\theta,M_p(r_A(\theta),r_B(\theta))}.$$

Comparing both expressions we obtain (1), so the proof of claim, and main lemma, is complete.

3   Proving the Main Theorems

Proof (Proof of Theorem 3) First assume that $f(2) > 1$. We claim that $\{0\}$ is the identity element with respect to $\oplus$.

Indeed, denote the identity element by S. If $S \neq \{0\}$ there exists $0 \neq a \in S$, and then by the star property $[0, a] \subseteq S$. But then we get from monotonicity that

$$[0, a] = [0, a] \oplus S \supseteq [0, a] \oplus [0, a] = f(2) [0, a] = [0, f(2) \cdot a].$$

Since $f(2) > 1$, this is obviously a contradiction. It follows that $\oplus$ satisfies all the assumptions of Lemma 1, so $\oplus = \tilde{+}_p$ for some $p > 0$.

Next, assume that $f(2) < 1$. For a star-shaped set $A \in \mathcal{S}_0^n$, define its "star polar" $A^* \in \mathcal{S}_0^n$ by the relation $r_{A^*}(\theta) = r_A(\theta)^{-1}$ for every direction $\theta \in S^{n-1}$. Define a new addition $\boxplus : \mathcal{S}_0^n \times \mathcal{S}_0^n \to \mathcal{S}_0^n$ by

$$A \boxplus B = \left(A^* \oplus B^*\right)^*.$$

Notice that $\boxplus$ is indeed an addition operation in the sense of Definition 1—if K is the identity element of $\oplus$ then K^* is the identity element of $\boxplus$. It is easy to check that $\boxplus$ is monotone. Finally, for every $A \in \mathcal{S}_0^n$ and $m \in \mathbb{N}$ we have

$$m \,\square\, A = \underbrace{A \boxplus A \boxplus \cdots \boxplus A}_{m \text{ times}} = \left(\underbrace{A^* \oplus A^* \oplus \cdots \oplus A^*}_{m \text{ times}}\right)^* = \left(f(m)A^*\right)^* = \frac{1}{f(m)} A,$$

so $\boxplus$ satisfy the homothety property with homothety function $g(m) = \frac{1}{f(m)}$. In particular $g(2) = \frac{1}{f(2)} > 1$, so by the previous case we have $\boxplus = \tilde{+}_p$ for some $p > 0$. But then for every $A, B \in \mathcal{S}_0^n$ we have

$$A \oplus B = \left(A^* \boxplus B^*\right)^* = \left(A^* \tilde{+}_p B^*\right)^* = A \tilde{+}_{-p} B,$$

so the theorem is proved in this case as well.

Finally, assume $f(2) = 1$, so $f \equiv 1$. If the identity element of $\oplus$ is $\{0\}$, then we have

$$A \oplus B \supseteq A \oplus \{0\} = A$$

$$A \oplus B \supseteq \{0\} \oplus B = B,$$

so $A \oplus B \supseteq A \cup B = A \tilde{+}_\infty B$. But the opposite inclusion is also true, since

$$A \oplus B \subseteq \left(A \tilde{+}_\infty B\right) \oplus \left(A \tilde{+}_\infty B\right) = A \tilde{+}_\infty B,$$

so we indeed have $\oplus = \tilde{+}_\infty$. If the identity element of $\oplus$ is $\mathbb{R}^n$, an almost identical argument proves that $\oplus = \tilde{+}_{-\infty}$. This finished the proof of the theorem.

Next, we want to prove Theorem 4. We assume the conditions of the theorem holds, and prove several claims reducing the theorem to the situation of Lemma 1.

Claim For every $A, B \in \mathcal{S}_0^n$ and every integer m we have $m \odot (A \cap B) = (m \odot A) \cap (m \odot B)$.

Proof One inclusion is immediate from monotonicity: $A \cap B \subseteq A$ implies $m \odot (A \cap B) \subseteq m \odot A$. Similarly $m \odot (A \cap B) \subseteq m \odot B$, so we see that indeed

$$m \odot (A \cap B) \subseteq (m \odot A) \cap (m \odot B).$$

For the second inclusion, by divisibility there exists $C \in \mathcal{S}_0^n$ such that

$$m \odot C = (m \odot A) \cap (m \odot B).$$

Since $m \odot C \subseteq m \odot A$, the strong monotonicity implies that $C \subseteq A$. Similarly $C \subseteq B$, and then $C \subseteq A \cap B$ so

$$(m \odot A) \cap (m \odot B) = m \odot C \subseteq m \odot (A \cap B).$$

This completes the proof.

For $\theta \in S^{n-1}$ and $c \in [0, \infty]$, let us write

$$H_{\theta,c} = \{x \in \mathbb{R}^n : \langle x, \theta \rangle \le c\} \in \mathcal{S}_0^n.$$

Claim For every $m \in \mathbb{N}$ there exists a number $f(m) \ge 1$ such that

$$m \odot H_{\theta,c} = f(m) H_{\theta,c} = H_{\theta, f(m)c}$$

for all $\theta \in S^{n-1}$ and $c \in (0, \infty)$.

Proof Note that by monotonicity we have

$$m \odot H_{\theta,c} \supseteq [(m-1) \odot \{0\}] \oplus H_{\theta,c} = H_{\theta,c},$$

and since $m \odot H_{\theta,c}$ is assumed to be convex it follows that $m \odot H_{\theta,c} = H_{\theta,\lambda c}$ for some $\lambda \ge 1$. Our goal is to prove that λ is independent of θ and c.

So, assume that $m \odot H_{\theta,c} = H_{\theta,\lambda c}$ and $m \odot H_{\eta,d} = H_{\eta,\mu d}$. Our goal is to prove that $\lambda = \mu$, and we may assume that $\theta \ne \eta$. This means that we can find a point $x_0 \in \mathbb{R}^n$ such that $\langle x_0, \theta \rangle = c$ and $\langle x_0, \eta \rangle = d$. If we define $A = (-\infty, x_0]$ to be the ray emanating from x_0 and passing through the origin, then

$$A = H_{\theta,c} \cap \mathbb{R}x_0 = H_{\eta,d} \cap \mathbb{R}x_0.$$

Now we apply the previous claim to $H_{\theta,c}$ and $\mathbb{R}x_0$ and see that

$$m \odot A = m \odot (H_{\theta,c} \cap \mathbb{R}x_0) = (m \odot H_{\theta,c}) \cap (m \odot \mathbb{R}x_0) = H_{\theta,\lambda c} \cap \mathbb{R}x_0 = (-\infty, \lambda x_0].$$

Notice that we used that fact that $\oplus$ preserves subspaces to deduce that $m \odot \mathbb{R}x_0 = \mathbb{R}x_0$.

But exactly the same reasoning shows us that

$$m \odot A = m \odot (H_{\eta,d} \cap \mathbb{R}x_0) = (m \odot H_{\eta,d}) \cap (m \odot \mathbb{R}x_0) = H_{\eta,\mu d} \cap \mathbb{R}x_0 = (-\infty, \mu x_0]$$

This shows that $\lambda = \mu$ as we wanted.

Claim For every $m \in \mathbb{N}$ we have

$$m \odot R_{\theta,c} = f(m) R_{\theta,c} = R_{\theta,f(m)c},$$

where $f(m)$ is the same constant from Claim 3.

Proof Since $m \odot R_{\theta,c} \supseteq R_{\theta,c}$, we must have $m \odot R_{\theta,c} = R_{\theta,d}$ for some $d \geq c$. All we need to show is that $d = f(m)c$.

On the one hand $H_{\theta,c} \subseteq R_{\theta,c}$, so by monotonicity

$$R_{\theta,d} = m \odot R_{\theta,c} \supseteq m \odot H_{\theta,c} = H_{\theta,f(m)c}.$$

Comparing radial functions in direction θ, we see that $d \geq f(m)c$.

On the other hand, for every $\varepsilon > 0$ we know that $H_{\theta,c+\varepsilon} \not\subseteq R_{\theta,c}$, and so by strong monotonicity

$$H_{\theta,f(m)(c+\varepsilon)} = m \odot H_{\theta,c} \not\subseteq m \odot R_{\theta,c} = R_{\theta,d}.$$

This means that for some direction $\eta \in S^{n-1}$ we must have $r_{H_{\theta,f(m)(c+\varepsilon)}}(\eta) > r_{R_{\theta,d}}(\eta)$. But $r_{R_{\theta,d}}(\eta) = \infty$ for all $\eta \neq \theta$, so we must have $\eta = \theta$ and

$$f(m) \cdot (c + \varepsilon) = r_{H_{\theta,f(m)(c+\varepsilon)}}(\theta) > r_{R_{\theta,d}}(\theta) = d.$$

Sending $\varepsilon \to 0$ we see that $d \leq f(m)c$, which completes the proof.

Proof (Proof of Theorem 4) All assumptions of Lemma 1 holds, as the only assumption of the lemma that wasn't assumed in the theorem is exactly Claim 3. Hence $\oplus = \tilde{+}_p$ for some $p > 0$ like we wanted.

4   Polynomiality of Volume

Remember the Minkowski addition has remarkable property, not shared by other p-additions when $p > 1$. For $A \in \mathcal{S}_0^n$ (and in particular $A \in \mathcal{K}_0^n$) let us denote by $|A| \in [0, \infty]$ the Lebesgue volume of A. Minkowski's theorem then states that for every convex bodies $K_1, K_2, \ldots, K_m \in \mathcal{K}_0^n$ the function $f : (0, \infty)^m \to \mathbb{R}$ defined by

$$f(\lambda_1, \lambda_2, \ldots, \lambda_m) = |\lambda_1 K_1 + \lambda_2 K_2 + \cdots + \lambda_m K_m|$$

is an homogeneous polynomial of degree n, with non-negative coefficients. This theorem allows the introduction of mixed volumes, a fundamental notion in convexity which will not be needed here.

Inspired by Minkowski's theorem, we define:

Definition 3 An additional $\oplus$ on $\mathcal{K}_0^n$ (resp. $\mathcal{S}_0^n$) is polynomial if for every $A, B \in \mathcal{K}_0^n$ (resp. $A, B \in \mathcal{S}_0^n$), the function

$$f(m, k) = |(m \odot A) \oplus (k \odot B)|$$

is a polynomial on $\mathbb{N}^2$.

Notice that the polynomiality property is weaker then Minkowski's theorem, as there are only two bodies and the polynomial is not assumed to be homogeneous. Still, it is not difficult to check that the p-addition is not polynomial for any $1 < p < \infty$, and so the following corollary of Theorem 2 was proved in [7]:

Corollary 1 Let $\oplus$ be a polynomial addition on $\mathcal{K}_0^n$ satisfying all properties of Theorem 2. Then $\oplus = +_1$ or $\oplus = +_\infty$.

Notice that the case $p = \infty$ is somewhat degenerate: we have $m \cdot_\infty A = A$ for all $A \in \mathcal{K}_0^n$ and $m \in \mathbb{N}$, so the function $|(m \odot A) \oplus (k \odot B)|$ is a constant function, which is a polynomial.

For star bodies, however, the situation is different. By integration in polar coordinates, we have the formula

$$|A| = \omega_n \cdot \int_{S^{n-1}} r_A(\theta)^n d\sigma(\theta),$$

where ω_n is the volume of the unit Euclidean ball, and σ is the Haar probability measure on S^{n-1}. It follows that for every $A, B \in \mathcal{S}_0^n$, every $m, k \in \mathbb{N}$ and every $p > 0$ one has

$$\left|(m \tilde{\cdot}_p A) \tilde{+}_p (k \tilde{\cdot}_p B)\right| = \omega_n \cdot \int_{S^{n-1}} (m \cdot r_A(\theta)^p + k \cdot r_B(\theta)^p)^{\frac{n}{p}} d\sigma(\theta),$$

and this expression is a polynomial in m and k whenever $\frac{n}{p}$ is an integer. In fact, in this case the polynomial is an homogeneous polynomial of degree $\frac{n}{p}$.

By taking A to be the unit ball and $B = \{0\}$ we see that the condition that $\frac{n}{p}$ is an integer is also a necessary condition for polynomiality. We summarize the discussion in the following corollary:

Corollary 2 *Assume* $\oplus : \mathcal{S}_0^n \times \mathcal{S}_0^n \to \mathcal{S}_0^n$ *is polynomial, and satisfy all the conditions of Theorem 4. Then there exists* $k \in \mathbb{N}$ *such that* $\oplus$ *is the* $\frac{n}{k}$*-radial sum, and for every* $A, B \in \mathcal{S}_0^n$ *the function*

$$f(m, k) = |(m \odot A) \oplus (k \odot B)|$$

is an homogeneous polynomial of degree k *with non-negative coefficients.*

Acknowledgements We would like to thank the referee for the careful review and the detailed comments.

Both authors are supported by ISF grant 826/13 and BSF grant 2012111. The second named author is also supported by the Adams Fellowship Program of the Israel Academy of Sciences and Humanities.

References

1. P. Erdős, On the distribution function of additive functions. Ann. Math. **47**(1), 1–20 (1946)
2. W.J. Firey, *p*-means of convex bodies. Math. Scand. **10**, 17–24 (1962)
3. R. Gardner, D. Hug, W. Weil, Operations between sets in geometry. J. Eur. Math. Soc. **15**(6), 2297–2352 (2013)
4. E. Howe, A new proof of Erdös's theorem on monotone multiplicative functions. Am. Math. Mon. **93**(8), 593–595 (1986)
5. E. Lutwak, The Brunn-Minkowski-Firey theory I: mixed volumes and the Minkowski problem. J. Differ. Geom. **38**(1), 131–150 (1993)
6. E. Lutwak, The Brunn-Minkowski-Firey theory II: affine and geominimal surface areas. Adv. Math. **118**(2), 244–294 (1996)
7. V. Milman, L. Rotem, Characterizing addition of convex sets by polynomiality of volume and by the homothety operation. Commun. Contemp. Math. **17**(3), 1450022 (2015)

On Mimicking Rademacher Sums in Tail Spaces

Krzysztof Oleszkiewicz

Abstract We establish upper and lower bounds for the L^1 distance from a Rademacher sum to the mth tail space on the discrete cube. The bounds are tight, up to the value of multiplicative constants.

2010 **Mathematics Subject Classification.** Primary: 60E15, 42C10

1 Introduction

Throughout the paper, $n > m \geq 2$ will be integers, and we will use the standard notation $[n] := \{1, 2, \ldots, n\}$. We will equip the discrete cube $\{-1, 1\}^n$ with the normalized counting (equivalently, uniform probability) measure $\mu_n = (\frac{1}{2}\delta_{-1} + \frac{1}{2}\delta_1)^{\otimes n}$. Let $\mathbb{E}$ denote the expectation with respect to this measure, and let $r_1, r_2, \ldots, r_n$ be the standard Rademacher functions on the discrete cube, i.e. the coordinate projections $r_j(x) = x_j$ for $x \in \{-1, 1\}^n$ and $j \in [n]$. Furthermore, for $A \subseteq [n]$, we define the Walsh functions by $w_A = \prod_{j \in A} r_j$, with $w_\emptyset \equiv 1$.

The Walsh functions $(w_A)_{A \subseteq [n]}$ form a complete orthonormal system in $L^2(\{-1, 1\}^n, \mu_n)$. Thus, every $f : \{-1, 1\}^n \to \mathbb{R}$ admits a unique Walsh-Fourier expansion $f = \sum_{A \subseteq [n]} \hat{f}(A) w_A$, with coefficients given by $\hat{f}(A) = \langle f, w_A \rangle = \mathbb{E}[f \cdot w_A]$. For $m \geq 2$, we will denote by $T^{>m} = T^{>m}_{\{-1,1\}^n}$ the linear span of $(w_A)_{A \subseteq [n]:|A|>m}$, which is called the mth tail space on the discrete cube. For $f : \{-1, 1\}^n \to \mathbb{R}$ we consider its L^1 distance to the mth tail space, $\mathrm{dist}_{L^1}\left(f, T^{>m}_{\{-1,1\}^n}\right) := \inf_{g \in T^{>m}} \mathbb{E}|f - g|$. By $T^{\leq m} = T^{\leq m}_{\{-1,1\}^n}$ we will denote the space of Walsh-Fourier chaoses of order not exceeding m, i.e. the linear span of $(w_A)_{A \subseteq [n]:|A| \leq m}$.

K. Oleszkiewicz (✉)

Institute of Mathematics, University of Warsaw, ul. Banacha 2, 02-097 Warsaw, Poland

e-mail: koles@mimuw.edu.pl

© Springer International Publishing AG 2017

B. Klartag, E. Milman (eds.), *Geometric Aspects of Functional Analysis*, Lecture Notes in Mathematics 2169, DOI 10.1007/978-3-319-45282-1_21

2 Main Result

The following theorem provides an answer to a question of Robert Bogucki, Piotr Nayar, and Michał Wojciechowski (2013, personal communication).

Theorem 2.1 *For $a_1 \geq a_2 \geq \ldots \geq a_n \geq a_{n+1} = 0$, let $S : \{-1, 1\}^n \to \mathbb{R}$ be defined by $S = a_1 r_1 + a_2 r_2 + \ldots + a_n r_n$. Furthermore, let $\alpha = \alpha \left((a_j)_{j=1}^n, m \right)$ be given by $\alpha := \min_{k \in [n]} \left((\sum_{j=1}^k a_j^2)^{1/2} + m a_{k+1} \right)$. Then*

$$\frac{1}{37} \cdot \alpha \leq \operatorname{dist}_{L^1} \left(S, T_{\{-1,1\}^n}^{>m} \right) \leq \frac{8}{\pi} \cdot \alpha.$$

In particular, $\operatorname{dist}_{L^1} \left(\sum_{j=1}^n r_j, T_{\{-1,1\}^n}^{>m} \right) \simeq \min(m, \sqrt{n})$, which is the case Bogucki, Nayar, and Wojciechowski were originally interested in.

We will need the following discrete cube dual counterpart to the classical Bernstein inequality.

Lemma 2.2 *For any integers $n > m \geq 2$ and any real numbers $a_1, a_2, \ldots, a_n$, there is a function $f : \{-1, 1\}^n \to \mathbb{R}$ with*

$$\mathbb{E}|f| \leq \frac{8m}{\pi} \cdot \max_{j \in [n]} |a_j|$$

and such that $\hat{f}(\{j\}) = a_j$ for $j \in [n]$, and $\hat{f}(A) = 0$ for all $A \subseteq \{1, 2, \ldots, n\}$ of cardinality $0, 2, 3, 4, \ldots, m$.

Proof By the homogeneity, we assume that $\max_{j \in [n]} |a_j| = 1$. For $m = 2$ it suffices to consider $f = \frac{1}{2} \prod_{j=1}^n (1 + a_j r_j) - \frac{1}{2} \prod_{j=1}^n (1 - a_j r_j)$. Indeed, by the triangle inequality, $\mathbb{E}|f| \leq 1 < 16/\pi$. Note that $\mathbb{E}|1 + a_j r_j| = \mathbb{E}[1 + a_j r_j] = 1$.

For general $m \geq 2$, let us consider a Fejér type function

$$\psi_m(x) = \sum_{k=1}^m k \sin kx + \sum_{k=1}^{m-1} (m - k) \sin \left((m + k)x \right),$$

or, equivalently,

$$\psi_m(x) = \sum_{k=1}^{2m-1} \min(k, 2m - k) \frac{e^{ikx} - e^{-ikx}}{2i} = \left(\sum_{l=-m+1}^{m-1} e^{ilx/2} \right)^2 \frac{e^{imx} - e^{-imx}}{2i}$$

$$= \left(\frac{e^{imx/2} - e^{-imx/2}}{e^{ix/2} - e^{-ix/2}} \right)^2 \sin mx = \sin mx \cdot \sin^2(mx/2) / \sin^2(x/2).$$

Since, clearly, $|\psi_m(x)| \leq \sum_{k=1}^{m} k + \sum_{k=1}^{m-1}(m-k) = m^2$,

$$\int_{-\pi}^{\pi} |\psi_m(x)| \, dx \leq \int_{-2/m}^{2/m} m^2 \, dx + 2 \int_{2/m}^{\pi} \frac{dx}{\sin^2(x/2)} = 4m + 4\cot(1/m) \leq 8m.$$

Using the orthogonality in $L^2([-\pi, \pi], dx)$, we have $\int_{-\pi}^{\pi} \psi_m(x) \sin x \, dx = \pi$, $\int_{-\pi}^{\pi} \psi_m(x) \, dx = 0$, and $\int_{-\pi}^{\pi} \psi_m(x) \sin^k x \, dx = 0$ for $2 \leq k \leq m$. For even k's the last equality is trivial since ψ_m is an odd function, and for odd k's, by the binomial formula, $\sin^k x = \left(\frac{e^{ix} - e^{-ix}}{2i}\right)^k$ can be expressed as a linear combination $\beta_{k,1} \sin x + \beta_{k,3} \sin 3x + \ldots + \beta_{k,k} \sin kx$. Thus,

$$\frac{1}{\pi} \int_{-\pi}^{\pi} \psi_m(x) \sin^k x \, dx = \beta_{k,1} + 3\beta_{k,3} + \ldots + k\beta_{k,k}$$

$$= \frac{d}{dx} (\beta_{k,1} \sin x + \beta_{k,3} \sin 3x + \ldots + \beta_{k,k} \sin kx)\Big|_{x=0} = \frac{d \sin^k x}{dx}\Big|_{x=0} = 0.$$

Choosing f defined by

$$f = \frac{1}{\pi} \int_{-\pi}^{\pi} \psi_m(x) \prod_{j=1}^{n} (1 + a_j r_j \sin x) \, dx$$

$$= \sum_{A \subseteq [n]} \frac{\prod_{j \in A} a_j}{\pi} \int_{-\pi}^{\pi} \psi_m(x) \sin^{|A|} x \, dx \cdot w_A \in \sum_{j=1}^{n} a_j r_j + T_{\{-1,1\}^n}^{>m},$$

we finish the proof. Indeed, $\mathbb{E}|f| \leq \frac{1}{\pi} \int_{-\pi}^{\pi} |\psi_m(x)| \, dx \leq 8m/\pi$ – recall that $\max_j |a_j| \leq 1$, so that, for every $x \in [-\pi, \pi]$, the independent random variables $(1 + a_j r_j \sin x)_{j=1}^{n}$ are nonnegative and have mean 1. $\square$

Proof of Theorem 2.1 The upper bound easily follows from Lemma 2.2. Indeed, for $k < n$, by the lemma, applied to the cube $\{-1, 1\}^{n-k}$ instead of $\{-1, 1\}^n$, we may find a Walsh-Fourier polynomial f in $r_{k+1}, r_{k+2}, \ldots, r_n$ such that $f - \sum_{j=k+1}^{n} a_j r_j \in T_{\{-1,1\}^n}^{>m}$ and $\mathbb{E}|f| \leq \frac{8m}{\pi} \max_{k+1 \leq j \leq n} |a_j| = \frac{8}{\pi} \cdot m a_{k+1}$. Thus

$$\text{dist}_{L^1}\left(S, T_{\{-1,1\}^n}^{>m}\right) \leq \mathbb{E}\left|S - \left(\sum_{j=k+1}^{n} a_j r_j - f\right)\right| = \mathbb{E}\left|\sum_{j=1}^{k} a_j r_j + f\right|$$

$$\leq \mathbb{E}\left|\sum_{j=1}^{k} a_j r_j\right| + \mathbb{E}|f| \leq \left\|\sum_{j=1}^{k} a_j r_j\right\|_{L^2} + \frac{8}{\pi} m a_{k+1}$$

$$= \left(\sum_{j=1}^{k} a_j^2\right)^{1/2} + \frac{8}{\pi} m a_{k+1}.$$

For $k = n$, we simply note that

$$\text{dist}_{L^1}\left(S, T^{>m}_{\{-1,1\}^n}\right) \leq E|S - 0| \leq \|S\|_{L^2} = \left(\sum_{j=1}^{n} a_j^2\right)^{1/2}.$$

Taking the minimum over $k \in [n]$, we deduce the upper bound.

To prove the lower bound, we introduce the following auxiliary functions:

$$W_m(t) = \sum_{l=0}^{\lfloor\frac{m-1}{2}\rfloor} \frac{(-1)^l t^{2l+1}}{(2l+1)!}, \quad R_m(t) = \sum_{l=\lfloor\frac{m+1}{2}\rfloor}^{\infty} \frac{(-1)^l t^{2l+1}}{(2l+1)!}.$$

Obviously, $W_m(t) + R_m(t) = \sin t$ and for $t \in [-m/6, m/6]$ we have

$$|R_m(t)| \leq \sum_{l=\lfloor\frac{m+1}{2}\rfloor}^{\infty} (e/6)^{2l+1} \leq 2^{-m},$$

since $k! \geq (k/e)^k$. Hence $|W_m(t)| \leq 2$ for $t \in [-m/6, m/6]$.
Let $f = S - g$ for some $g \in T^{>m}_{\{-1,1\}^n}$. If $a_1 \geq ma_2$, then

$$\mathbb{E}|f| \geq \mathbb{E}[fr_1] = a_1 \geq \frac{1}{2}\left((a_1^2)^{1/2} + ma_2\right) \geq \alpha/2.$$

If $a_1 < ma_2$, then let κ denote the largest $k \in [n-1]$ for which $\sum_{j=1}^{k} a_j^2 < m^2 a_{k+1}^2$ (note that $\sum_{j=1}^{k} a_j^2 - m^2 a_{k+1}^2$ increases in k), so that $\sum_{j=1}^{\kappa} a_j^2 < m^2 a_{\kappa+1}^2$, and thus $\kappa < m^2$, and

$$\sum_{j=1}^{\kappa+1} a_j^2 \geq m^2 a_{\kappa+2}^2. \tag{1}$$

Note that $\sum_{j=1}^{\kappa} a_j \leq \sqrt{\kappa}\left(\sum_{j=1}^{\kappa} a_j^2\right)^{1/2} < m\left(\sum_{j=1}^{\kappa} a_j^2\right)^{1/2}$. For $l \in [\kappa]$, let $b_l = \frac{a_l}{6}\left(\sum_{j=1}^{\kappa} a_j^2\right)^{-1/2}$, so that $\left|\sum_{j=1}^{\kappa} b_j r_j\right| \leq \sum_{j=1}^{\kappa} b_j \leq m/6$. Also, $\sum_{j=1}^{\kappa} b_j^2 = 1/36$, so that, for all $j \in [\kappa]$, we have $b_j \in [0, 1/6]$, and thus $\cos b_j \geq e^{-b_j^2}$, and, for all $l \in [\kappa]$,

$$\mathbb{E}r_l \sin\left(\sum_{j=1}^{\kappa} b_j r_j\right) = \operatorname{Im} \mathbb{E}r_l e^{i\sum_{j=1}^{\kappa} b_j r_j} = \operatorname{Im}\left(\mathbb{E}[r_l e^{ib_l r_l}] \cdot \prod_{j \in [\kappa]\setminus\{l\}} \mathbb{E}[e^{ib_j r_j}]\right)$$

$$= \sin b_l \cdot \prod_{j \in [\kappa]\setminus\{l\}} \cos b_j = \tan b_l \cdot \prod_{j \in [\kappa]} \cos b_j \geq b_l e^{-\sum_{j=1}^{\kappa} b_j^2}$$

$$= e^{-1/36} \cdot b_l.$$

Now we are in a position to finish the proof of the lower bound. We have

$$2\mathbb{E}|f| \geq \mathbb{E}f\, W_m\left(\sum_{j=1}^{\kappa} b_j r_j\right) = \mathbb{E}(S - g)W_m\left(\sum_{j=1}^{\kappa} b_j r_j\right)$$

$$= \mathbb{E}\left(\sum_{l=1}^{\kappa} a_l r_l\right) W_m\left(\sum_{j=1}^{\kappa} b_j r_j\right) = \sum_{l=1}^{\kappa} a_l \cdot \mathbb{E}r_l W_m\left(\sum_{j=1}^{\kappa} b_j r_j\right).$$

The second equality follows from the fact that both g and $\sum_{j=\kappa+1}^{n} a_j r_j$ are orthogonal to $W_m\left(\sum_{j=1}^{\kappa} b_j r_j\right)$ in $L^2\left(\{-1, 1\}^n, \mu_n\right)$. Indeed, $\deg W_m \leq m$, so that $W_m\left(\sum_{j=1}^{\kappa} b_j r_j\right)$ is a Walsh-Fourier chaos of order not exceeding m in variables $r_1, r_2, \ldots, r_\kappa$. Since

$$\mathbb{E}r_l W_m(\sum_{j=1}^{\kappa} b_j r_j) = \mathbb{E}r_l \sin(\sum_{j=1}^{\kappa} b_j r_j) - \mathbb{E}r_l R_m(\sum_{j=1}^{\kappa} b_j r_j) \geq e^{-1/36} b_l - 2^{-m},$$

we arrive at

$$2\mathbb{E}|f| \geq e^{-1/36} \sum_{l=1}^{\kappa} a_l b_l - 2^{-m} \sum_{l=1}^{\kappa} a_l = \frac{e^{-1/36}}{6}\left(\sum_{l=1}^{\kappa} a_l^2\right)^{1/2} - 2^{-m} \sum_{l=1}^{\kappa} a_l$$

$$\geq \left(\frac{e^{-1/36}}{6} - m \cdot 2^{-m}\right)\left(\sum_{l=1}^{\kappa} a_l^2\right)^{1/2}.$$

We have $\frac{e^{-1/36}}{6} - 20 \cdot 2^{-20} > 2/13$, so that, for $m \geq 20$, $\mathbb{E}|f| \geq \frac{1}{13}\left(\sum_{l=1}^{\kappa} a_l^2\right)^{1/2}$. Finally, recall that $\alpha \leq \left(\sum_{l=1}^{\kappa+1} a_l^2\right)^{1/2} + ma_{\kappa+2}$ and, by (1), the last expression can be bounded from above by $2\left(\sum_{l=1}^{\kappa+1} a_l^2\right)^{1/2} \leq 2\left(2\sum_{l=1}^{\kappa} a_l^2\right)^{1/2}$, so that $\mathbb{E}|f| \geq \frac{\alpha}{13 \cdot 2\sqrt{2}} \geq \alpha/37$, for $m \geq 20$.

The remaining case, $2 \leq m < 20$, is much easier:

$$\alpha \leq a_1 + ma_2 \leq (m+1)a_1 = (m+1)\mathbb{E}[fr_1] \leq 20\mathbb{E}|f| \leq 37\mathbb{E}|f|. \quad \square$$

Certainly, with some additional effort, the numerical constants can be improved, at the cost of clarity.

Unfortunately, the outlined method does not seem to extend to a more general situation. Even for a chaos f of order 2, it does not seem to yield bounds for $\mathrm{dist}_{L^1}\left(f, T^{>m}_{\{-1,1\}^n}\right)$.

3   Gaussian Counterpart

Motivated by a question of the referee, we will briefly discuss a Gaussian counterpart of the main result. Let γ_n denote the standard Gaussian probability measure on $\mathbb{R}^n$, i.e. $d\gamma_n(x) = (2\pi)^{-n/2}e^{-|x|^2/2}dx$. On the real line, the Hermite polynomials $(H_c)_{c=0}^{\infty}$ form a natural orthogonal basis of the Hilbert space $L^2(\mathbb{R}, \gamma_1)$. In $L^2(\mathbb{R}^n, \gamma_n)$, the same role is played by their tensor products $(\mathbf{H_c})_{c \in \{0,1,2,\dots\}^n}$, where $\mathbf{H_c}(x) = \prod_{j=1}^n H_{c_j}(x_j)$, for $x = (x_1, x_2, \dots, x_n) \in \mathbb{R}^n$. For a multi-index $\mathbf{c} = (c_1, c_2, \dots, c_n)$, let $|\mathbf{c}| = \sum_{j=1}^n c_j$. For a positive integer m, it is natural to express $L^2(\mathbb{R}^n, \gamma_n)$ as $T^{\leq m}_{(\mathbb{R}^n, \gamma_n)} \oplus T^{>m}_{(\mathbb{R}^n, \gamma_n)}$, where $T^{\leq m}_{(\mathbb{R}^n, \gamma_n)}$ is a finite-dimensional (thus closed) linear span of $(\mathbf{H_c})_{|\mathbf{c}| \leq m}$, identical with $\{P \in \mathbb{R}[x_1, x_2, \dots, x_n] : \deg P \leq m\}$, and $T^{>m}_{(\mathbb{R}^n, \gamma_n)}$ is the $L^2(\mathbb{R}^n, \gamma_n)$-closure of the linear span of $(\mathbf{H_c})_{|\mathbf{c}| > m}$. This decomposition is closely related to the one we discussed on the discrete cube (see section "Introduction"). In fact, it may be obtained from it by a CLT-type limit transition. Also, for every nonnegative integer k, the linear span of $(\mathbf{H_c})_{|\mathbf{c}|=k}$ is the eigenspace associated with the eigenvalue k for the standard (Ornstein-Uhlenbeck) heat semigroup generator on $L^2(\mathbb{R}^n, \gamma_n)$, just as the linear span of $(w_A)_{|A|=k}$ is the eigenspace associated with the eigenvalue k for the standard heat semigroup generator on the discrete cube. Therefore, it seems interesting that, in contrast to Theorem 2.1, we have the following proposition.

Proposition 3.1 *For all positive integers n and m, and every $f \in L^2(\mathbb{R}^n, \gamma_n)$ with $\int_{\mathbb{R}^n} f \, d\gamma_n = 0$, for every $\varepsilon > 0$, there is a polynomial Q belonging to the linear span of $(\mathbf{H_c})_{|\mathbf{c}| > m}$ such that $\int_{\mathbb{R}^n} |f(x) - Q(x)| \, d\gamma_n(x) < \varepsilon$. Thus, $\mathrm{dist}_{L^1(\mathbb{R}^n, \gamma_n)}\left(f, T^{>m}_{(\mathbb{R}^n, \gamma_n)}\right) = 0$, in particular for $f(x) = \sum_{j=1}^n a_j x_j$, where $a_1, a_2, \dots, a_n$ are arbitrary real numbers.*

Proof Assume that the main assertion is not true, i.e., for some positive integers m and n, there exist $\varepsilon > 0$ and a square-integrable mean-zero f such that $U := \{g \in L^2(\mathbb{R}^n, \gamma_n) : \int_{\mathbb{R}^n} |f - g| \, d\gamma_n < \varepsilon\}$ is disjoint with the linear span of $(\mathbf{H_c})_{|\mathbf{c}| > m}$. One easily checks that U is an open convex subset of $L^2(\mathbb{R}^n, \gamma_n)$, so that, by the geometric Hahn-Banach (Mazur's) theorem and by the Riesz representation theorem, there exists a function $P \in L^2(\mathbb{R}^n, \gamma_n)$ such that $\int_{\mathbb{R}^n} P(x)\mathbf{H}_{\mathbf{c}}(x)\,d\gamma_n(x) = 0$, for every $\mathbf{c}$ with $|\mathbf{c}| > m$, and $\int_{\mathbb{R}^n} P(x)g(x)\,d\gamma_n(x) > 0$ for every $g \in U$. Since P is orthogonal to the span of $(\mathbf{H}_{\mathbf{c}})_{|\mathbf{c}|>m}$, it is also orthogonal to its L^2-closure, $T^{>m}_{(\mathbb{R}^n, \gamma_n)}$, and thus belongs to its orthogonal complement, $T^{\leq m}_{(\mathbb{R}^n, \gamma_n)}$. We have proved that P is a polynomial of degree not exceeding m. Since $f \in U$ and $\int_{\mathbb{R}^n} f\,d\gamma_n = 0$, we know that P cannot be a constant polynomial and thus is unbounded on $\mathbb{R}^n$. For $M > 0$, let $A_M = \{x \in \mathbb{R}^n : |P(x)| > M\}$. Let $h_M(x) = \frac{\varepsilon}{2\gamma_n(A_M)}\mathrm{sgn}(P(x))\mathbf{1}_{A_M}(x)$, so that $g_M = f - h_M$ belongs to U. Therefore, $\int_{\mathbb{R}^n} P(x)g_M(x)\,d\gamma_n(x)$ is positive, and $\int_{\mathbb{R}^n} P(x)f(x)\,d\gamma_n(x) > \int_{\mathbb{R}^n} P(x)h_M(x)\,d\gamma_n(x) \geq \varepsilon M/2$. By letting M tend to infinity, we obtain a contradiction.

Remark 3.2 In Proposition 3.1, the assumption $f \in L^2(\mathbb{R}^n, \gamma_n)$ can be easily weakened to $f \in L^1(\mathbb{R}^n, \gamma_n)$. It suffices to note that every mean-zero function is an L^1-limit of mean-zero square-integrable functions.

Acknowledgements I would like to thank the anonymous referee for a stimulating question about the Gaussian case. Research supported by NCN grant DEC-2012/05/B/ST1/00412.

Stability for Borell-Brascamp-Lieb Inequalities

Andrea Rossi and Paolo Salani

Abstract We study stability issues for the so-called Borell-Brascamp-Lieb inequalities, proving that when near equality is realized, the involved functions must be L^1-close to be p-concave and to coincide up to homotheties of their graphs.

1 Introduction

The aim of this paper is to study the stability of the so-called *Borell-Brascamp-Lieb inequality* (BBL inequality below), which we recall hereafter.

Proposition 1.1 (BBL Inequality) *Let* $0 < \lambda < 1, -\frac{1}{n} \leq p \leq +\infty, 0 \leq f, g, h \in L^1(\mathbb{R}^n)$ *and assume the following holds*

$$h((1 - \lambda)x + \lambda y) \geq \mathcal{M}_p(f(x), g(y); \lambda) \tag{1}$$

for every $x, y \in \mathbb{R}^n$. *Then*

$$\int_{\mathbb{R}^n} h \, dx \geq \mathcal{M}_{\frac{p}{np+1}} \left(\int_{\mathbb{R}^n} f \, dx, \int_{\mathbb{R}^n} g \, dx ; \lambda \right). \tag{2}$$

Here the number $p/(np+1)$ has to be interpreted in the obvious way in the extremal cases (i.e. it is equal to $-\infty$ when $p = -1/n$ and to $1/n$ when $p = +\infty$) and the quantity $\mathcal{M}_q(a, b; \lambda)$ represents the (λ-weighted) q-*mean* of two nonnegative numbers a and b, that is $\mathcal{M}_q(a, b; \lambda) = 0$ if $ab = 0$ for every $q \in \mathbb{R} \cup \{\pm\infty\}$ and

$$\mathcal{M}_q(a, b; \lambda) = \begin{cases} \max\{a, b\} & q = +\infty, \\ [(1 - \lambda)a^q + \lambda b^q]^{\frac{1}{q}} & 0 \neq q \in \mathbb{R}, \\ a^{1-\lambda}b^{\lambda} & q = 0, \\ \min\{a, b\} & q = -\infty, \end{cases} \quad \text{if } ab > 0. \tag{3}$$

A. Rossi • P. Salani (✉)
Dipartimento di Matematica "Ulisse Dini", Università degli Studi di Firenze, Viale Morgagni 67/A, 50134 Firenze, Italy
e-mail: andrea.rossi@unifi.it; paolo.salani@unifi.it

© Springer International Publishing AG 2017
B. Klartag, E. Milman (eds.), *Geometric Aspects of Functional Analysis*,
Lecture Notes in Mathematics 2169, DOI 10.1007/978-3-319-45282-1_22

The BBL inequality was first proved (in a slightly different form) for $p > 0$ by Henstock and Macbeath (with $n = 1$) in [22] and by Dinghas in [11]. Then it was generalized by Brascamp and Lieb in [6] and by Borell in [4]. The case $p = 0$ is usually known as *Prékopa-Leindler inequality*, as it was previously proved by Prékopa [25] and Leindler [24] (later rediscovered by Brascamp and Lieb in [5]).

In this paper we deal only with the case $p > 0$ and are particularly interested in the equality conditions of BBL, that are discussed in [13] (see Theoreme 12 therein). To avoid triviality, if not otherwise explicitly declared, we will assume throughout the paper that $f, g \in L^1(\mathbb{R}^n)$ are nonnegative compactly supported functions [with supports supp (f) and supp (g)] such that

$$F = \int_{\mathbb{R}^n} f \, dx > 0 \quad \text{and} \quad G = \int_{\mathbb{R}^n} g \, dx > 0 \, .$$

Let us restate a version of the BBL inequality including its equality condition in the case

$$p = \frac{1}{s} > 0 \, ,$$

adopting a slightly different notation.

Proposition 1.2 *Let $s > 0$ and f, g be as said above. Let $\lambda \in (0, 1)$ and h be a nonnegative function belonging to $L^1(\mathbb{R}^n)$ such that*

$$h((1 - \lambda)x + \lambda y) \geqslant \left((1 - \lambda)f(x)^{1/s} + \lambda g(y)^{1/s}\right)^s \tag{4}$$

for every $x \in$ supp (f), $y \in$ supp (g).

Then

$$\int_{\mathbb{R}^n} h \, dx \geqslant \mathcal{M}_{\frac{1}{n+s}}(F, G; \lambda) \, . \tag{5}$$

Moreover equality holds in (5) if and only if there exists a nonnegative concave function φ such that

$$\varphi(x)^s = a_1 f(b_1 x - \bar{x}_1) = a_2 g(b_2 x - \bar{x}_2) = a_3 h(b_3 x - \bar{x}_3) \quad a.e. \ x \in \mathbb{R}^n \, , \tag{6}$$

for some $\bar{x}_1, \bar{x}_2, \bar{x}_3 \in \mathbb{R}^n$ and suitable $a_i, b_i > 0$ for $i = 1, 2, 3$.

Notice that, given f and g, the smallest function satisfying (4) (hence the smallest function to which Proposition 1.2 possibly applies to) is their p-Minkowksi sum (or (p, λ)-supremal convolution), defined as follows (for $p = \frac{1}{s}$)

$$h_{s,\lambda}(z) = \sup \left\{ \left((1 - \lambda)f(x)^{1/s} + \lambda g(y)^{1/s}\right)^s : z = (1 - \lambda)x + \lambda y \right\} \tag{7}$$

for $z \in (1 - \lambda)$ supp $(f) + \lambda$ supp (g) and $h_{s,\lambda}(z) = 0$ if $z \notin (1 - \lambda)$ supp $(f) + \lambda$ supp (g).

When dealing with a rigid inequality, a natural question arises about the stability of the equality case; here the question at hand is the following: if we are close to equality in (5), must the functions f, g and h be close (in some suitable sense) to satisfy (6)?

The investigation of stability issues in the case $p = 0$ was started by Ball and Böröczky in [2, 3] and new related results are in [7]. The general case $p > 0$ has been very recently faced in [19]. But the results of [19], as well as the quoted results for $p = 0$, hold only in the restricted class of p-concave functions, hence answering only a half of the question. Let us recall here the definition of p-concave function: a nonnegative function u is p-concave for some $p \in \mathbb{R} \cup \{\pm\infty\}$ if

$$u((1 - \lambda)x + \lambda y) \geq \mathcal{M}_p(u(x), u(y); \lambda) \quad \text{for every } x, y \in \mathbb{R}^n \text{ and every } \lambda \in (0, 1).$$

Roughly speaking, u is p-concave if it has convex support Ω and: (1) u^p is concave in Ω for $p > 0$; (2) $\log u$ is concave in Ω for $p = 0$; (3) u^p is convex in Ω for $p < 0$; (4) u is quasi-concave, i.e. all its superlevel sets are convex, for $p = -\infty$; (5) u is a positive constant in Ω, for $p = +\infty$.

Here we want to remove this restriction, proving that near equality in (5) is possible if and only if the involved functions are close to coincide up to homotheties of their graphs and they are also nearly p-concave, in a suitable sense. But before stating our main result in detail, we need to introduce some notation: for $s > 0$, we say that two functions $v, \hat{v} : \mathbb{R}^n \to [0, +\infty)$ are s-equivalent if there exist $\mu_v > 0$ and $\bar{x} \in \mathbb{R}^n$ such that

$$\hat{v}(x) = \mu_v^s \, v \left(\frac{x - \bar{x}}{\mu_v} \right) \qquad \text{a.e. } x \in \mathbb{R}^n. \tag{8}$$

Now we are ready to state our main result, which regards the case $s = 1/p \in \mathbb{N}$. Later (see Sect. 4) we will extend the result to the case $0 < s \in \mathbb{Q}$ in Corollary 4.3 and finally (see Corollary 5.1 in Sect. 5) we will give a slightly weaker version, valid for every $s > 0$.

Theorem 1.3 *Let f, g, h as in Proposition 1.2 with*

$$0 < s \in \mathbb{N}.$$

Assume that

$$\int_{\mathbb{R}^n} h \, dx \leq \mathcal{M}_{\frac{1}{n+s}} (F, G; \lambda) + \varepsilon \tag{9}$$

for some $\varepsilon > 0$ small enough.

Then there exist a $\frac{1}{s}$-concave function $u : \mathbb{R}^n \longrightarrow [0, +\infty)$ and two functions $\hat{f}$ and $\hat{g}$, s-equivalent to f and g in the sense of (8) [with suitable μ_f and μ_g given *in (46)] such that the following hold:*

$$u \geq \hat{f} \qquad u \geq \hat{g}, \tag{10}$$

$$\int_{\mathbb{R}^n} (u - \hat{f}) \, dx + \int_{\mathbb{R}^n} (u - \hat{g}) \, dx \leq C_{n+s} \left(\frac{\varepsilon}{\mathcal{M}_{\frac{1}{n+s}}(F, G; \lambda)} \right), \tag{11}$$

where $C_{n+s}(\eta)$ is an infinitesimal function for $\eta \longrightarrow 0$ [whose explicit expression is given later, see (15)].

Notice that the function u is bounded, hence as a byproduct of the proof we obtain that the functions f and g have to be bounded as well (see Remark 3.1).

The proof of the above theorem is based on a proof of the BBL inequality due to Klartag [23], which directly connects the BBL inequality to the Brunn-Minkowski inequality, and the consequent application of a recent stability result for the Brunn-Minkowski inequality by Figalli and Jerison [15], which does not require any convexity assumption of the involved sets. Indeed [15] is the first paper, at our knowledge, investigating on stability issues for the Brunn-Minkowski inequality outside the realm of convex bodies. Noticeably, Figalli and Jerison ask therein for a functional counterpart of their result, pointing out that *"at the moment some stability estimates are known for the Prékopa-Leindler inequality only in one dimension or for some special class of functions [2, 3], and a general stability result would be an important direction of future investigations."* Since BBL inequality is the functional counterpart of the Brunn-Minkowksi inequality (for any $p > 0$ as much as for $p = 0$), this paper can be considered a first answer to the question by Figalli and Jerison.

The paper is organized as follows. The Brunn-Minkowski inequality and the stability result of [15] are recalled in Sect. 2, where we also discuss the equivalence between the Brunn-Minkowski and the BBL inequality. In Sect. 3 we prove Theorem 1.3. Finally Sect. 4 contains the already mentioned generalization to the case of rational s, namely Corollary 4.3, while Sect. 5 is devoted to Corollary 5.1, where we prove a stability for every $s > 0$ under a suitable normalization for $\int f$ and $\int g$. The paper ends with an Appendix where we give the proofs of some easy technical lemmas for the reader's convenience.

2 Preliminaries

2.1 Notation

Throughout the paper the symbol $|\cdot|$ is used to denote different things and we hope this is not going to cause confusion. In particular: for a real number a we denote by $|a|$ its absolute value, as usual; for a vector $x = (x_1, \ldots, x_m) \in \mathbb{R}^m$ we denote by $|x|$

its euclidean norm, that is $|x| = \sqrt{x_1^2 + \cdots + x_m^2}$; for a set $A \subset \mathbb{R}^m$ we denote by $|A|$ its (m-dimensional) Lebesgue measure or, sometimes, its outer measure if A is not measurable.

The support set of a nonnegative function $f : \mathbb{R}^m \to [0, +\infty)$ is denoted by supp (f), that is supp $(f) = \overline{\{x \in \mathbb{R}^m : f(x) > 0\}}$.

Let $\lambda \in (0, 1)$, the Minkowski convex combination (of coefficient λ) of two nonempty sets $A, B \subseteq \mathbb{R}^n$ is given by

$$(1 - \lambda)A + \lambda B = \{(1 - \lambda)a + \lambda b : a \in A, \ b \in B\}.$$

2.2 About the Brunn-Minkowski Inequality

The classical form of the Brunn-Minkowski inequality (BM in the following) regards only convex bodies and it is at the core of the related theory (see [26]). Its validity has been extended later to the class of measurable sets and we refer to the beautiful paper by Gardner [18] for a throughout presentation of BM inequality, its history and its intriguing relationships with many other important geometric and analytic inequalities. Let us now recall it (in its general form).

Proposition 2.1 (Brunn-Minkowski Inequality) *Given $\lambda \in (0, 1)$, let $A, B \subseteq \mathbb{R}^n$ be nonempty measurable sets. Then*

$$|(1 - \lambda)A + \lambda B|^{1/n} \geqslant (1 - \lambda) |A|^{1/n} + \lambda |B|^{1/n} \tag{12}$$

(where $| \cdot |$ possibly means outer measure if $(1 - \lambda)A + \lambda B$ is not measurable).

In addition, if $|A|, |B| > 0$, then equality in (12) holds if and only if there exist a convex set $K \subseteq \mathbb{R}^n$, $v_1, v_2 \in \mathbb{R}^n$ and $\lambda_1, \lambda_2 > 0$ such that

$$\lambda_1 A + v_1 \subseteq K, \quad \lambda_2 B + v_2 \subseteq K, \quad |K \setminus (\lambda_1 A + v_1)| = |K \setminus (\lambda_2 B + v_2)| = 0. \tag{13}$$

We remark that equality holds in (12) if and only if the involved sets are convex (up to a null measure set) and homothetic.

The stability of BM inequality was first investigated only in the class of convex sets, see for instance [12, 14, 16, 17, 20, 27]. Very recently Christ [9, 10] started the investigation without convexity assumptions, and its qualitative results have been made quantitative and sharpened by Figalli and Jerison in [15]; here is their result, for $n \geqslant 2$.

Proposition 2.2 *Let $n \geqslant 2$, and $A, B \subset \mathbb{R}^n$ be measurable sets with $|A| = |B| = 1$. Let $\lambda \in (0, 1)$, set $\tau = \min\{\lambda, 1 - \lambda\}$ and $S = (1 - \lambda)A + \lambda B$. If*

$$|S| \leqslant 1 + \delta \tag{14}$$

for some $\delta \leq e^{-M_n(\tau)}$, then there exists a convex $K \subset \mathbb{R}^n$ such that, up to a translation,

$$A, B \subseteq K \quad \text{and} \quad |K \setminus A| + |K \setminus B| \leq \tau^{-N_n} \delta^{\sigma_n(\tau)}.$$

The constant N_n can be explicitly computed and we can take

$$M_n(\tau) = \frac{2^{3^{n+2}} n^{3^n} |\log \tau|^{3^n}}{\tau^{3^n}}, \qquad \sigma_n(\tau) = \frac{\tau^{3^n}}{2^{3^{n+1}} n^{3^n} |\log \tau|^{3^n}}.$$

Remark 2.3 As already said, the proof of our main result is based on Proposition 2.2 and now we can give the explicit expression of the infinitesimal function C_{n+s} of Theorem 1.3:

$$C_{n+s}(\eta) = \frac{\eta^{\sigma_{n+s}(\tau)}}{\omega_s \, \tau^{N_{n+s}}}, \tag{15}$$

where ω_s denotes the measure of the unit ball in $\mathbb{R}^s$.

Next, for further use, we rewrite Proposition 2.2 without the normalization constraint about the measures of the involved sets A and B.

Corollary 2.4 *Let $n \geq 2$ and $A, B \subset \mathbb{R}^n$ be measurable sets with $|A|, |B| \in (0, +\infty)$. Let $\lambda \in (0, 1)$, set $\tau = \min\{\lambda, 1 - \lambda\}$ and $S = (1 - \lambda)A + \lambda B$. If*

$$\frac{|S| - \left[(1 - \lambda)|A|^{1/n} + \lambda |B|^{1/n}\right]^n}{\left[(1 - \lambda)|A|^{1/n} + \lambda |B|^{1/n}\right]^n} \leq \delta \tag{16}$$

for some $\delta \leq e^{-M_n(\tau)}$, then there exist a convex $K \subset \mathbb{R}^n$ and two homothetic copies $\tilde{A}$ and $\tilde{B}$ of A and B such that

$$\tilde{A}, \tilde{B} \subseteq K \quad \text{and} \quad |K \setminus \tilde{A}| + |K \setminus \tilde{B}| \leq \tau^{-N_n} \delta^{\sigma_n(\tau)}.$$

Proof The proof is standard and we give it just for the sake of completeness. First we set

$$\tilde{A} = \frac{A}{|A|^{1/n}}, \qquad \tilde{B} = \frac{B}{|B|^{1/n}}$$

so that $|\tilde{A}| = |\tilde{B}| = 1$. Then we define

$$\tilde{S} := \mu\tilde{A} + (1 - \mu)\tilde{B} \quad \text{with} \quad \mu = \frac{(1 - \lambda)|A|^{1/n}}{(1 - \lambda)|A|^{1/n} + \lambda |B|^{1/n}},$$

and observe that $|\tilde{S}| \geq 1$ by the Brunn-Minkowski inequality. It is easily seen that

$$\tilde{S} = \frac{S}{(1-\lambda)|A|^{1/n} + \lambda|B|^{1/n}}.$$

Now we see that the hypothesis (14) holds for $\tilde{A}, \tilde{B}, \tilde{S}$, indeed

$$|\tilde{S}| - 1 = \frac{|S| - \left[(1-\lambda)|A|^{1/n} + \lambda|B|^{1/n}\right]^n}{\left[(1-\lambda)|A|^{1/n} + \lambda|B|^{1/n}\right]^n} \leq \delta,$$

by (16). Finally Proposition 2.2 applied to $\tilde{A}, \tilde{B}$ and $\tilde{S}$ implies the result and this concludes the proof. □

2.3 The Equivalence Between BBL and BM Inequalities

The equivalence between the two inequalities is well known and it becomes apparent as soon as one notices that the (p, λ)-supremal convolution defined in (7) corresponds to the Minkowski linear combinations of the graphs of f^p and g^p. In particular, for $p = 1$, (2) coincides with (12) where $A = \{(x, t) \in \mathbb{R}^{n+1} : 0 \leq t \leq f(x)\}$ and $B = \{(x, t) \in \mathbb{R}^{n+1} : 0 \leq t \leq g(x)\}$.

To be precise, that Proposition 1.1 implies (12) is easily seen by applying (2) to the case $f = \chi_A$, $g = \chi_B$, $h = \chi_{(1-\lambda)A+\lambda B}$, $p = +\infty$. The opposite implication can be proved in several ways; hereafter we present a proof due to Klartag [23], which is particularly useful for our goals.

To begin, given two integers $n, s > 0$, let $f : \mathbb{R}^n \longrightarrow [0, +\infty)$ be an integrable function with nonempty support (to avoid the trivial case in which f is identically zero). Following Klartag's notations and ideas [23] (see also [1]), we associate with f the nonempty measurable set

$$K_{f,s} = \left\{(x, y) \in \mathbb{R}^{n+s} = \mathbb{R}^n \times \mathbb{R}^s : x \in \text{supp}(f), |y| \leq f(x)^{1/s}\right\}, \qquad (17)$$

where obviously $x \in \mathbb{R}^n$ and $y \in \mathbb{R}^s$. In other words, $K_{f,s}$ is the subset of $\mathbb{R}^{n+s}$ obtained as union of the s-dimensional closed balls of center $(x, 0)$ and radius $f(x)^{1/s}$, for x belonging to the support of f, or, if you prefer, the set in $\mathbb{R}^{n+s}$ obtained by rotating with respect to $y = 0$ the $(n + 1)$-dimensional set $\{(x, y) \in \mathbb{R}^{n+s} : 0 \leq y_1 \leq f(x)^{1/s}, y_2 = \cdots = y_s = 0\}$.

We observe that $K_{f,s}$ is convex if and only if f is $(1/s)$-concave [that is for us a function f having compact convex support such that $f^{1/s}$ is concave on supp (f)]. If supp (f) is compact, then $K_{f,s}$ is bounded if and only if f is bounded.

Moreover, thanks to Fubini's Theorem, it holds

$$|K_{f,s}| = \int_{\text{supp}(f)} \omega_s \cdot \left(f(x)^{1/s} \right)^s \, dx = \omega_s \int_{\mathbb{R}^n} f(x) \, dx. \tag{18}$$

In this way, the integral of f coincides, up to the constant ω_s, with the volume of $K_{f,s}$. Now we will use this simple identity to prove Proposition 1.2 as a direct application of the BM inequality.

Although of course the set $K_{f,s}$ depends heavily on s, for simplicity from now on we will remove the subindex s and just write K_f for $K_{f,s}$.

Let us start with the simplest case, when $p = 1/s$ with s positive integer.

Proposition 2.5 (BBL, Case $1/p = s \in \mathbb{N}$) *Let n, s be positive integers, $\lambda \in (0,1)$ and $f, g, h : \mathbb{R}^n \longrightarrow [0, +\infty)$ be integrable functions, with $\int f > 0$ and $\int g > 0$. Assume that for any $x_0 \in \text{supp}(f)$, $x_1 \in \text{supp}(g)$*

$$h\left((1-\lambda)x_0 + \lambda x_1\right) \geq \left[(1-\lambda)f(x_0)^{1/s} + \lambda g(x_1)^{1/s}\right]^s. \tag{19}$$

Then

$$\left(\int_{\mathbb{R}^n} h \, dx \right)^{\frac{1}{n+s}} \geq (1-\lambda) \left(\int_{\mathbb{R}^n} f \, dx \right)^{\frac{1}{n+s}} + \lambda \left(\int_{\mathbb{R}^n} g \, dx \right)^{\frac{1}{n+s}}. \tag{20}$$

Proof Since the integrals of f and g are positive, the sets K_f and K_g have positive measure. Let Ω_λ be the Minkowski convex combination (with coefficient λ) of $\Omega_0 = \text{supp}(f)$ and $\Omega_1 = \text{supp}(g)$. Now consider the function $h_{s,\lambda}$ as defined by (7); to simplify the notation, we will denote $h_{s,\lambda}$ by h_λ from now on. First notice that the support of h_λ is Ω_λ. Then it is easily seen that

$$K_{h_\lambda} = (1-\lambda)K_f + \lambda K_g. \tag{21}$$

Moreover, since $h \geq h_\lambda$ by assumption (19), we have

$$K_h \supseteq K_{h_\lambda}. \tag{22}$$

By applying Proposition 2.1 to K_{h_λ}, K_f, K_g we get

$$|K_h|^{\frac{1}{n+s}} \geq |K_{h_\lambda}|^{\frac{1}{n+s}} \geq (1-\lambda) |K_f|^{\frac{1}{n+s}} + \lambda |K_g|^{\frac{1}{n+s}}, \tag{23}$$

where $|K_{h_\lambda}|$ possibly means the outer measure of the set K_{h_λ}.

Finally (18) yields

$$|K_h| = \omega_s \int_{\mathbb{R}^n} h \, dx, \qquad |K_f| = \omega_s \int_{\mathbb{R}^n} f \, dx, \qquad |K_g| = \omega_s \int_{\mathbb{R}^n} g \, dx,$$

thus dividing (23) by $\omega_s^{\frac{1}{n+s}}$ we get (20).   □

Next we show how it is possible to generalize Proposition 2.5 to a positive rational index s. The idea is to apply again the Brunn-Minkowski inequality to sets that generalize those of the type (17). What follows is a slight variant of the proof of Theorem 2.1 in [23].

The case of a positive rational index s requires the following definition. Given $f : \mathbb{R}^n \longrightarrow [0, +\infty)$ integrable and a positive integer q (it will be the denominator of the rational s) we consider the auxiliary function $\tilde{f} : \mathbb{R}^{nq} \longrightarrow [0, +\infty)$ defined as

$$\tilde{f}(x) = \tilde{f}(x_1, \ldots, x_q) = \prod_{j=1}^{q} f(x_j), \tag{24}$$

where $x = (x_1, \ldots, x_q) \in (\mathbb{R}^n)^q$. We observe that, by construction,

$$\int_{\mathbb{R}^{nq}} \tilde{f}\, dx = \left(\int_{\mathbb{R}^n} f\, dx \right)^q; \tag{25}$$

moreover $\quad \operatorname{supp} \tilde{f} = (\operatorname{supp} f) \times \ldots \times (\operatorname{supp} f) = (\operatorname{supp} f)^q$.

As just done, from now on we write A^q to indicate the Cartesian product of q copies of a set A.

Remark 2.6 Let A, B be nonempty sets, $q > 0$ be an integer, μ a real. Clearly

$$(A + B)^q = A^q + B^q, \qquad (\mu A)^q = \mu A^q.$$

To compare products of real numbers of the type (24) the following lemma is useful. It's a consequence of Hölder's inequality (see [21], Theorem 10) for families of real numbers (in our case for two sets of q positive numbers).

Lemma 2.7 *Given an integer $q > 0$, let $\{a_1, \ldots, a_q\}$, $\{b_1, \ldots, b_q\}$ be two sets of q real numbers. Then*

$$\left| \prod_{j=1}^{q} a_j \right| + \left| \prod_{j=1}^{q} b_j \right| \leq \left[\prod_{j=1}^{q} \left(|a_j|^q + |b_j|^q \right) \right]^{1/q}.$$

From this lemma we deduce the following.

Corollary 2.8 *Let $\lambda \in (0, 1)$, $s = \frac{p}{q}$ with integers $p, q > 0$.*
Given $f, g : \mathbb{R}^n \longrightarrow [0, +\infty)$, $x_1, \ldots, x_q, x_1', \ldots, x_q' \in \mathbb{R}^n$, it holds

$$(1 - \lambda) \prod_{j=1}^{q} f(x_j)^{1/p} + \lambda \prod_{j=1}^{q} g(x_j')^{1/p} \leq \prod_{j=1}^{q} \left[(1 - \lambda) f(x_j)^{1/s} + \lambda g(x_j')^{1/s} \right]^{1/q}.$$

Proof Observing that

$$(1 - \lambda) \prod_{j=1}^{q} f(x_j)^{1/p} + \lambda \prod_{j=1}^{q} g(x_j')^{1/p} = \prod_{j=1}^{q} (1 - \lambda)^{1/q} f(x_j)^{1/p} + \prod_{j=1}^{q} \lambda^{1/q} g(x_j')^{1/p},$$

the result follows directly from Lemma 2.7 applied to $\{a_1, \ldots, a_q\}$, $\{b_1, \ldots, b_q\}$ with

$$a_j = (1 - \lambda)^{1/q} f(x_j)^{1/p}, \qquad b_j = \lambda^{1/q} g(x_j')^{1/p}, \qquad j = 1, \ldots, q.$$

$\square$

Let

$$s = \frac{p}{q}$$

with integers $p, q > 0$ that we can assume are coprime.

Given an integrable function $f : \mathbb{R}^n \longrightarrow [0, +\infty)$ not identically zero, we define the nonempty measurable subset of $\mathbb{R}^{nq+p}$

$$W_{f,s} = K_{\tilde{f},p} = \{(x, y) \in (\mathbb{R}^n)^q \times \mathbb{R}^p : x \in \text{supp}(\tilde{f}), |y| \leq \tilde{f}(x)^{1/p}\} \qquad (26)$$

$$= \left\{ (x_1, \ldots, x_q, y) \in (\mathbb{R}^n)^q \times \mathbb{R}^p : x_j \in \text{supp}(f) \; \forall j = 1, \ldots, q, \quad |y| \leq \prod_{j=1}^{q} f(x_j)^{1/p} \right\}.$$

We notice that this definition naturally generalizes (17), since in the case of an integer $s > 0$ it holds $s = p$, $q = 1$, so in this case $\tilde{f} = f$ and $W_{f,s} = K_f$.

As for $K_{f,s}$, for simplicity we will remove systematically the subindex s and write W_f in place of $W_{f,s}$ if there is no possibility of confusion. Clearly

$$|W_f| = \int_{\text{supp}(\tilde{f})} \omega_p \cdot (\tilde{f}(x)^{1/p})^p \, dx = \omega_p \int_{\mathbb{R}^{nq}} \tilde{f}(x) \, dx = \omega_p \left(\int_{\mathbb{R}^n} f(x) \, dx \right)^q \qquad (27)$$

where the last equality is given by (25).

Moreover we see that W_f is convex if and only if $\tilde{f}$ is $\frac{1}{p}$-concave (that is, if and only if f is $\frac{1}{s}$-concave, see Lemma 4.1 later on). Next we set

$$W = (1 - \lambda) W_f + \lambda W_g. \qquad (28)$$

Finally, we notice that, by (21), we have

$$W = K_{\tilde{h}_{p,\lambda},p},$$

where $\tilde{h}_{p,\lambda}$ is the $(1/p, \lambda)$-supremal convolution of $\tilde{f}$ and $\tilde{g}$ as defined in (7). In other words, W is the set made by the elements $(z, y) \in (\mathbb{R}^n)^q \times \mathbb{R}^p$ such that $z \in (1 - \lambda) \operatorname{supp}(\tilde{f}) + \lambda \operatorname{supp}(\tilde{g})$ and

$$
\begin{aligned}
|y| \leq \sup \{ &(1 - \lambda) \tilde{f}(x)^{1/p} + \lambda \tilde{g}(x')^{1/p} : \\
&z = (1 - \lambda)x + \lambda x', x \in \operatorname{supp}(\tilde{f}), x' \in \operatorname{supp}(\tilde{g}) \}.
\end{aligned}
\tag{29}
$$

Lemma 2.9 *With the notations introduced above, it holds*

$$
W \subseteq W_{h_\lambda} \subseteq W_h,
$$

where h_λ is the $(1/s, \lambda)$-supremal convolution of f, g, and h is as in Proposition 1.2.

Proof The second inclusion is obvious, since $h \geq h_\lambda$ by assumption (4). Regarding the other inclusion, first we notice that (26) and Remark 2.6 yield

$$
\begin{aligned}
W_{h_\lambda} &= \{(z, y) \in (\mathbb{R}^n)^q \times \mathbb{R}^p : z \in \operatorname{supp}(\widetilde{h_\lambda}), \ |y| \leq \widetilde{h_\lambda}(z)^{1/p}\} \\
&= \{(z, y) \in (\mathbb{R}^n)^q \times \mathbb{R}^p : z \in ((1 - \lambda) \operatorname{supp}(f) + \lambda \operatorname{supp}(g))^q, \ |y| \leq \widetilde{h_\lambda}(z)^{1/p}\} \\
&= \{(z, y) \in (\mathbb{R}^n)^q \times \mathbb{R}^p : z \in (1 - \lambda) \operatorname{supp}(\tilde{f}) + \lambda \operatorname{supp}(\tilde{g}), \ |y| \leq \widetilde{h_\lambda}(z)^{1/p}\},
\end{aligned}
$$

where $\widetilde{h_\lambda}$ is the function associated to h_λ by (24). To conclude it is sufficient to compare this with the condition given by (29).

For every $z \in (1 - \lambda) \operatorname{supp}(\tilde{f}) + \lambda \operatorname{supp}(\tilde{g})$ consider

$$
\sup \{(1 - \lambda) \tilde{f}(x)^{1/p} + \lambda \tilde{g}(x')^{1/p}\} = \sup \left\{ (1 - \lambda) \prod_{j=1}^{q} f(x_j)^{1/p} + \lambda \prod_{j=1}^{q} g(x'_j)^{1/p} \right\},
$$

where the supremum is made with respect to $x \in \operatorname{supp}(\tilde{f})$, $x' \in \operatorname{supp}(\tilde{g})$ such that $z = (1 - \lambda)x + \lambda x'$. Corollary 2.8 then implies

$$
\begin{aligned}
\sup \{(1 - \lambda) \tilde{f}(x)^{1/p} + \lambda \tilde{g}(x')^{1/p}\} &\leq \sup \left\{ \prod_{j=1}^{q} [(1 - \lambda) f(x_j)^{1/s} + \lambda g(x'_j)^{1/s}]^{1/q} \right\} \\
&\leq \prod_{j=1}^{q} \left\{ \sup [(1 - \lambda) f(x_j)^{1/s} + \lambda g(x'_j)^{1/s}]^{1/q} \right\} \\
&= \prod_{j=1}^{q} \left\{ h_\lambda \left((1 - \lambda)x_j + \lambda x'_j \right)^{1/qs} \right\} \\
&= \widetilde{h_\lambda} \left((1 - \lambda)x + \lambda x' \right)^{1/p} = \widetilde{h_\lambda}(z)^{1/p},
\end{aligned}
$$

having used the definition (24) in the penultimate equality. Therefore if

$$|y| \le \sup \left\{ (1-\lambda)\tilde{f}(x)^{1/p} + \lambda\tilde{g}(x')^{1/p} \right\},$$

that is if $(z, y) \in W$ by (29), then

$$|y| \le \widetilde{h_\lambda}(z)^{1/p},$$

i.e. $(z, y) \in W_{h_\lambda}$. This concludes the proof.   □

We are ready to prove the following version of the Borell-Brascamp-Lieb inequality, which holds for any positive real index s (and in fact also for $s = 0$).

Proposition 2.10 (BBL for $p > 0$) *Let $s > 0$, $\lambda \in (0, 1)$, let $n > 0$ be integer. Given $f, g, h : \mathbb{R}^n \longrightarrow [0, +\infty)$ integrable such that $\int f > 0$ and $\int g > 0$, assume that for any $x_0 \in \text{supp}(f)$, $x_1 \in \text{supp}(g)$*   ·

$$h\left((1-\lambda)x_0 + \lambda x_1\right) \ge \left[(1-\lambda)f(x_0)^{1/s} + \lambda g(x_1)^{1/s}\right]^s. \tag{30}$$

Then

$$\left(\int_{\mathbb{R}^n} h \, dx\right)^{\frac{1}{n+s}} \ge (1-\lambda)\left(\int_{\mathbb{R}^n} f \, dx\right)^{\frac{1}{n+s}} + \lambda\left(\int_{\mathbb{R}^n} g \, dx\right)^{\frac{1}{n+s}}. \tag{31}$$

Proof Assume first that $s > 0$ is rational and let $s = \frac{p}{q}$ with p, q coprime positive integers. Thanks to (28) we can apply Proposition 2.1 to W_f, W_g (that are nonempty measurable subsets of $\mathbb{R}^{nq+p}$), so

$$|W|^{\frac{1}{nq+p}} \ge (1-\lambda)\left|W_f\right|^{\frac{1}{nq+p}} + \lambda\left|W_g\right|^{\frac{1}{nq+p}},$$

where $|W|$ possibly means the outer measure of the set W. On the other hand Lemma 2.9 implies $|W_h| \ge |W|$, thus

$$|W_h|^{\frac{1}{nq+p}} \ge (1-\lambda)\left|W_f\right|^{\frac{1}{nq+p}} + \lambda\left|W_g\right|^{\frac{1}{nq+p}}.$$

Finally the latter inequality with the identity (27) is equivalent to

$$\omega_p^{\frac{1}{nq+p}}\left(\int_{\mathbb{R}^n} h \, dx\right)^{\frac{q}{nq+p}} \ge \omega_p^{\frac{1}{nq+p}}\left[(1-\lambda)\left(\int_{\mathbb{R}^n} f \, dx\right)^{\frac{q}{nq+p}} + \lambda\left(\int_{\mathbb{R}^n} g \, dx\right)^{\frac{q}{nq+p}}\right].$$

Dividing by $\omega_p^{\frac{1}{nq+p}}$ we get (31), since

$$\frac{q}{nq+p} = \frac{q}{q(n+s)} = \frac{1}{n+s}$$

is exactly the required index. The case of a real $s > 0$ (and also $s = 0$) follows by a standard approximation argument. $\square$

3 The Proof of Theorem 1.3

The idea is to apply the result of Figalli-Jerison, more precisely Corollary 2.4, to the sets K_{h_λ}, K_f, K_g, and then translate the result in terms of the involved functions. We remember that with h_λ we denote the function $h_{s,\lambda}$ given by (7).

We also recall that we set $F = \int f$ and $G = \int g$.

Thanks to (18), assumption (9) is equivalent to

$$\omega_s^{-1} |K_h| \leq \omega_s^{-1} \left[(1 - \lambda) |K_f|^{\frac{1}{n+s}} + \lambda |K_g|^{\frac{1}{n+s}} \right]^{n+s} + \varepsilon,$$

which, by (22), implies

$$|K_{h_\lambda}| \leq \left[(1 - \lambda) |K_f|^{\frac{1}{n+s}} + \lambda |K_g|^{\frac{1}{n+s}} \right]^{n+s} + \varepsilon \omega_s. \tag{32}$$

If ε is small enough, by virtue of (21) we can apply Corollary 2.4 to the sets K_{h_λ}, K_f, K_g and from (32) we obtain that they satisfy assumption (16) with

$$\delta = \frac{\varepsilon \omega_s}{\mathcal{M}_{\frac{1}{n+s}} (|K_f|, |K_g|; \lambda)} = \frac{\varepsilon}{\mathcal{M}_{\frac{1}{n+s}} (F, G; \lambda)}. \tag{33}$$

Then, if $\delta \leq e^{-M_{n+s}(\tau)}$, there exist a convex $K \subset \mathbb{R}^{n+s}$ and two homothetic copies $\hat{K}_f$ and $\hat{K}_g$ of K_f and K_g such that $|\hat{K}_f| = |\hat{K}_g| = 1$ and

$$\left(\hat{K}_f \cup \hat{K}_g \right) \subseteq K \tag{34}$$

and

$$\left| K \setminus \hat{K}_f \right| + \left| K \setminus \hat{K}_g \right| \leq \tau^{-N_{n+s}} \left(\frac{\varepsilon}{\mathcal{M}_{\frac{1}{n+s}} (F, G; \lambda)} \right)^{\sigma_{n+s}(\tau)}. \tag{35}$$

Remark 3.1 Since $|\hat{K}_f| = |\hat{K}_g| = 1$, (35) implies that the convex set K has finite positive measure. Then it is bounded (since convex), whence (34) yields the boundedness of K_f and K_g which in turn implies the boundedness of the functions f and g. For simplicity, we can assume the convex K is compact (possibly substituting it with its closure).

In what follows, we indicate with $(x,y) \in \mathbb{R}^n \times \mathbb{R}^s$ an element of $\mathbb{R}^{n+s}$. When we say [see just before (34)] that $\hat{K}_f$ and $\hat{K}_g$ are homothetic copies of K_f and K_g, we mean that there exist $z_0 = (x_0, y_0) \in \mathbb{R}^{n+s}$ and $z_1 = (x_1, y_1) \in \mathbb{R}^{n+s}$ such that

$$\hat{K}_f = |K_f|^{-\frac{1}{n+s}} \left(K_f - z_0\right) \quad \text{and} \quad \hat{K}_g = |K_g|^{-\frac{1}{n+s}} \left(K_g - z_1\right) . \tag{36}$$

Clearly, without loss of generality we can take $z_0 = 0$.

To conclude the proof, we want now to show that, up to a suitable symmetrization, we can take $y_1 = 0$ (i.e. the translation of the homothetic copy $\hat{K}_g$ of K_g is horizontal) and that the convex set K given by Figalli and Jerison can be taken of the type K_u for some $\frac{1}{s}$-concave function u.

For this, let us introduce the following Steiner type symmetrization in $\mathbb{R}^{n+s}$ with respect to the n-dimensional hyperspace $y = 0$ (see for instance [8]). Let C be a bounded measurable set in $\mathbb{R}^{n+s}$, for every $\bar{x} \in \mathbb{R}^n$ we set

$$C(\bar{x}) = \{y \in \mathbb{R}^s : (\bar{x}, y) \in C\}$$

and

$$r_C(\bar{x}) = \left(\omega_s^{-1}|C(\bar{x})|\right)^{1/s} . \tag{37}$$

Then we define the S-symmetrand of C as follows

$$S(C) = \left\{(\bar{x}, y) \in \mathbb{R}^{n+s} : C \cap \{x = \bar{x}\} \neq \emptyset, |y| \leq r_C(\bar{x})\right\} , \tag{38}$$

i.e. $S(C)$ is obtained as union of the s-dimensional closed balls of center $(\bar{x}, 0)$ and radius $r_C(\bar{x})$, for $\bar{x} \in \mathbb{R}^n$ such that $C \cap \{x = \bar{x}\}$ is nonempty. Thus, fixed $\bar{x}$, the (s-dimensional) measure of the corresponding section of $S(C)$ is

$$\mathcal{H}^s(S(C) \cap \{x = \bar{x}\}) = \omega_s r_C(\bar{x})^s = |C(\bar{x})| . \tag{39}$$

We describe the main properties of S-symmetrization, for bounded measurable subsets of $\mathbb{R}^{n+s}$:

(i) if $C_1 \subseteq C_2$ then $S(C_1) \subseteq S(C_2)$ (obvious by definition);
(ii) $|C| = |S(C)|$ (consequence of (39) and Fubini's Theorem) so the S-symmetrization is measure preserving;
(iii) if C is convex then $S(C)$ is convex (the proof is based on the BM inequality in $\mathbb{R}^s$ and, for the sake of completeness, is given in the Appendix).

Now we symmetrize $K, \hat{K}_f, \hat{K}_g$ [and then replace them with $S(K), S(\hat{K}_f), S(\hat{K}_g)$].
Clearly

$$S(\hat{K}_f) = \hat{K}_f, \tag{40}$$

$$S(\hat{K}_g) = S\left(|K_g|^{-\frac{1}{n+s}}(K_g - (x_1, y_1))\right) = |K_g|^{-\frac{1}{n+s}}(K_g - (x_1, 0)). \tag{41}$$

Moreover, (iii) implies that $S(K)$ is convex and by (i) and (34) we have

$$(S(\hat{K}_f) \cup S(\hat{K}_g)) \subseteq S(K). \tag{42}$$

The latter, (35) and Fubini's theorem imply

$$\left|S(K) \setminus S(\hat{K}_f)\right| + \left|S(K) \setminus S(\hat{K}_g)\right| \leqslant \tau^{-N_{n+s}}\left(\frac{\varepsilon}{\mathcal{M}_{\frac{1}{n+s}}(F, G; \lambda)}\right)^{\sigma_{n+s}(\tau)}. \tag{43}$$

Finally we notice that $S(K)$ is a compact convex set of the desired form.

Remark 3.2 Consider the set K_u associated to a function $u : \mathbb{R}^n \to [0, +\infty)$ by (17)
and let $\bar{x} \in \mathbb{R}^n, \bar{z} = (\bar{x}, 0) \in \mathbb{R}^{n+s}, \mu > 0$ and

$$H = \mu\left(K_u - \bar{z}\right).$$

Then

$$H = K_v$$

[the set associated to v by (17)] where

$$v(x) = \mu^s u\left(\frac{x - \bar{x}}{\mu}\right). \tag{44}$$

From the previous remarks, we see that the sets $S(\hat{K}_f)$ and $S(\hat{K}_g)$ are in fact
associated via (17) to two functions $\hat{f}$ and $\hat{g}$, such that

$$S(\hat{K}_f) = K_{\hat{f}}, \quad S(\hat{K}_g) = K_{\hat{g}}, \tag{45}$$

and $\hat{f}$ and $\hat{g}$ are s-equivalent to f and g respectively, in the sense of (8) with

$$\mu_f = (\omega_s F)^{\frac{-1}{n+s}}, \qquad \mu_g = (\omega_s G)^{\frac{-1}{n+s}}. \tag{46}$$

We notice that the support sets Ω_0 and Ω_1 of $\hat{f}$ and $\hat{g}$ are given by

$$\Omega_0 = \{x \in \mathbb{R}^n : (x,0) \in S(\hat{K}_f)\}, \qquad \Omega_1 = \{x \in \mathbb{R}^n : (x,0) \in S(\hat{K}_g)\}$$

and that they are in fact homothetic copies of the support sets of the original functions f and g.

Now we want to find a $\frac{1}{s}$-concave function u such that $S(K)$ is associated to u via (17). We define $u : \mathbb{R}^n \longrightarrow [0, +\infty)$ as follows

$$u(x) = \begin{cases} r_K(x)^s & \text{if } (x,0) \in S(K), \\ 0 & \text{otherwise}, \end{cases}$$

and prove that

$$K_u = S(K). \tag{47}$$

First notice that

$$\text{supp}\,(u) = \{x \in \mathbb{R}^n : (x,0) \in S(K)\}. \tag{48}$$

Indeed we have $\{z \in \mathbb{R}^n : u(z) > 0\} \subseteq \{x \in \mathbb{R}^n : (x,0) \in S(K)\}$, whence $\text{supp}\,(u) = \overline{\{z \in \mathbb{R}^n : u(z) > 0\}} \subseteq \{x \in \mathbb{R}^n : (x,0) \in S(K)\}$, since the latter is closed. Vice versa let x such that $(x,0) \in S(K)$.

If $r_K(x) > 0$ [see (37)] then $x \in \text{supp}\,(u)$ obviously. Otherwise suppose $r_K(x) = 0$, then, by the convexity of $S(K)$ and the fact that $S(K)$ is not contained in $\{y = 0\}$, evidently

$$[(U \setminus \{x\}) \cap \{z \in \mathbb{R}^n : r_K(z) > 0\}] \neq \emptyset$$

for every neighborhood U of x, i.e. $x \in \text{supp}\,(u)$.

By the definition of u and (17), using (48), we get

$$\begin{aligned} K_u &= \{(x,y) \in \mathbb{R}^n \times \mathbb{R}^s : x \in \text{supp}\,(u), |y| \leq u(x)^{1/s}\} \\ &= \{(x,y) \in \mathbb{R}^n \times \mathbb{R}^s : (x,0) \in S(K), |y| \leq u(x)^{1/s}\} \\ &= \{(x,y) \in \mathbb{R}^n \times \mathbb{R}^s : (x,0) \in S(K), |y| \leq r_K(x)\} = S(K). \end{aligned}$$

Therefore we have shown (47) and from the convexity of K follows that u is a $\frac{1}{s}$-concave function. Being $K_u \supseteq \left(K_{\hat{f}} \cup K_{\hat{g}}\right)$, clearly

$$\text{supp}\,(u) \supseteq (\Omega_0 \cup \Omega_1), \qquad u \geq \hat{f} \text{ in } \Omega_0, \qquad u \geq \hat{g} \text{ in } \Omega_1.$$

The final estimate can be deduced from (43). Indeed, thanks to (18), we get

$$\left| K_u \setminus K_{\hat{f}} \right| = |K_u| - \left| K_{\hat{f}} \right| = \omega_s \int_{\mathbb{R}^n} (u - \hat{f}) \, dx,$$

and the same equality holds for $\left| K_u \setminus K_{\hat{g}} \right|$. So (43) becomes

$$\int_{\mathbb{R}^n} (u - \hat{f}) \, dx + \int_{\mathbb{R}^n} (u - \hat{g}) \, dx \leq \omega_s^{-1} \tau^{-N_n+s} \left(\frac{\varepsilon}{\mathcal{M}_{\frac{1}{n+s}}(F, G; \lambda)} \right)^{\sigma_{n+s}(\tau)},$$

that is the desired result.

4 A Generalization to the Case s Positive Rational

We explain how Theorem 1.3 can be generalized to a positive rational index s. Given $f : \mathbb{R}^n \longrightarrow [0, +\infty)$ and an integer $q > 0$, we consider the auxiliary function $\tilde{f} : \mathbb{R}^{nq} \longrightarrow [0, +\infty)$ given by (24), i.e.

$$\tilde{f}(x) = \tilde{f}(x_1, \ldots, x_q) = \prod_{j=1}^{q} f(x_j),$$

with $x = (x_1, \ldots, x_q) \in (\mathbb{R}^n)^q$. Clearly f is bounded if and only if $\tilde{f}$ is bounded. We study further properties of functions of type (24).

Lemma 4.1 *Given an integer $q > 0$, and a real $t > 0$ let $\tilde{u} : \mathbb{R}^{nq} \longrightarrow [0, +\infty)$ be a function of the type (24). Then $\tilde{u}$ is t-concave if and only if the function $u : \mathbb{R}^n \longrightarrow [0, +\infty)$ is (qt)-concave.*

Proof Suppose first that $\tilde{u}^t$ is concave. Fixed $\lambda \in (0, 1)$, $x, x' \in \mathbb{R}^n$, we consider the element of $\mathbb{R}^{nq}$ which has all the q components identical to $(1 - \lambda)x + \lambda x'$. From hypothesis it holds

$$\tilde{u}^t \left((1 - \lambda)x + \lambda x', \ldots, (1 - \lambda)x + \lambda x' \right) \geq (1 - \lambda)\tilde{u}^t(x, \ldots, x) + \lambda \tilde{u}^t(x', \ldots, x'),$$

i.e. [thanks to (24)]

$$u^{qt} \left((1 - \lambda)x + \lambda x' \right) \geq (1 - \lambda)u^{qt}(x) + \lambda u^{qt}(x').$$

Thus u^{qt} is concave.

Vice versa assume that u^{qt} is concave, and fix $\lambda \in (0, 1)$, $x = (x_1, \ldots, x_q)$, $x' = (x_1', \ldots, x_q') \in (\mathbb{R}^n)^q$. We have

$$\tilde{u}^t \left((1-\lambda)x + \lambda x'\right) = \prod_{j=1}^{q} u^t \left((1-\lambda)x_j + \lambda x_j'\right) = \prod_{j=1}^{q} \left[u^{qt} \left((1-\lambda)x_j + \lambda x_j'\right)\right]^{1/q}$$

$$\geq \prod_{j=1}^{q} \left[(1-\lambda)u^{qt}(x_j) + \lambda u^{qt}(x_j')\right]^{1/q} \geq \prod_{j=1}^{q} (1-\lambda)^{1/q} u^t(x_j)$$

$$+ \prod_{j=1}^{q} \lambda^{1/q} u^t(x_j') = (1-\lambda) \prod_{j=1}^{q} u^t(x_j) + \lambda \prod_{j=1}^{q} u^t(x_j')$$

$$= (1-\lambda)\tilde{u}^t(x) + \lambda \tilde{u}^t(x'),$$

where the first inequality holds by concavity of u^{qt}, while in the second one we have used Lemma 2.7 with $a_j = (1-\lambda)^{1/q} u^t(x_j)$, $b_j = \lambda^{1/q} u^t(x_j')$. Hence u^t is concave. $\square$

Lemma 4.2 *Let $q > 0$ integer and $u \geq f \geq 0$ in $\mathbb{R}^n$. Then*

$$\tilde{u} - \tilde{f} \geq \widetilde{u - f}.$$

Proof The proof is by induction on the integer $q \geq 1$. The case $q = 1$ is trivial, because in such case $\tilde{u} = u$, $\tilde{f} = f$, $\widetilde{u - f} = u - f$. For the inductive step assume that the result is true until the index q, and denote with $\tilde{\tilde{u}}, \tilde{\tilde{f}}, \widetilde{\widetilde{u - f}}$ the respective functions of index $q + 1$. By the definition (24)

$$\left(\tilde{\tilde{u}} - \tilde{\tilde{f}}\right)(x_1, \ldots, x_{q+1}) = \tilde{u}(x_1, \ldots, x_q)u(x_{q+1}) - \tilde{f}(x_1, \ldots, x_q)f(x_{q+1}),$$

$$\widetilde{\widetilde{u - f}}(x_1, \ldots, x_{q+1}) = \widetilde{u - f}(x_1, \ldots, x_q) \cdot (u - f)(x_{q+1}).$$

These two equalities imply

$$\left(\tilde{\tilde{u}} - \tilde{\tilde{f}}\right)(x_1, \ldots, x_{q+1})$$

$$= \widetilde{\widetilde{u - f}}(x_1, \ldots, x_{q+1}) - \widetilde{u - f}(x_1, \ldots, x_q) \cdot \left[u(x_{q+1}) - f(x_{q+1})\right]$$

$$+ \tilde{u}(x_1, \ldots, x_q)u(x_{q+1}) - \tilde{f}(x_1, \ldots, x_q)f(x_{q+1})$$

$$\geq \widetilde{\widetilde{u - f}}(x_1, \ldots, x_{q+1}) - \left(\tilde{u} - \tilde{f}\right)(x_1, \ldots, x_q)\left[u(x_{q+1}) - f(x_{q+1})\right]$$

$$+ \tilde{u}(x_1, \ldots, x_q)u(x_{q+1}) - \tilde{f}(x_1, \ldots, x_q)f(x_{q+1})$$

$$= \widetilde{u - f}(x_1, \ldots, x_{q+1}) + f(x_{q+1}) \left[\tilde{u}(x_1, \ldots, x_q) - \tilde{f}(x_1, \ldots, x_q) \right]$$
$$+ \tilde{f}(x_1, \ldots, x_q) \left[u(x_{q+1}) - f(x_{q+1}) \right]$$
$$\geq \widetilde{u - f}(x_1, \ldots, x_{q+1}),$$

having used the inductive hypothesis and the assumption $u \geq f \geq 0$. □

Corollary 4.3 *Given an integer $n > 0$, $\lambda \in (0, 1)$, $s = \frac{p}{q}$ with p, q positive integers, let $f, g \in L^1(\mathbb{R}^n)$ be nonnegative compactly supported functions such that*

$$F = \int_{\mathbb{R}^n} f \, dx > 0 \quad and \quad G = \int_{\mathbb{R}^n} g \, dx > 0.$$

Let $h : \mathbb{R}^n \longrightarrow [0, +\infty)$ satisfy assumption (19) and suppose there exists $\varepsilon > 0$ small enough such that

$$\left(\int_{\mathbb{R}^n} h \, dx \right)^q \leq \left[\mathcal{M}_{\frac{1}{n+s}} (F, G; \lambda) \right]^q + \varepsilon. \tag{49}$$

Then there exist a $\frac{1}{p}$-concave function $u' : \mathbb{R}^{nq} \longrightarrow [0, +\infty)$ and two functions $\hat{f}, \hat{g} : \mathbb{R}^{nq} \longrightarrow [0, +\infty)$, p-equivalent to $\tilde{f}$ and $\tilde{g}$ [given by (24)] in the sense of (8) with

$$\mu_{\tilde{f}} = \omega_p^{\frac{-1}{nq+p}} F^{\frac{-1}{n+s}}, \qquad \mu_{\tilde{g}} = \omega_p^{\frac{-1}{nq+p}} G^{\frac{-1}{n+s}},$$

such that the following hold:

$$u' \geq \hat{f}, \qquad u' \geq \hat{g},$$

and

$$\int_{\mathbb{R}^{nq}} (u' - \hat{f}) dx + \int_{\mathbb{R}^{nq}} (u' - \hat{g}) \, dx \leq C_{nq+p} \left(\frac{\varepsilon}{\mathcal{M}_{\frac{1}{nq+p}} (F^q, G^q; \lambda)} \right). \tag{50}$$

Proof We can assume $h = h_\lambda$. Since f and g are nonnegative compactly supported functions belonging to $L^1(\mathbb{R}^n)$, thus by (24) $\tilde{f}, \tilde{g}$ are nonnegative compactly supported functions belonging to $L^1(\mathbb{R}^{nq})$. The assumption (49) is equivalent, considering the corresponding functions $\tilde{f}, \tilde{g}, \tilde{h} : \mathbb{R}^{nq} \longrightarrow [0, +\infty)$ and using (25), to

$$\int_{\mathbb{R}^{nq}} \tilde{h}\, dx \leq \left[(1-\lambda) \left(\int_{\mathbb{R}^{nq}} \tilde{f}\, dx \right)^{\frac{1}{nq+qs}} + \lambda \left(\int_{\mathbb{R}^{nq}} \tilde{g}\, dx \right)^{\frac{1}{nq+qs}} \right]^{nq+qs} + \varepsilon$$

i.e.
$$\int_{\mathbb{R}^{nq}} \tilde{h}\, dx \leq \mathcal{M}_{\frac{1}{nq+p}} (F^q, G^q ; \lambda) + \varepsilon. \tag{51}$$

We notice that the index $qs = p$ is integer, while nq is exactly the dimension of the space in which $\tilde{f}, \tilde{g}, \tilde{h}$ are defined. To apply Theorem 1.3, we have to verify that $\tilde{f}, \tilde{g}, \tilde{h}$ satisfy the corresponding inequality (19) of index qs. Given $x_1, \ldots, x_q \in \mathrm{supp}(f)$, $x_1', \ldots, x_q' \in \mathrm{supp}(g)$, let $x = (x_1, \ldots, x_q)$, $x' = (x_1', \ldots, x_q') \in (\mathbb{R}^n)^q$. By hypothesis, we know that f, g, h satisfy (19), in particular for every $j = 1, \ldots, q$

$$h \left((1-\lambda)x_j + \lambda x_j' \right) \geq \left[(1-\lambda)f(x_j)^{1/s} + \lambda g(x_j')^{1/s} \right]^s.$$

This implies

$$\prod_{j=1}^{q} h \left((1-\lambda)x_j + \lambda x_j' \right) \geq \left[\prod_{j=1}^{q} \left[(1-\lambda)f(x_j)^{1/s} + \lambda g(x_j')^{1/s} \right] \right]^s$$

$$\geq \left[(1-\lambda) \left(\prod_{j=1}^{q} f(x_j) \right)^{1/qs} + \lambda \left(\prod_{j=1}^{q} g(x_j') \right)^{1/qs} \right]^{qs}, \tag{52}$$

where the last inequality is due to Corollary 2.8. By definition of (24), (52) means that for every $x \in \mathrm{supp}(\tilde{f})$, $x' \in \mathrm{supp}(\tilde{g})$ we have

$$\tilde{h} \left((1-\lambda)x + \lambda x' \right) \geq \left[(1-\lambda)\tilde{f}(x)^{1/qs} + \lambda \tilde{g}(x')^{1/qs} \right]^{qs},$$

i.e. the functions $\tilde{f}, \tilde{g}, \tilde{h} : \mathbb{R}^{nq} \longrightarrow [0, +\infty)$ satisfy the hypothesis (19) with the required index qs. Therefore we can apply Theorem 1.3 and conclude that there exist a $\frac{1}{p}$-concave function $u' : \mathbb{R}^{nq} \longrightarrow [0, +\infty)$ and two functions $\hat{f}, \hat{g}$, p-equivalent to $\tilde{f}$ and $\tilde{g}$, with the required properties. The estimate (11), applied to (51), implies

$$\int_{\mathbb{R}^{nq}} (u' - \hat{f})\, dx + \int_{\mathbb{R}^{nq}} (u' - \hat{g})\, dx \leq C_{nq+p} \left(\frac{\varepsilon}{\mathcal{M}_{\frac{1}{nq+p}} (F^q, G^q ; \lambda)} \right).$$

□

Remark 4.4 Assume $F = G$ and, for simplicity, suppose that $\hat{f} = \tilde{f}$, $\hat{g} = \tilde{g}$ in Corollary 4.3 (as it is true up to a p-equivalence). Moreover assume that the $\frac{1}{p}$-concave function $u' : \mathbb{R}^{nq} \longrightarrow [0, +\infty)$, given by Corollary 4.3, is of the type (24), i.e. $u' = \tilde{u}$ where $u : \mathbb{R}^n \longrightarrow [0, +\infty)$ has to be $\frac{1}{s}$-concave by Lemma 4.1. In this case Corollary 4.3 assumes a simpler statement, which naturally extends the result of Theorem 1.3. Indeed (50), thanks to Lemma 4.2, becomes

$$\int_{\mathbb{R}^{nq}} \widetilde{u-f}\, dx + \int_{\mathbb{R}^{nq}} \widetilde{u-g}\, dx \leqslant C_{nq+p} \left(\frac{\varepsilon}{\mathcal{M}_{\frac{1}{nq+p}}(F^q, G^q\,;\lambda)} \right), \qquad \text{i.e.}$$

$$\left[\int_{\mathbb{R}^n} (u-f)\, dx \right]^q + \left[\int_{\mathbb{R}^n} (u-g)\, dx \right]^q \leqslant C_{nq+p} \left(\frac{\varepsilon}{\mathcal{M}_{\frac{1}{nq+p}}(F^q, G^q\,;\lambda)} \right).$$

$$(53)$$

Unfortunately the function u' constructed in Theorem 1.3 is not necessarily of the desired form, that is in general we can not find a function $u : \mathbb{R}^n \longrightarrow [0, +\infty)$ such that $u' = \tilde{u}$ (a counterexample can be explicitly given). Then our proof can not be easily extended to the general case $s \in \mathbb{Q}$ to get (53).

5   A Stability for $s > 0$

To complete the paper, we give a (weaker) version of our main stability result Theorem 1.3 which works for an arbitrary real index $s > 0$. For this, let us denote by $[s]$ the integer part of s, i.e. the largest integer not greater than s. Obviously $[s] + 1 > s \geqslant [s]$, whereby (by the monotonicity of p-means with respect to p, i.e. $\mathcal{M}_p(a, b; \lambda) \leqslant \mathcal{M}_q(a, b; \lambda)$ if $p \leqslant q$) for every $a, b \geqslant 0$, $\lambda \in (0, 1)$

$$\left[(1 - \lambda)a^{\frac{1}{s}} + \lambda b^{\frac{1}{s}} \right]^s \geqslant \left[(1 - \lambda)a^{\frac{1}{[s]+1}} + \lambda b^{\frac{1}{[s]+1}} \right]^{[s]+1}, \qquad (54)$$

$$\left[(1 - \lambda)a^{\frac{1}{n+s}} + \lambda b^{\frac{1}{n+s}} \right]^{n+s} \geqslant \left[(1 - \lambda)a^{\frac{1}{n+[s]+1}} + \lambda b^{\frac{1}{n+[s]+1}} \right]^{n+[s]+1}. \qquad (55)$$

We arrive to the following corollary for every index $s > 0$.

Corollary 5.1 *Given $s > 0$, $\lambda \in (0, 1)$, let $f, g : \mathbb{R}^n \longrightarrow [0, +\infty)$ be integrable functions such that*

$$\int_{\mathbb{R}^n} f\, dx = \int_{\mathbb{R}^n} g\, dx = 1. \qquad (56)$$

Assume $h : \mathbb{R}^n \longrightarrow [0, +\infty)$ satisfies assumption (30) and there exists $\varepsilon > 0$ small enough such that

$$\int_{\mathbb{R}^n} h\, dx \leqslant 1 + \varepsilon. \qquad (57)$$

Then there exist a $\frac{1}{[s]+1}$-concave function $u : \mathbb{R}^n \longrightarrow [0, +\infty)$ and two functions $\hat{f}$ and $\hat{g}$, $([s]+1)$-equivalent to f and g in the sense of (44) (with $\mu_f = \mu_g = (\omega_{[s]+1})^{\frac{-1}{n+[s]+1}}$) such that

$$u \geq \hat{f}, \qquad u \geq \hat{g},$$

and

$$\int_{\mathbb{R}^n} (u - \hat{f}) \, dx + \int_{\mathbb{R}^n} (u - \hat{g}) \, dx \leq C_{n+[s]+1}(\varepsilon).$$

Proof We notice that the assumption (30) (i.e. the hypothesis of BBL of index $\frac{1}{s}$), through (54), implies that for every $x_0 \in \mathrm{supp}(f)$, $x_1 \in \mathrm{supp}(g)$

$$h\left((1-\lambda)x_0 + \lambda x_1\right) \geq \left[(1-\lambda)f(x_0)^{\frac{1}{[s]+1}} + \lambda g(x_1)^{\frac{1}{[s]+1}}\right]^{[s]+1},$$

i.e. the corresponding hypothesis of BBL for the index $\frac{1}{[s]+1}$. Therefore, thanks to the assumptions (56) and (57), it holds $\int h \leq 1 + \varepsilon = \mathcal{M}_{\frac{1}{n+[s]+1}}(\int f, \int g; \lambda) + \varepsilon$, so we can apply directly Theorem 1.3 using the integer $[s]+1$ as index. This concludes the proof. □

Remark 5.2 If we don't use the normalization (56) and want to write a result for generic unrelated $F = \int f$ and $G = \int g$, we can notice that assumption (57) should be replaced by

$$\int_{\mathbb{R}^n} h \, dx \leq \mathcal{M}_{\frac{1}{n+[s]+1}}(F, G; \lambda) + \varepsilon.$$

On the other hand, thanks to assumption (30), we can apply Proposition 2.10 and obtain

$$\int_{\mathbb{R}^n} h \, dx \geq \mathcal{M}_{\frac{1}{n+s}}(F, G; \lambda).$$

Then we would have

$$\mathcal{M}_{\frac{1}{n+s}}(F, G; \lambda) \leq \mathcal{M}_{\frac{1}{n+[s]+1}}(F, G; \lambda) + \varepsilon.$$

The latter inequality is possible only if F and G are close to each others, thanks to the stability of the monotonicity property of p-means, which states

$$\mathcal{M}_{\frac{1}{n+[s]+1}}(F, G; \lambda) \leq \mathcal{M}_{\frac{1}{n+s}}(F, G; \lambda),$$

with equality if and only if $F = G$. In this sense the normalization (56) cannot be completely avoided and the result obtained in Corollary 5.1 is weaker than what desired. Indeed notice in particular that it does not coincide with Theorem 1.3 even in the case when s is integer, since $[s] + 1 > s$ in that case as well.

Acknowledgements The second author has been partially supported by INdAM in the framework of a GNAMPA project, and by MIUR in the framework of a PRIN 2013 project and a FIR 2013 project.

Appendix

Here we show that the S-symmetrization, introduced in Remark 3.1, preserves the convexity of the involved set (that is the property (iii) therein).

We use the notations of Remark 3.1, in particular we refer to (37) and (38), and remember that C is a bounded measurable set in $\mathbb{R}^{n+s}$. We need the following preliminary result, based on the Brunn-Minkowski inequality in $\mathbb{R}^s$.

Lemma 5.3 *If C is a bounded convex set in $\mathbb{R}^{n+s}$, then for every $t \in (0, 1)$ and every $x_0, x_1 \in \mathbb{R}^n$ such that $C(x_0)$ and $C(x_1)$ are nonempty, it holds*

$$(1 - t)r_C(x_0) + tr_C(x_1) \leq r_C((1 - t)x_0 + tx_1). \tag{58}$$

Proof By definition of (37)

$$r_C(x_0) = \omega_s^{-1/s}|C(x_0)|^{1/s}, \quad r_C(x_1) = \omega_s^{-1/s}|C(x_1)|^{1/s},$$

thus

$$(1 - t)r_C(x_0) + tr_C(x_1) = \omega_s^{-1/s}\left[(1 - t)|C(x_0)|^{1/s} + t|C(x_1)|^{1/s}\right]. \tag{59}$$

Since C is convex, we notice that $C(x_0), C(x_1)$ are (nonempty) convex sets in $\mathbb{R}^s$ such that

$$(1 - t)C(x_0) + tC(x_1) \subseteq C((1 - t)x_0 + tx_1). \tag{60}$$

Applying BM inequality (i.e. Proposition 2.1) to the sets $C(x_0), C(x_1) \subset \mathbb{R}^s$, (59) implies

$$(1 - t)r_C(x_0) + tr_C(x_1) \leq \omega_s^{-1/s}|(1 - t)C(x_0) + tC(x_1)|^{1/s}$$
$$\leq \omega_s^{-1/s}|C((1 - t)x_0 + tx_1)|^{1/s} = r_C((1 - t)x_0 + tx_1),$$

where in the last inequality we use (60). $\qquad\qquad\square$

Proposition 5.4 *If C is convex then S(C) is convex.*

Proof Let $t \in (0,1)$, and let $P = (x_0, y_0), Q = (x_1, y_1)$ be two distinct points belonging to $S(C)$, i.e. $C(x_0), C(x_1)$ are nonempty sets and

$$|y_0| \leq r_C(x_0), \qquad |y_1| \leq r_C(x_1). \tag{61}$$

We prove that

$$(1-t)P + tQ = ((1-t)x_0 + tx_1, (1-t)y_0 + ty_1) \in S(C).$$

By assumptions and (60) the set $C((1-t)x_0 + tx_1)$ is nonempty. Furthermore by the triangle inequality, (61) and Lemma 5.3 we obtain

$$|(1-t)y_0 + ty_1| \leq (1-t)|y_0| + t|y_1| \leq (1-t)r_C(x_0) + tr_C(x_1) \leq r_C((1-t)x_0 + tx_1).$$

Then $(1-t)P + tQ \in S(C)$, i.e. $S(C)$ is convex. $\qquad\square$

References

1. S. Artstein, B Klartag, V. Milman, The Santaló point of a function, and a functional form of the Santaló inequality. Mathematika **51**(1–2), 33–48 (2004)
2. K.M. Ball, K.J. Böröczky, Stability of the Prékopa-Leindler inequality. Mathematika **56**(2), 339–356 (2010)
3. K.M. Ball, K.J. Böröczky, Stability of some versions of the Prékopa-Leindler inequality. Monatsh. Math. **163**(1), 1–14 (2011)
4. C. Borell, Convex set functions in d-space. Period. Math. Hung. **6**(2), 111–136 (1975)
5. H.J. Brascamp, E.H. Lieb, Some inequalities for Gaussian measures and the long-range order of one-dimensional plasma, in *Functional Integration and Its Applications*, ed. by A.M. Arthurs (Clarendon Press, Oxford, 1975), pp. 1–14
6. H.J. Brascamp, E.H. Lieb, On extensions of the Brunn-Minkowski and Prékopa-Leindler theorems, including inequalities for log concave functions, and with an application to the diffusion equation. J. Funct. Anal. **22**(4), 366–389 (1976)
7. D. Bucur, I. Fragalà, Lower bounds for the Prékopa-Leindler deficit by some distances modulo translations. J. Convex Anal. **21**(1), 289–305 (2014)
8. Y.D. Burago, V.A. Zalgaller, *Geometric Inequalities* (Springer, Berlin, 1988)
9. M. Christ, Near equality in the two-dimensional Brunn-Minkowski inequality. Preprint 2012, arXiv:1206.1965v2
10. M. Christ, Near equality in the Brunn-Minkowski inequality. Preprint 2012, arXiv:1207.5062v1
11. A. Dinghas, Uber eine Klasse superadditiver Mengenfunktionale von Brunn-Minkowski-Lusternikschem Typus. Math. Z. **68**, 111–125 (1957)
12. V.I. Diskant, Stability of the solution of a Minkowski equation. Sibirsk. Mat. Ž. **14**, 669–673, 696 (1973) [Russian]
13. S. Dubuc, Critères de convexité et inégalités intégrales. Ann. Inst. Fourier **27**(1), x, 135–165 (1977) [French. English summary]
14. R. Eldan, B. Klartag, Dimensionality and the stability of the Brunn-Minkowski inequality. Ann. Sc. Norm. Super. Pisa Cl. Sci. **XIII**(5), 975–1007 (2014)

15. A. Figalli, D. Jerison, Quantitative stability for the Brunn-Minkowski inequality. Preprint 2014, arXiv:1502.06513v1
16. A. Figalli, F. Maggi, A. Pratelli, A refined Brunn-Minkowski inequality for convex sets. Ann. Inst. Henri Poincare **26**, 2511–2519 (2009)
17. A. Figalli, F. Maggi, A. Pratelli, A mass transportation approach to quantitative isoperimetric inequality. Invent. Math. **182**(1), 167–211 (2010)
18. R.J. Gardner, The Brunn-Minkowski inequality. Bull. Am. Math. Soc. **39**(3), 355–405 (2002)
19. D. Ghilli, P. Salani, Quantitative Borell-Brascamp-Lieb inequalities for power concave functions. Preprint (2015)
20. H. Groemer, On the Brunn-Minkowski theorem. Geom. Dedicata **27**(3), 357–371 (1988)
21. G. Hardy, J.E. Littlewood, G. Pólya, Inequalities (Cambridge University Press, Cambridge, 1934)
22. R. Henstock, A.M. Macbeath, On the measure of sum sets, I. The theorems of Brunn, Minkowski and Lusternik. Proc. Lond. Math. Soc. 3(3), 182–194 (1953)
23. B. Klartag, Marginals of geometric inequalities, in *Geometric Aspects of Functional Analysis.* Lecture Notes in Mathematics, vol. 1910 (Springer, Berlin, 2007), pp. 133–166
24. L. Leindler, On a certain converse of Hölder's inequality, II. Acta Sci. Math. **33**(3–4), 217–223 (1972)
25. A. Prékopa, Logarithmic concave measures with application to stochastic programming. Acta Sci. Math. **32**, 301–316 (1971)
26. R. Schneider, *Convex Bodies: The Brunn-Minkowski Theory.* Encyclopedia of Mathematics and Its Applications, vol. 44 (Cambridge University Press, Cambridge, 1993)
27. A. Segal, Remark on stability of Brunn-Minkowski and isoperimetric inequalities for convex bodies, in *Geometric Aspects of Functional Analysis.* Lecture Notes in Mathematics, vol. 2050 (Springer, Heidelberg, 2012), pp. 381–391

LECTURE NOTES IN MATHEMATICS

Editors in Chief: J.-M. Morel, B. Teissier;

Editorial Policy

1. Lecture Notes aim to report new developments in all areas of mathematics and their applications – quickly, informally and at a high level. Mathematical texts analysing new developments in modelling and numerical simulation are welcome.

   Manuscripts should be reasonably self-contained and rounded off. Thus they may, and often will, present not only results of the author but also related work by other people. They may be based on specialised lecture courses. Furthermore, the manuscripts should provide sufficient motivation, examples and applications. This clearly distinguishes Lecture Notes from journal articles or technical reports which normally are very concise. Articles intended for a journal but too long to be accepted by most journals, usually do not have this "lecture notes" character. For similar reasons it is unusual for doctoral theses to be accepted for the Lecture Notes series, though habilitation theses may be appropriate.

2. Besides monographs, multi-author manuscripts resulting from SUMMER SCHOOLS or similar INTENSIVE COURSES are welcome, provided their objective was held to present an active mathematical topic to an audience at the beginning or intermediate graduate level (a list of participants should be provided).

   The resulting manuscript should not be just a collection of course notes, but should require advance planning and coordination among the main lecturers. The subject matter should dictate the structure of the book. This structure should be motivated and explained in a scientific introduction, and the notation, references, index and formulation of results should be, if possible, unified by the editors. Each contribution should have an abstract and an introduction referring to the other contributions. In other words, more preparatory work must go into a multi-authored volume than simply assembling a disparate collection of papers, communicated at the event.

3. Manuscripts should be submitted either online at www.editorialmanager.com/lnm to Springer's mathematics editorial in Heidelberg, or electronically to one of the series editors. Authors should be aware that incomplete or insufficiently close-to-final manuscripts almost always result in longer refereeing times and nevertheless unclear referees' recommendations, making further refereeing of a final draft necessary. The strict minimum amount of material that will be considered should include a detailed outline describing the planned contents of each chapter, a bibliography and several sample chapters. Parallel submission of a manuscript to another publisher while under consideration for LNM is not acceptable and can lead to rejection.

4. In general, **monographs** will be sent out to at least 2 external referees for evaluation.

   A final decision to publish can be made only on the basis of the complete manuscript, however a refereeing process leading to a preliminary decision can be based on a pre-final or incomplete manuscript.

   Volume Editors of **multi-author works** are expected to arrange for the refereeing, to the usual scientific standards, of the individual contributions. If the resulting reports can be forwarded to the LNM Editorial Board, this is very helpful. If no reports are forwarded or if other questions remain unclear in respect of homogeneity etc, the series editors may wish to consult external referees for an overall evaluation of the volume.

5. Manuscripts should in general be submitted in English. Final manuscripts should contain at least 100 pages of mathematical text and should always include
   - a table of contents;
   - an informative introduction, with adequate motivation and perhaps some historical remarks: it should be accessible to a reader not intimately familiar with the topic treated;
   - a subject index: as a rule this is genuinely helpful for the reader.
   - For evaluation purposes, manuscripts should be submitted as pdf files.

6. Careful preparation of the manuscripts will help keep production time short besides ensuring satisfactory appearance of the finished book in print and online. After acceptance of the manuscript authors will be asked to prepare the final LaTeX source files (see LaTeX templates online: https://www.springer.com/gb/authors-editors/book-authors-editors/manuscriptpreparation/5636) plus the corresponding pdf- or zipped ps-file. The LaTeX source files are essential for producing the full-text online version of the book, see http://link.springer.com/bookseries/304 for the existing online volumes of LNM). The technical production of a Lecture Notes volume takes approximately 12 weeks. Additional instructions, if necessary, are available on request from lnm@springer.com.

7. Authors receive a total of 30 free copies of their volume and free access to their book on SpringerLink, but no royalties. They are entitled to a discount of 33.3 % on the price of Springer books purchased for their personal use, if ordering directly from Springer.

8. Commitment to publish is made by a *Publishing Agreement*; contributing authors of multiauthor books are requested to sign a *Consent to Publish form*. Springer-Verlag registers the copyright for each volume. Authors are free to reuse material contained in their LNM volumes in later publications: a brief written (or e-mail) request for formal permission is sufficient.

Addresses:
Professor Jean-Michel Morel, CMLA, École Normale Supérieure de Cachan, France
E-mail: moreljeanmichel@gmail.com

Professor Bernard Teissier, Equipe Géométrie et Dynamique,
Institut de Mathématiques de Jussieu – Paris Rive Gauche, Paris, France
E-mail: bernard.teissier@imj-prg.fr

Springer: Ute McCrory, Mathematics, Heidelberg, Germany,
E-mail: lnm@springer.com

Printed in the United States
By Bookmasters